1978

CO-AWZ-743

Concepts in mammalian embryoge

3 0301 00076796 8

CONCEPTS IN
MAMMALIAN
EMBRYOGENESIS

Cell Monograph Series
Benjamin Lewin, series editor

1. *Concepts in Mammalian Embryogenesis*, edited by Michael I.
Sherman

CONCEPTS IN MAMMALIAN EMBRYOGENESIS

Edited by
Michael I. Sherman

LIBRARY
College of St. Francis
JOLIET, ILL.

The MIT Press
Cambridge, Massachusetts, and London, England

Copyright © 1977 by
The Massachusetts Institute of Technology

All rights reserved. No part of this book may be reproduced in any form or by any means, electronic or mechanical, including photocopying, recording, or by any information storage and retrieval system, without permission in writing from the publisher.

This book was set in VIP Times Roman by Creative Composition Inc., printed on R & E Book by Murray Printing Company and bound in Holliston Roxite by Murray Printing Company in the United States of America

Library of Congress Cataloging in Publication Data
  Main entry under title:

Concepts in mammalian embryogenesis.

  (Cell monograph series; 1)
  Includes bibliographies and index.
  1. Embryology—Mammals. I. Sherman, Michael I.
II. Series.
QL959.C6   599'.03'3   77–4144
ISBN 0–262–19158–X

CONTENTS

SERIES PREFACE

The Cell Monograph Series is intended to provide timely accounts of topics important in molecular and cellular biology. It shares this purpose with the reviews published in Cell, but with the difference that in a book it is possible to take a much broader approach and to review many more aspects of a subject. Each volume in the series will be focused on a coherent theme and will attempt to provide a complete account of its topic. The books will be concerned with topical subjects that might be considered to lie at the leading edge of research. In some instances these will be subjects where a great deal of detailed information has been obtained and a review of the state of the art is useful; in other cases the subject will be one that is currently coming to fruition, so that a book bringing together recent advances may serve an important function for both those within the field and those working in other areas. This latter approach has been taken in this volume, the first, which discusses early mammalian embryogenesis. Here the potential of current research is evident: an understanding of the remarkable processes responsible for the development of a fertilized egg into an individual organism.

Benjamin Lewin

CONTRIBUTORS

David J. Adler
Department of Molecular Biology
Roswell Park Memorial Institute
Buffalo, New York

Anton Berns
Laboratory of Biochemistry
University of Nijmegen
Nijmegen, The Netherlands

W. David Billington
Department of Pathology
The Medical School
University of Bristol
Bristol, United Kingdom

Verne M. Chapman
Department of Molecular Biology
Roswell Park Memorial Institute
Buffalo, New York

Christopher F. Graham
Department of Zoology
University of Oxford;
Oxford, United Kingdom

Rudolf Jaenisch
Heinrich-Pette-Institut fur
Experimentelle Virologie
und Immunologie
Hamburg, Federal Republic of
Germany

Eric J. Jenkinson
Department of Pathology
The Medical School
University of Bristol
Bristol, United Kingdom

Cole Manes
La Jolla Research Foundation
La Jolla, California

Virginia E. Papaioannou
Department of Zoology
University of Oxford
Oxford, United Kingdom

Janet Rossant
Department of Zoology
University of Oxford
Oxford, United Kingdom

Michael I. Sherman
Roche Institute of Molecular
Biology
Nutley, New Jersey

Jonathan Van Blerkom
Department of Molecular, Cellular
and Developmental Biology
University of Colorado
Boulder, Colorado

John D. West
Department of Molecular Biology
Roswell Park Memorial Institute
Buffalo, New York

Linda R. Wudl
Roche Institute of Molecular
Biology
Nutley, New Jersey

GLOSSARY

Term	Synonyms[1]	Definition
definitive endoderm		The endoderm of the late fetus and the adult, e.g., respiratory and alimentary epithelium.
ectoplacental cone (EPC)		The thickened part of the trophoblast layer at the embryonic pole. Cells at the outside border of this layer give rise to secondary giant trophoblast cells. The remaining cells eventually constitute part of the placental disc.
embryonal carcinoma (EC)	primitive teratocarcinoma cell (PTC)	The relatively undifferentiated stem cell of teratocarcinomas. Embryonal carcinoma cell lines may be *pluripotent*, i.e., capable of giving rise to a variety of differentiated cell types, or *nullipotent*, i.e., unable to differentiate to other cell types.
endoderm carcinoma		Carcinomas whose cells resemble either the parietal or the visceral endoderm of the early embryo. This term is used generally if the distinction between visceral and parietal endoderm has not been, or cannot be, made.
equivalent gestation day (EGD)		The age of the embryo had it been left in utero. This does not necessarily indicate that the embryo has reached an equivalent developmental stage.

[1]These synonyms have not been used in this monograph.

Term	Synonyms[1]	Definition
first day of pregnancy, second day of pregnancy, etc.	day 0, day 1, etc.; day 1/2, day 1 1/2, etc.	The first day of pregnancy is the day of observation of the sperm plug after mating. It begins at the time of copulation (approximately midnight for the mouse) and extends for 24 hours, when the second day begins.
inner cell mass (ICM)	epiblast	The cluster of cells in the blastocyst surrounded by the trophectoderm layer.
parietal endoderm	distal endoderm; parietal yolk sac	The single layer of cells, derived from the primitive endoderm (q.v.), which grows adherent to the inner wall of the trophoblast layer, forming the outer boundary of the yolk sac cavity from the early egg cylinder stage.
parietal endoderm carcinoma	parietal yolk sac (PYS) carcinoma	Tumors consisting of cells bearing a close resemblance to cells of the parietal endoderm layer of the conceptus.
primitive endoderm		The single layer of cells lining the blastocoel cavity of the late blastocyst; the progenitor cells of the parietal and visceral endoderm.
T complex	T locus	A series of genetic loci in the mouse characterized by a variety of recessive *t* or dominant *T* mutations.
terato-carcinoma		A tumor containing embryonal carcinoma cells as its malignant element. Other cell types,

[1]These synonyms have not been used in this monograph.

Term	Synonyms[1]	Definition
		generally nonmalignant, are often found in the tumor as well.
teratoma		(1) Specifically, a benign tumor containing one or more cell types but lacking embryonal carcinoma (q.v.). (2) Generally, this term is used if it is not known whether the tumor is malignant or benign.
T region		The structural elements of chromosome 17 in the mouse, extending from the locus characterized by the mutant T (Brachyury) to the locus characterized by the mutant tf (tufted).
trophectoderm	trophoblast	The outside single layer of cells in the preimplantation blastocyst.
trophoblast		(1) Initially, the outer layer of cells in the peri- and postimplantation embryo; these cells anchor the conceptus to the uterine wall. (2) Later, the fetally derived component of the placenta, consisting initially of EPC and giant cells and subsequently constituting the intermediate layers (labyrinth, spongiotrophoblast) of the placental disc.
visceral endoderm	proximal endoderm	(1) Initially, the layer of endoderm cells enclosing the extraembryonic and embryonic ectoderm of the egg cylinder. (Others have applied this term only to the endoderm cells overlaying the extraembryonic

[1]These synonyms have not been used in this monograph.

Term	Synonyms[1]	Definition
		ectoderm. The term has been used here in a broader sense because there is no other convenient name for the endoderm of the egg cylinder; the fate of the cells overlying the embryonic ectoderm is unclear.)[2] (2) Later in development, the outermost layer of the yolk sac.
yolk sac	visceral yolk sac	The saclike structure of the mid-gestation embryo which forms the outer boundary of the exocoelom. It consists of an outer layer of endoderm cells (visceral endoderm) and an inner mesodermal layer.

[1]These synonyms have not been used in this monograph.

[2]See G. D. Snell and L. C. Stevens (1966), Early embryology, *in* Biology of the Laboratory Mouse, E. L. Green, ed. (New York: McGraw-Hill), pp. 205–245.

INTRODUCTION   Michael I. Sherman

The roots of early mammalian embryology extend back through hundreds of years (Bodemer, 1971). Despite the use of relatively unsophisticated techniques, a number of investigators late in the last century and early in the present one offered remarkable descriptions of the developing mammalian embryo. As Rossant and Papaioannou point out in chapter 1, these investigators put forward several proposals to describe cell lineage in the early mammalian embryo. The elimination of the incorrect alternatives has taken many years and a variety of technical developments. Although we are close to having a detailed fate map of cells up to the late egg-cylinder stage in rodents, it is perhaps surprising that some of the important pieces of the puzzle have fallen into place only in the past few years (chapter 1).

Despite the early start on investigations into mammalian embryology, almost nothing was known about the biochemistry of the early mammalian embryo fifteen years ago. When such studies were begun, primarily in the laboratories of Biggers and Brinster (see, e.g., Brinster, 1971; Biggers, 1971), they were largely related to efforts to develop a suitable medium for preimplantation embryo culture. Because mammalian embryos are small and not available in large numbers, progress in delineating their biochemical and molecular biological properties was slow compared to achievements with frog and sea-urchin embryos, wherein investigators had begun to introduce the concepts of gene amplification, activation of previously stored, stable messages, and delayed expression of paternal genes. It could only be assumed, therefore, that rules of development in embryos of lower species would apply to mammalian conceptuses as well. Comparative studies are possible, however, with recently developed microtechniques (see chapters 2 and 3), and such studies have revealed that these generalizations were not always warranted.

Most of the studies to be described in this monograph deal with the mouse embryo. An obvious disadvantage of using the mouse for embryological studies, namely its small size, is outweighed by the following advantages: mice are relatively inexpensive to maintain and breed well in captivity; females can be superovulated so that they will produce up to fifty normally developing early embryos; the genetics of the mouse are well established; and culture systems are available for the study of preimplantation (Whitten, 1971) and postimplantation (Sherman, 1975) mouse embryos. Furthermore, the existence of mutants affecting specific stages of development (chapters 3 and 4) and the availability of strains of mice with a high incidence of teratocarcinomas (tumors containing cells with marked similarities to early embryonic cells; see chapter 7) have led to some exciting experimental approaches. Despite the fact that the mouse embryo is the central character in this book, references are made to other mammalian species as appropriate.

An effort has been made to standardize terminology throughout this monograph, and the contributors all consented to use in their chapters the terms defined in the glossary. (The reader should note, however, that the terminology used here by the authors may not be that which they use in articles published elsewhere.) This effort reflects the view that standardization of terms is needed to eliminate the confusion that has arisen in the past over the use of different terminology by different investigators.

Several volumes dealing with various aspects of mammalian embryology have been published in the last six years (Blandau, 1971; Raspé, 1971; Balls and Wild, 1975; Markert and Papaconstantinou, 1975; Sherman and Solter, 1975; Elliott and O'Connor, 1976). Rather than to collect a number of articles describing the recent research of several authors, as most of these volumes have done, the purpose of this monograph is to provide the reader with a thorough overview of a few selected topics on mammalian embryogenesis that are of current interest and interrelate well with each other.

One of the underlying themes in all chapters of this monograph is the programming of events in early embryogenesis. The activation of "new" genes, which begins early in embryogenesis, complements preexisting gene action (beginning prior to fertilization) and helps to provide the developing embryo with the machinery necessary for DNA, RNA, and protein synthesis and with enzymes required for intermediary metabolism ("maintenance" gene products). The synthesis of "differentiative" gene products, those that irreversibly route cells along different developmental pathways, must also be taking place prior to, or at, implantation. Of the new gene products that have been analyzed during embryogenesis (RNA and total proteins, described in chapter 2; paternal gene products, described in chapter 3; cell surface constituents, described in chapter 5), a number are involved in maintenance functions. Others are characteristic of the differentiated functions that various cell types assume. We know nothing definite, however, about genes that *initiate* differentiation in early embryonic cells. Though it is possible that the products of some of these genes are already known, they have not been associated with their roles in the control of development.

Recent biological studies have revealed that a relationship exists between the position of a cell relative to other cells in the embryo and its ultimate fate (see chapter 1). However, just as neurobiologists have been struggling to understand how electrical impulses in the brain are ultimately translated into behavior, developmental embryologists have yet to explain how the microenvironment can trigger a cell to embark on a specific course of differentiation. One difficulty in investigating this problem at the molecular level is that only a small amount of tissue is available for such studies. Utilizing the kinds of microanalytic techniques described in chapters 2 and 3 and studying similar phenomena in teratocarcinomas, where large numbers of cells can be obtained, might compensate for this problem. For example, Graham has reviewed evidence that aggregates of certain murine embryonal carcinoma cells can give rise to an outer layer of endoderm in vitro analogous to the production of primitive endoderm by the outermost cells of the inner cell mass of the rodent embryo (see chapter 7). The value of the teratocarcinoma system in studying this differentiative event and others that occur in embryogenesis is dependent upon how faithfully tumor cells recapitulate embryonic processes. Such relationships between embryos and teratocarcinomas are thoroughly considered in chapter 7.

This monograph includes a chapter on T mutations in the mouse because this system has the potential for contributing to our knowledge of the mechanisms involved in development of the mammalian embryo. According to one popular theory, the T complex consists of a series of developmental genes encoding surface constituents essential for cell–cell recognition and interaction at particular stages in embryogenesis. This concept is not universally accepted (see chapter 4); nevertheless, a better understanding of how T mutations cause developmental arrest should add to our knowledge of how embryonic programming is controlled.

Recent immunological, histochemical, and biochemical analyses of embryonic cell surfaces unrelated to studies of T complex mutations have also aroused much interest. Embryonic cell surface membranes undergo a number of changes in the first half of gestation (see chapter

5). Because sensitive analytic techniques are now available for studying cell surface components, these constituents should be valuable markers of genetic programming during embryogenesis.

Finally, in chapter 6 Jaenisch and Berns consider the possible role of viruses in the control of embryonic development. Although there is no direct evidence that viruses have such a role, the presence of endogenous viruslike particles in early embryos and the ability of exogenous viruses to integrate into the genome of early embryonic cells suggest that the oncogene hypothesis merits further attention.

In summary, the chapters in this monograph demonstrate that mammalian embryologists are now equipped with morphological, biochemical, and molecular biological markers of early embryonic development and differentiation. Model systems are available for testing ideas of genetic and epigenetic control mechanisms. Proper use of these resources could revolutionize our understanding of mammalian embryogenesis within the next decade.

References

BALLS, M., and WILD, A. E., eds.
(1975).
The Early Development of Mammals.
London: Cambridge University Press.

BIGGERS, J. D. (1971).
Metabolism of mouse embryos. J.
Reprod. Fertil. Suppl. *14*, 41–54.

BLANDAU, R. J., ed. (1971).
The Biology of the Blastocyst.
Chicago: University of Chicago Press.

BODEMER, C. W. (1971).
The biology of the blastocyst in his-
torical perspective. *In* The Biology of
the Blastocyst, R. J. Blandau, ed.
(Chicago: University of Chicago
Press), pp. 1–25.

BRINSTER, R. L. (1971).
Mammalian embryo metabolism. *In*
The Biology of the Blastocyst, R. J.
Blandau, ed. (Chicago: University of
Chicago Press), pp. 303–318.

ELLIOTT, K., and O'CONNOR, M.,
eds. (1976).
Determination during Embryogenesis.
Amsterdam: Associated Scientific
Publishers.

MARKERT, C. L., and
PAPACONSTANTINOU, J., eds.
(1975).
The Developmental Biology of
Reproduction. New York: Academic
Press.

RASPÉ, G., ed. (1971).
Schering Symposium on Intrinsic
Factors in Early Mammalian
Development. Advances in the
Biosciences, *6*. Oxford: Pergamon
Press.

SHERMAN, M. I. (1975).
The culture of cells derived from
mouse blastocysts. Cell *5*, 345–349.

SHERMAN, M. I., and SOLTER, D.,
eds. (1975).
Teratomas and Differentiation. New
York: Academic Press.

WHITTEN, W. K. (1971).
Nutrient requirements for the culture
of preimplantation embryos in vitro.
Advan. Biosciences *6*, 129–141.

CONCEPTS IN MAMMALIAN EMBRYOGENESIS

1 THE BIOLOGY OF EMBRYOGENESIS

J. Rossant and V. E. Papaioannou

1.1. Introduction

The potential of the mammalian embryo for elucidating problems in developmental biology has been realized in recent years with the increasing use of experimental methods (see reviews by Graham, 1971; Herbert and Graham, 1974; Mintz, 1974; Gardner, 1975; Gardner and Papaioannou, 1975; Wilson and Stern, 1975; Gardner and Rossant, 1976). Previously, our knowledge of early mammalian embryogenesis was based on descriptive studies of living and fixed, sectioned embryos, and this inevitably resulted in a static picture of a dynamic process. Now, however, we can utilize experimental approaches to supplement and test this necessarily incomplete morphological picture. This review summarizes ideas of cell fate in embryogenesis based on descriptive studies and highlights the regions of controversy. It also considers how recent experimental techniques, mainly using mouse embryos, have resolved some of these controversies. Armed with a revised fate map of the embryo, we shall then describe studies aimed at elucidating how and when cell determination occurs in development.

1.2 Descriptive Studies

1.2.1 Cleavage Stages and Blastocyst Formation

The first accurate account of early cleavage and blastocyst formation in mammals was probably van Benenden's description of the early development of the rabbit (1875). Careful studies of several different species (e.g., Rauber, 1875; Heape, 1886; Robinson, 1892; Duval, 1895; Assheton, 1898; Jenkinson, 1900; Huber, 1915) followed his work. All authors seemed to agree that cleavage of the mammalian egg resulted in a blastocyst or blastula consisting of an outer layer of cells, the trophoblast (now termed trophectoderm), enclosing a group of cells, the epiblast (now called the inner cell mass, or ICM), which is located at one end of the blastocoelic cavity. The hypoblast (primitive endoderm, i.e., the progenitor of the parietal and visceral endoderm layers), a monolayer of cells, is found on the inner surface of the epiblast and trophoblast. At this stage the epiblast can be called the primitive ecto-

J. R. is a Beit Memorial Junior Research Fellow. V. E. P. is supported by the Cancer Research Campaign.

derm (figure 1.1). The early authors disagreed, however, on the lineage relationships of these cells. There were at least five different theories put forward to explain development of the observed structure (figure 1.2).

Robinson (1892) believed that the ICM, when it first appears, is totally endodermal and that the trophectoderm at the opposite pole of the blastocyst invaginates and proliferates to form the ectoderm. The intermediate stage on which he based his hypothesis has not been observed by other workers, and it is almost certainly an artefact [figure 1.2a(ii)].

Duval's interpretation, based on study of the bat, the mouse, and other species (Duval, 1891, 1895), is similar in some ways to Robinson's. The whole of the primitive ectoderm derives from an invagination of the trophectoderm after the first inner cells have spread over the inner surface of the trophectoderm to form a hypothetical "stade didermique" [figure 1.2b(ii)] consisting of endoderm and trophectoderm only. Neither Duval himself nor other workers observed this hypothetical stage.

Assheton based his hypothesis on studies of the rabbit (1894), goat (1908), and sheep (1898). He proposed a hypoblast origin of the trophoblast and believed that the blastocoel formed by a split between hypoblast cells caused by faster division of those cells in an outside position. The epiblast, first identified as large paler-staining cells in the morula, became enclosed by hypoblast as the blastocyst formed. Assheton suggested that nutritional, mechanical, and positional factors were important in determining which cells would become hypoblast.

Huber, looking at the rat (1915), and Sobotta (1903), studying the mouse, could see no distinction between cells of the ICM and trophectoderm in the early blastocyst and suggested that the trophoblast, which later overlies the ICM, derives from proliferation of the ICM or embryonic disc [figure 1.2d(ii)]. The endoderm was also thought to develop from the embryonic disc.

Jenkinson (1900), with other workers (Selenka, 1883; Cristiani, 1892), concluded that cytoplasmic regionalization of extranuclear units determining ICM and trophectoderm cell characteristics existed in the egg, so that these tissues had separate cell lineages from the earliest stages. The primitive endoderm was believed to arise from the ICM.

This brief survey reveals how similar series of specimens can be interpreted in many different ways and serves to emphasize how difficult it is to establish cell lineage relationships from descriptive studies

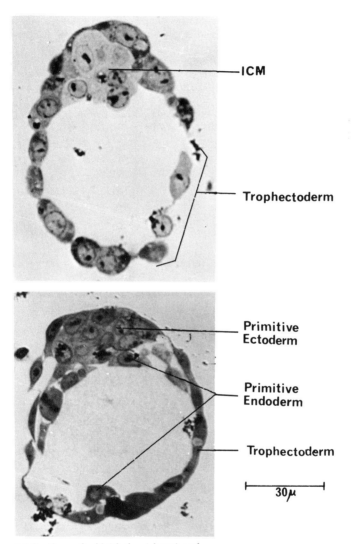

ICM

Trophectoderm

Primitive
Ectoderm

Primitive
Endoderm

Trophectoderm

├─────30μ─────┤

1.1 Sections of mid-4th-day (above) and
mid-5th-day (below) mouse blastocysts
showing delamination of primitive
endoderm from the ICM.

alone. Since these studies appeared at the turn of the century, further
descriptive studies have been made with the light microscope (Lewis
and Wright, 1935; Reinius, 1965; Dickson, 1966; Snell and Stevens,
1966; Rugh, 1968; Theiler, 1972) and with the electron microscope.

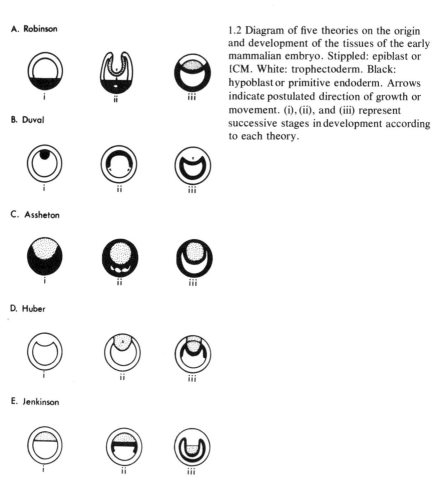

A. Robinson

B. Duval

C. Assheton

D. Huber

E. Jenkinson

1.2 Diagram of five theories on the origin and development of the tissues of the early mammalian embryo. Stippled: epiblast or ICM. White: trophectoderm. Black: hypoblast or primitive endoderm. Arrows indicate postulated direction of growth or movement. (i), (ii), and (iii) represent successive stages in development according to each theory.

Electron-microscopic (EM) studies, particularly in the rat (Enders and Schalfke, 1967; Schalfke and Enders, 1967) and the mouse (Potts and Wilson, 1967; Calarco and Brown, 1969; Enders, 1971; Nadijcka and Hillman, 1974) have shown that ICM, primitive ectoderm, and trophectoderm cells are morphologically very similar, although trophectoderm cells tend to contain more intracellular inclusions (Enders, 1971) and are attached to each other by tight junctional complexes (Enders and Schalfke, 1965; Hastings and Enders, 1975; Ducibella et al., 1975). Scanning EM studies indicate that the surfaces of ICM and trophectoderm cells also differ (Calarco and Epstein, 1973). Endoderm cells can be distinguished from both primitive ectoderm and trophectoderm by the presence of large amounts of rough endoplasmic reticulum (Enders and Schalfke, 1965; Enders, 1971). These differences, which are only clearly revealed at the EM level, do not provide any further information about the relationships between the different cell populations, but they serve to show that the three cell populations are morphologically distinct by the late blastocyst stage. These differences are confirmed, at least as far as ICM and trophectoderm cells are concerned, by recent biochemical studies in which two-dimensional electrophoretic analysis revealed five "spots" specific to ICM and four specific to trophectoderm (Van Blerkom et al., 1976; see chapter 2).

Another line of descriptive study has been pursued by Dalcq, Mulnard, and their co-workers (Dalcq and Jones-Seaton, 1949; Jones-Seaton, 1950; Dalcq, 1957; Mulnard, 1955, 1961, 1965) and more recently by others (Rodé et al., 1968; Cerisola and Izquierdo, 1969; Solter et al., 1973; Sherman, 1972; Izquierdo and Ortiz, 1975). This is the investigation of histochemical and enzyme activity differences between cells of early embryos. For example, Mulnard (1955) studied acid phosphatase activity in rat and mouse embryos and found that while the endoderm and trophectoderm of the blastocyst were both negative, the primitive ectoderm was positive. He concluded that the endoderm derived not from the ICM but from the trophectoderm, as Dalcq and Jones-Seaton (1949) had suggested.

1.2.2 Egg-Cylinder Formation

By the late blastocyst stage, three distinct cell populations are apparent: primitive endoderm, primitive ectoderm, and trophectoderm. In some rodents, including the rat (Huber, 1915), the mouse (Jenkinson, 1900, 1913), and the guinea pig (Selenka, 1884), an ingrowth of the embryonic region at about the time of implantation results in the formation of the egg cylinder and the so-called "inversion of germ layers" (Fraser, 1883; Cristiani, 1892; Snell and Stevens, 1966) (figure 1.3b). In some other mammalian species, e.g., the rabbit (Assheton, 1894; van Benenden, 1875), Tarsius (Hubrecht, 1902), and sheep (Assheton, 1898), no such invagination occurs, and the trophectoderm over the primitive ectoderm eventually disappears, exposing the embryonic knob to the uterine environment (figure 1.3a). In such species the embryo is gradually reenclosed by the amniotic folds (Mossman, 1937).

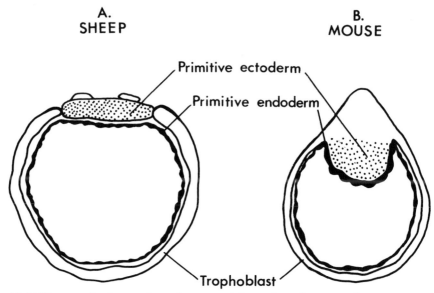

1.3 (a) Diagram of a sheep embryo after most of the trophectoderm overlying the primitive ectoderm has disappeared. (b) Diagram of mouse egg cylinder illustrating the "inversion of germ layers."

Little experimental information is available on cell fate in these species, however, and we shall confine further discussion to the mouse and rat, which have been intensively studied.

On the sixth or seventh day of gestation in the mouse and the seventh day in the rat, a division of the ectoderm into embryonic and extraembryonic regions is apparent (figure 1.4). Considerable difference of opinion has existed over the derivation of the extraembryonic ectoderm. Some workers believed that it develops as a proliferation of the trophectoderm cells at the embryonic pole of the blastocyst, which pushes the embryonic ectoderm into the blastocoel (Jenkinson, 1900, 1913; Jolly and Férester-Tadie, 1936). Robinson (1904) stated, however, that the embryonic ectoderm is formed from the ICM, although it is continuous with the overlying trophoblast. This idea has been perpetuated by later workers (Rugh, 1968; Theiler, 1972). Dalcq (1957) and Mulnard (1955) also postulated an ICM origin for the extraembryonic ectoderm based on similarities in acid phosphatase activity.

There has been almost unanimous agreement that the trophectoderm

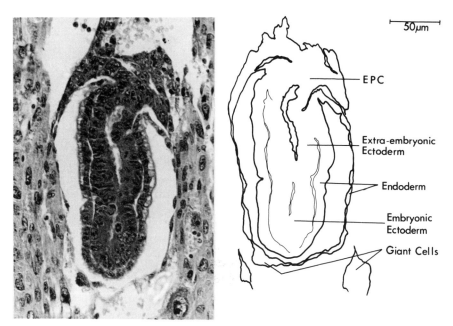

1.4 Photomicrograph and diagram of a sagittal section of a mid-7th-day mouse egg cylinder. A constriction in the middle of the cylinder marks the boundary between embryonic and extraembryonic ectoderm.

away from the embryonic region, the mural trophectoderm, transforms into the primary trophoblast giant cells (Duval, 1891; Jenkinson, 1902; Dickson, 1963, 1966) and that the ectoplacental cone (EPC) arises from proliferation of the trophectoderm overlying the ICM, the polar trophectoderm. Later the outer cells of the ectoplacental cone transform into the secondary trophoblast giant cells. Giant cell transformation involves the cessation of cell division and the endoreduplication of the DNA (Zybina, 1961, 1963a,b; Zybina and Tikhomirova, 1963; Andreeva, 1964; Zybina and Mos'yan, 1967; Hunt and Avery, 1971; Barlow and Sherman, 1972; Sherman et al., 1972) and can result in cells with as much as 1024–2048 times the haploid DNA content (Zybina, 1970; Nagl, 1972).

It is also agreed that the primitive endoderm of the blastocyst proliferates and spreads over the inner surface of the trophoblast, forming a layer of parietal endoderm as well as a visceral endoderm layer covering the egg cylinder.

1.2.3 Primitive Streak and Head Process

The next important event in embryogenesis is the appearance of mesoderm, which apparently arises from the posterior embryonic ectoderm
(Huber, 1915; Jolly and Férester-Tadié, 1936; Snell and Stevens, 1966)
and spreads anteriorly between the ectoderm and endoderm. The region of mesoderm proliferation is known as the primitive streak (figure
1.5), and its appearance establishes the anterior-posterior axis of the
embryo. It has occasionally been suggested that at least part of the
mesoderm is derived from the primitive endoderm (Sobotta, 1911; Bonnevie, 1950). At about the time that primitive streak formation occurs,
the amniotic folds begin to cut off the embryonic region from the ectoplacental cavity.

The head process arises at the anterior end of the primitive streak
(Jolly and Férester-Tadié, 1936; Snell and Stevens, 1966) and appears
to displace the visceral endoderm in the anterior midline of the embryo. Opinions vary about the relative roles of primitive endoderm and
the head process in forming the definitive gut endoderm of the fetus.
For example, Jenkinson (1913) states that the primitive endoderm of
the blastocyst is the tissue from which alimentary canal, yolk sac, and
allantois are all derived. However, Jolly and Férester-Tadié (1936), in a
careful histological study of head process formation in the rat and
mouse, state that the head process cells replace the primitive endoderm
cells and that the anterior part of the gut and the notochord derive
from this "nouvel hypoblaste." They suggest that the primitive endoderm plays only an accessory role in the formation of the embryo by
contributing to the hind gut and to the extraembryonic endoderm of the
yolk sac. If their description of events is correct, it follows that the
embryonic ectoderm of the late blastocyst and early egg cylinder
contains the potential for producing all three germ layers (ectoderm,
mesoderm, and endoderm) of the embryo.

1.2.4 Problems Arising from Descriptive Studies

Descriptive studies of early embryogenesis cannot answer the following
questions of cell fate and lineage:
What is the fate of the ICM and trophectoderm from the mature
blastocyst prior to endoderm formation?

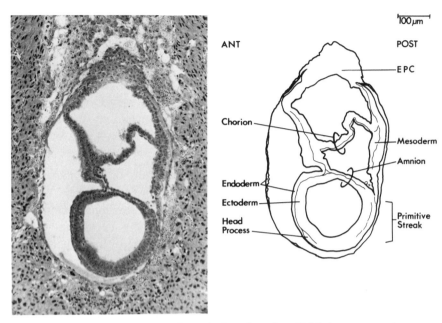

1.5 Photomicrograph and diagram of a sagittal section of a mid-8th-day mouse embryo. Appearance of the primitive streak establishes the anterior-posterior axis of the embryo.

1. Does the ICM contribute to overlying trophoblast (Huber)?
2. Is the primitive endoderm derived from the ICM (Jenkinson) or the trophectoderm (Assheton)?
3. Does the trophectoderm contribute to the later embryonic ectoderm (Duval)?
4. Is the extraembryonic ectoderm derived from the ICM (Robinson, Mulnard, Rugh) or the trophectoderm (Jenkinson)?

What is the fate of the primitive endoderm and primitive ectoderm of the late blastocyst?

1. Does the primitive endoderm give rise to yolk sac and definitive endoderm (Jenkinson)?
2. Does the embryonic ectoderm later give rise to definitive endoderm (Jolly and Férester-Tadié) as well as to mesoderm and ectoderm?

In order to answer these questions experimentally, techniques are required for determining the "prospective fate" of a given cell or tissue in an embryo. *The "prospective fate" of a cell is defined as what*

*happens to it in normal undisturbed development, as opposed to its
"prospective potency," which refers to the full range of developmental
performances of which a cell is capable under any circumstances*
(Weiss, 1939).

1.3 Experimental Studies

1.3.1 Cell Fate

Prospective fate of cells can only be traced by study of cells marked
so that their spatial relations within the embryo are undisturbed.
Such studies have been made using induced or spontaneous genetic
mosaicism (for reviews see Nesbitt and Gartler, 1971; Porter and
Rivers, 1975), cytoplasmic patterning in the living egg (Conklin, 1905;
Davidson, 1968), or vital dyes (Vogt, 1925; Weston, 1967). These methods have not proved easily applicable to mammalian embryos. However, development of experimental methods for producing chimeric
mouse embryos (Nicholas and Hall, 1940; Tarkowski, 1961; Mintz,
1962; Gardner, 1968) allows cell lineage relationships to be determined
more accurately, provided that the cells to be studied are placed in
their normal position in the embryo.

Chimeric mice consisting of a mixture of cells of two genotypes may
be made by aggregating cleavage-stage embryos of different genotypes (Tarkowski, 1961, 1965; Mintz, 1962, 1964a,b, 1965, 1971) or
by injecting entire ICMs or isolated ICM cells into blastocysts
(Gardner, 1968, 1971, 1974). With the use of genetic markers such as
isozymal variants (Mintz and Baker, 1967; Chapman et al., 1971;
Condamine et al., 1971), chromosome markers (Mystkowska and
Tarkowski, 1968; Ford et al., 1975) or coat color markers (Mintz, 1965,
1967; Tarkowski, 1964, 1965; Gardner, 1968, 1971; Gardner and Lyon,
1971; McLaren and Bowman, 1969; Nesbitt, 1974), it is possible to
trace the progeny of the two cell populations in resulting mice. The
technique of chimera production is useful for studying many sorts of
developmental and immunological problems (see Mintz, 1974). Intraspecific chimeras have also been produced in the rat (Mayer and Fritz,
1973), the rabbit (Gardner and Munro, 1973), and possibly the sheep
(Pighills et al., 1968), but little work on cell fate has been done in these
species. Limited development has also been achieved after aggregation
of morulae of different species, namely the rat and mouse (Mulnard,

1973; Zeilmaker, 1973; Stern, 1975) and the bank vole and mouse (Mystkowska, 1975). Further development of interspecific chimeras has resulted after injection of rat ICMs into mouse blastocysts (Gardner and Johnson, 1973, 1975). Both inter- and intraspecific chimeras have been used to investigate cell fate. The following sections discuss how the production of experimental chimeras and other manipulative techniques have been used to answer the questions posed by descriptive analyses of early embryogenesis.

Fate of the ICM and trophectoderm from the fourth-day mouse blastocyst

Injection of isolated fourth-day ICMs or single fourth-day ICM cells into fourth-day blastocysts can result in single fetuses with widespread chimerism (Gardner, 1968, 1971, 1975). However, the injected ICM cells never contribute to the EPC or trophoblast giant cells (Gardner, 1975). Confirmation of these results has been obtained by injecting rat ICMs into mouse blastocysts where individual cells of rat or mouse origin can be detected in resulting conceptuses by immunofluorescence (Gardner and Johnson, 1973, 1975). These results suggest that the EPC and giant cells are derived from the trophectoderm of the blastocyst (Jenkinson) rather than by proliferation of the ICM (Huber). Because the reciprocal experiment in which trophectoderm cells were injected into mouse blastocysts failed to produce any sort of chimerism (Gardner, 1971), the fate of the ICM and trophectoderm cannot be firmly established by this technique. An alternative approach is to "reconstitute" blastocysts by inserting ICMs of one genotype into microsurgically isolated trophectodermal vesicles of another (Gardner et al., 1973). Analysis of resulting conceptuses at early somite stages revealed that the EPC and giant cells were entirely of the trophectoderm genotype and that a major part of the embryonic plus extraembryonic membranes fraction was of the ICM genotype. However, a trophectoderm contribution was also detected in the embryonic plus membranes fraction in a majority of conceptuses. In the light of descriptive studies, it was possible that this trophectoderm contribution was to the primitive endoderm (Mulnard, 1955; Dalcq, 1957) and/or to the extraembryonic ectoderm (Jenkinson, 1902).

First, considering the primitive endoderm, there is now evidence to suggest that the proposed trophectoderm origin is incorrect (Mulnard, 1974). Vesicles of pure trophectoderm can be prepared by various methods (Gardner, 1971, 1972; Sherman, 1975a; Snow, 1973a,b), but no histological evidence of endoderm formation was found in such structures either in culture (Ansell and Snow, 1975; Sherman, 1975a) or when transferred to the uterus (Gardner, 1972; Snow 1973a,b). Rat ICMs isolated from the blastocyst before endoderm formation contribute to the primitive endoderm as well as to ectoderm when injected into mouse blastocysts (Gardner and Johnson, 1973, 1975). Also, when fourth-day ICMs are dissected microsurgically from mouse blastocysts and injected into empty zonae, the outer cells show the ultrastructural characteristics of endoderm (see Enders, 1971) after one or two days in the oviduct (Rossant, 1975b). Similarly, the outer cells of ICMs isolated from blastocysts immunosurgically show morphological characteristics of endoderm after 24 hr in culture (Solter and Knowles, 1975). Thus the weight of evidence suggests that the primitive endoderm derives from the ICM and not from the trophectoderm.

The second possible explanation for the results of the reconstituted blastocyst experiments is that the extraembryonic ectoderm is derived from the trophectoderm. In postimplantation development the extraembryonic ectoderm forms the ectoderm of the chorion, which is eventually separated from the embryonic region by the coalescence of the amniotic folds. The chorion later forms a close union with the EPC and the allantois (figure 1.6), forming the chorioallantoic placenta (Duval, 1891; Jenkinson, 1902; Snell and Stevens, 1966). In analysis of reconstituted blastocysts, the chorion was included whenever possible with the embryonic plus membranes fraction. This fact could explain the trophectoderm contribution to this fraction.

More direct evidence for this suggestion is provided by the experiments on rat/mouse chimeras. When rat ICMs are injected into mouse blastocysts (Gardner and Johnson, 1973, 1975) or aggregated with mouse morulae (Rossant, 1976a), the progeny of the rat cells may be widespread in endoderm and embryonic ectoderm, but they are never found in the extraembryonic ectoderm. Occasionally, in the injection experiments, the rat ICM failed to aggregate with the host mouse ICM and formed a separate egg cylinder inside the mouse trophoblast, a twinning situation which also sometimes occurs when a mouse ICM is injected into a mouse blastocyst (V. E. Papaioannou and R. L. Gardner, unpublished). In this situation the whole of the endoderm and embryonic ectoderm of the egg cylinder formed from the rat ICM was of rat origin, but the EPC and extraembryonic ectoderm consisted of mouse

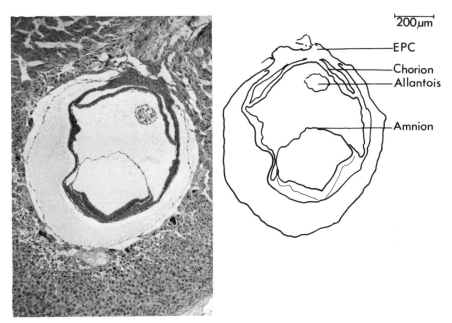

1.6 Photomicrograph and diagram of a frontal section of a mid-9th-day mouse embryo through the neural folds, showing the fusion of the chorion and the EPC and the approach of the allantois, which arises more posteriorly.

cells (Gardner and Johnson, 1975). This is strong evidence for a trophectoderm origin of the extraembryonic ectoderm, but it remains to be confirmed by injecting rat ICMs into mouse trophectodermal vesicles where no interference from mouse ICM cells could occur.

Fate of the primitive endoderm and ectoderm from the fifth-day mouse blastocyst

To follow the fate of fifth-day endoderm and ectoderm cells, one would ideally like to inject genetically marked cells into their proper position in fifth-day blastocysts. However, fifth-day blastocysts fail to reimplant and develop when transferred to the uterus (A. K. Tarkowski, personal communication) and the only possible experiment is to inject fifth-day cells into fourth-day blastocysts. In preliminary experiments, Gardner and Papaioannou (1975) found that when either fragments of endoderm or single endoderm cells (recognized by their rough appearance in disaggregates) were injected into blastocysts, chimerism was nearly always restricted to the yolk sac and placenta. No indication of a contribution to the presumed endodermal derivatives of the fetus (gut, liver, lungs, etc.) was detected by electrophoretic separation of iso-zyme markers. Ectoderm cells, on the other hand, contributed to many tissues, including the gut and its derivatives. This finding strongly sug-gests that the primitive endoderm of the blastocyst does not contribute in a major way to the definitive endoderm of the fetus. However, the limits of detection in this system (about 5 to 7 percent) do not preclude a minor contribution, such as a contribution to the hind gut, as sug-gested by Jolly and Ferester-Tadié (1936). It seems likely that the fate of most of the primitive endoderm from the fifth-day blastocyst is to form the visceral endoderm of the yolk sac (Gardner and Papaioannou, 1975), although this is not yet proven because the yolk sac contains mesoderm as well as endoderm tissues and the experiments described did not distinguish between these two tissues.

The suggestion that the definitive endoderm of the fetus is derived in the most part, if not entirely, from the embryonic ectoderm is con-firmed by study of the potency of later isolated embryonic ectoderm and endoderm tissues in ectopic sites (section 1.3.2).

Revised fate map
The manipulation experiments discussed have enabled the fate of cells in embryogenesis to be determined more critically than descriptive studies alone would allow. A plan of cell lineage relationships con-structed in the light of these studies is presented in figure 1.7. This plan requires further confirmatory experiments, but it seems likely to be valid in most essential features. Studies on cell potency can also pro-vide indirect evidence on cell fate.

1.3.2 Cell Potency

With cell fate in early embryogenesis more firmly established, it is possible to study cell potency, i.e., the development of a cell outside its normal environment, and to investigate the mechanisms by which differences between cells are produced. *When a cell's potency is equal to its fate, a cell is said to be determined.* There are various methods for assessing whether embryonic cells are determined (Weiss, 1939; Huxley and de Beer, 1934; Needham, 1942; Hadorn, 1965; Gehring,

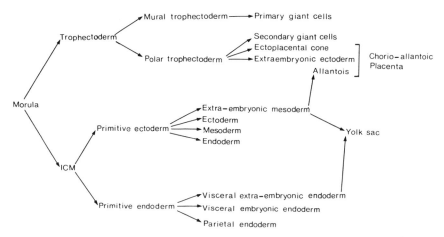

1.7 Tentative revised plan of cell lineage relationships in early mouse embryogenesis.

1972; Nöthiger, 1972; Holtzer et al., 1972), but in embryological experiments the most common approach is to study the potency of cells in isolation or after transplantation to a different embryonic site. Ideally, the potency of a particular cell type should be tested under a variety of different conditions before it is concluded that such cells are determined (Needham, 1942).

Potency of cells from the fourth-day blastocyst
Polar and mural trophectoderm   Trophectoderm can first be distinguished as a distinct tissue by the fourth day. Two populations of cells can be described by their position at this time: the polar trophectoderm overlying the ICM and the mural trophectoderm surrounding the blastocoelic cavity. Vesicles of mural trophectoderm microsurgically isolated from the mouse blastocyst never produce ICM derivatives when transferred to the uterus, although they will implant and form trophoblast giant cells (Gardner, 1972). Also, mural trophectoderm cells, when injected into the blastocoel of intact blastocysts, do not integrate with the host ICM or contribute to its derivatives (Gardner, 1971). Thus, when isolated or when transplanted to a different embryonic site, the trophectoderm cells of the fourth-day blastocyst seem to represent a determined cell population. Nothing is known about the potency of polar trophectoderm because it is difficult to separate this tissue from the ICM without damage. However, it should be mentioned

that the properties of polar and mural trophectoderm appear to be similar because mural trophectoderm can substitute for polar trophectoderm in reconstituted blastocysts (Gardner et al., 1973).

ICM   The ICM cells on the fourth day resemble the relatively "undifferentiated" morula cells in several ways (Gardner, 1972; Rossant, 1975a):

1. Neither ICMs nor early morulae show any signs of cavitation or fluid accumulation (Gardner, 1972).
2. ICMs and morulae cannot induce decidualization of a receptive uterus (McLaren and Michie, 1956; Gardner, 1972; Kaufman and Gardner, 1974).
3. ICMs aggregate readily in culture (Gardner, 1971, 1972), as do morulae (Tarkowski, 1961; Mintz, 1962).

The morphological distinctions characteristic of the blastocyst stage could therefore coincide with the determination of the trophectoderm (or at least the mural trophectoderm) alone, leaving the ICM unrestricted in its potency, rather than with the determination of both cell populations. The potency of ICM cells has been tested in various ways. In one series of experiments, microsurgically isolated mouse (Rossant, 1975a) and rat (Rossant, 1976a) ICMs were aggregated with 8-cell mouse embryos in culture. Morphologically normal blastocysts developed, and these gave rise to chimeric fetuses when transferred to the mouse uterus. However, no contribution of ICM cell progeny could be detected in the trophectoderm derivatives either by electrophoretic analysis of isozymal differences in mouse/mouse chimeras (Rossant, 1975a) or by the more sensitive immunofluorescent analysis of rat/mouse chimeras (Rossant, 1976a).

In further experiments, aggregated pairs of fourth-day ICMs were inserted into empty zonae and transferred to the oviduct (Rossant, 1975b). The cells survived and divided in this environment for one to two days, but no morphological evidence of trophectoderm formation was observed and no contribution to trophectoderm derivatives occurred when these ICMs were recovered from the oviduct and injected into host blastocysts. Thus, in the situations tested, the potency of the ICM cells is equal to their fate and they can be considered to be determined. By the 64-cell, mature blastocyst stage, therefore, both ICM and trophectoderm cells appear to be committed to separate developmental pathways.

Mechanism and time of determination   Much speculation and experimentation has been directed at the problem of how ICM versus trophectoderm determination is established in the early embryo. The question has been reviewed extensively elsewhere (Graham, 1971; Herbert and Graham, 1974; Wilson and Stern, 1975; Gardner and Rossant, 1976); we shall only outline the two major hypotheses and review the more critical experiments for and against them.

Many early workers sought differences between early blastomeres that could be followed through to either ICM or trophectoderm cells, in a way analogous to the cytoplasmic localization found in eggs of lower vertebrates and invertebrates (reviewed by Wilson, 1925; Davidson, 1968). For example, van Benenden (1875, 1880) suggested that one of the blastomeres at the 2-cell stage in the rabbit was larger and clearer than the other and that this blastomere gave rise to the trophectoderm, whereas the other gave rise to the ICM. Assheton (1898) concluded that in the sheep, differences between inside and outside cells can be detected at the early morula stage, the inside cells being the progenitors of the embryonic ectoderm. Heuser and Streeter (1928) thought that the progenitors of the ICM and trophectoderm were separate after the first cleavage in the pig. However, other workers were unable to detect any differences between early blastomeres [e.g., Heape (1886) in the mole and rabbit], and to the present day, no one has detected distinct morphological differentiation within the living egg or early cleavage stages (Austin, 1961; Wilson and Stern, 1975). EM studies have also failed to reveal any segregation of cytoplasmic organelles or other structures in early cleavage (Sotelo and Porter, 1959; Enders and Schlafke, 1965; Calarco and Brown, 1969).

On the basis of histochemical evidence, Dalcq (1957) proposed that ICM and trophectoderm determination was brought about by segregation of cytoplasmic factors in the fertilized egg during cleavage. Dalcq (1957) and Mulnard (1955, 1961) reported that two areas of cytoplasm, dorsal and ventral, could be distinguished in fixed rat and mouse eggs. By the 8-cell stage, segregation of these areas of cytoplasm had occurred, so that some cells contained only "dorsal" cytoplasm and gave rise to the ICM whereas others contained only "ventral" cytoplasm and produced trophectoderm. More recent studies have not confirmed these observations (Rodé et al., 1968; Cerisola and Izquierdo, 1969; Solter et al., 1973; Mulnard, 1974). Direct evidence against Dalcq's hypothesis has come from the study of mouse blastomeres isolated at the 8-cell stage. If segregation of dorsal and ventral cytoplasm has occurred by this stage and is determinative, these isolated blastomeres should produce only ICM or trophectoderm cells. In the mouse, such isolated blastomeres do not seem capable of producing a complete embryo in vivo (Rossant, 1976b), but they can contribute to both ICM and trophectoderm derivatives when combined with blastomeres of a different genotype (Kelly, 1975). These results are clearly incompatible with Dalcq's hypothesis. One could, however, still postulate a mechanism of segregation similar to that envisaged by Heape (1886), whereby factors originally present in all blastomeres (he suggests yolk particles) would pass specifically to inside cells at the blastocyst stage, thus making them different from outside cells.

The second hypothesis of ICM and trophectoderm determination is the epigenetic "inside-outside" hypothesis (Mintz, 1964a,b, 1965; Tarkowski, 1965; Tarkowski and Wroblewska, 1967; Graham, 1971), which suggests that cell position during cleavage is important in determination. Blastomeres that happen to remain on the outside of the embryo during cleavage form trophectoderm, but those that become completely enclosed by other blastomeres form the ICM. The most direct observations in support of this hypothesis have come from experiments in which alterations of blastomere position during cleavage have been shown to affect determination. Various experiments have been performed using ^{3}H-thymidine (Mintz, 1965; Garner and McLaren, 1974), strain differences in natural cytoplasmic granularity (Mintz, 1964a, 1965), and silicone oil droplets (Stern and Wilson, 1972; Wilson et al., 1972) as markers, in which morphologically normal blastocysts were formed from aggregates of morulae, despite little apparent cell sorting. Thus, cells that were on the outside of the morulae before aggregation could contribute to the ICM of the composite blastocyst if they became enclosed by the aggregation process. However, all these experiments are interpreted in terms of morphology at the blastocyst stage, which may not be indicative of postimplantation potential (Adams, 1965; Rossant, 1976b; Snow, 1976). It is possible that cell position could be altered without altering cell potency so that, e.g., an "outside" cell could be trapped in the ICM without becoming a functional ICM cell.

Probably the best results in favor of the "inside-outside" hypothesis are those of Hillman et al. (1972), who used genetically marked as well as ^{3}H-thymidine–labeled cells to follow their postimplantation development in blastomere and morula aggregates. They found that cells placed in outside positions tended to contribute to trophectoderm and, more importantly, to postimplantation trophoblast tissues. Similarly, inside cells tended to contribute to the ICM and its postimplantation derivatives. But even these results are not conclusive, because cells in the "wrong" position might still have been eliminated in the postimplantation embryo. However, the weight of evidence suggests that cell position plays an important role in cell determination. It is still not clear when ICM and trophectoderm determination occurs (Gardner and Rossant, 1976), since little is known about cell potency between the 8-cell and blastocyst stages.

Snow (1973a,b) has suggested that determination occurs as early as the 16-cell stage. He found that development of the ICM was suppressed when 2- or 8-cell mouse embryos were cultured to the blastocyst stage in the presence of low concentrations of ^{3}H-thymidine. Blastocyst cell number was also reduced, and this reduction seemed to begin at the 16- to 32-cell stage. Snow suggested that prospective ICM cells are selectively killed at the 16-cell stage and that ICM determination therefore occurs early in cleavage. However, he does not present clear evidence for actual cell death at the 16-cell stage, and the data are consistent with a retardation in cell division beginning at this stage. This is a known effect of ^{3}H-thymidine (Wimber, 1964). If the embryo then cavitated at the normal time in development despite a lower cell number, the "inside-outside" hypothesis would predict formation of blastocysts with selectively reduced ICM cell numbers. For example, a blastocyst containing 30 cells would be expected to contain only three to five inside cells (Barlow et al., 1972), which might be too few to produce a functional ICM following implantation (Sherman, 1975a; Snow, 1976; Rossant, 1976b).

Theoretical considerations suggest that determination occurs later in cleavage. At the 16-cell stage in normal embryos, there is, on average, only one inside cell, yet by the blastocyst stage (64 cells) there is an average of 11 (Barlow et al., 1972). Therefore, if the inside cell at the 16-cell stage is the only one giving rise to the ICM of the blastocyst, it must divide about twice as fast as the outside cells. However, inside cells divide only slightly faster than outside ones at this stage (Barlow et al., 1972), so that evidence favors late determination or progressive determination, with cells becoming determined as they become enclosed. These two possibilities cannot be distinguished until more is known of the potency of individual blastomeres after the 8-cell stage.

Also, little is known about the actual molecular mechanisms involved in producing ICM/trophectoderm determination. If the "inside-outside" hypothesis is correct, what are the factors that make the microenvironment of the inside cells different from that of the outside cells? Various suggestions have been made (see discussion following Gardner and Rossant, 1976), including the importance of cell junctions (Hastings and Enders, 1975; Ducibella et al., 1975; Ducibella and Anderson, 1975). It has been shown that focal tight junctions first appear at the morula stage (Calarco and Brown, 1969; Hastings and Enders, 1975; Ducibella et al., 1975). Formation of such junctions might produce a different ionic and molecular environment inside the morula and thus affect cell determination (McMahon, 1974). No proof of such a hypothesis has yet been found.

Potency of cells from the fifth-day blastocyst
Primitive endoderm and ectoderm   The potency of endoderm and ectoderm cells from the fifth-day mouse blastocyst has not yet been extensively studied. Endoderm cells appear morphologically differentiated (Enders, 1971), so it is possible that they are determined and can never produce ectoderm derivatives. Conversely, the primitive ectoderm cells, which do not differ morphologically from fourth-day ICM cells (Enders, 1971), might retain the capacity to produce primitive endoderm after the original separation of the two layers.
Mechanism and time of determination   Various pieces of evidence suggest that cell position may be important in the formation of primitive endoderm. As discussed in section 1.3.1, isolated fourth-day mouse ICMs inserted into zonae and placed in the oviduct (Rossant, 1975a) or grown in culture following immunosurgical isolation (Solter and Knowles, 1975) form a complete peripheral monolayer of endoderm cells. Further evidence for a position effect is provided by rat/mouse chimeras. When rat ICMs are injected into mouse blastocysts, they usually spread over the surface of the host ICM, in the position where endoderm would normally form (Gardner and Johnson, 1975). In analysis of chimeras later in development, endodermal chimerism is dominant. This seems to be a true position effect since rat ICMs can occasionally form both endoderm and embryonic ectoderm of separate egg cylinders inside the mouse trophoblast (see section 1.3.1) (Gardner

and Johnson, 1975). Thus it appears that cells that are on the outside of the ICM and are exposed to the blastocoelic, oviducal, or culture fluid become determined to form endoderm. More experiments are being performed to ensure that this is a true position effect and not selective migration of predetermined endoderm cells to the outside of the ICM. As with ICM/trophectoderm determination, nothing is known about the exact timing of determination of endoderm and ectoderm or about the molecular mechanisms involved.

Potency of cells from the postimplantation embryo
Endoderm and ectoderm   So far, transfer of tissues to ectopic sites has been the only method used to study the potency of postimplantation cells. Grobstein (1952) mechanically removed the primitive endoderm from the embryonic portion of pre-head process mouse embryos (c. seventh day) and, after a short culture period, transferred the remaining ectodermal fragments to the anterior chamber of the eye. In this site, the cells differentiated into cornified epithelium, stromal tissue, and gut epithelium suggesting that seventh-day ectoderm has the potential to form ectodermal, mesodermal, and endodermal derivatives. Recently, these results have been confirmed and extended by use of enzymatic separation of tissues from rat egg cylinders (Levak-Švajger et al., 1969). Renal homografts of primitive ectoderm isolated from bilayered rat embryos contained gut derivatives as well as ectodermal and mesodermal tissues (Levak-Švajger and Švajger, 1971), as found by Grobstein in the mouse. Pure endodermal grafts failed to proliferate. At the primitive streak stage, ectoderm isolated with attached mesoderm produced gut derivatives, ectodermal, and mesodermal tissues, but by the head-fold stage, isolated ectoderm no longer produced gut endoderm (Levak-Švajger and Švajger, 1974). At this stage, endoderm, when isolated with overlying mesoderm, differentiated into gut derivatives. Similar results have been obtained in the mouse (Diwan and Stevens, 1976).

These experiments on the potency of ectoderm and endoderm suggest that the definitive gut endoderm arises from the primitive ectoderm. The potential of ectoderm to form gut endoderm is lost by the head-fold stage. This is compatible with the suggestion of Jolly and Férester-Tadié (1936) that most, if not all, of the definitive endoderm of the embryo arises from the head process, a structure that develops between the primitive streak and head-fold stages. Such an origin for the definitive endoderm would then be analogous to the situation in the chick, for which there is considerable evidence that definitive endoderm also arises from the epiblast or ectoderm at the head process stage (Hunt, 1937a,b; Nicolet, 1967, 1970; Vakaet, 1962). The normal fate of the primitive endoderm from pre-head-fold stages cannot be established by these experiments since isolated endoderm failed to proliferate, but the blastocyst injection experiments in the mouse (Gardner and Papaioannou, 1975) suggest that it either degenerates (Jolly and Férester-Tadié, 1936) or forms extraembryonic endoderm. **Trophoblast** The primary trophoblast giant cells derived from the mural trophectoderm are indistinguishable in nuclear volume and phagocytic activity from the secondary giant cells that develop later from the ectoplacental cone (Zybina and Tikhomirova, 1963). The secondary giant cells undergo a process of fragmentation of the nucleus, which could represent segregation of the chromosomes into diploid nuclear units (Zybina and Tikhomirova, 1963; Zybina and Mos'yan, 1967). However, this process is related to the time of degeneration of the giant cells and there is no evidence that any giant cells ever return to a diploid, mitotically active state.

The potential to form giant cells is retained by the diploid derivatives of the polar trophectoderm. When isolated at early postimplantation stages, EPCs produce large numbers of giant cells either in ectopic sites (Grobstein, 1950; Simmons and Russell, 1962; Simmons and Weintraub, 1965; Billington, 1965; Clarke, 1969; Hunt and Avery, 1976) or in culture (Jenkinson and Billington, 1974; Sherman, 1975b). In addition, sixth- and seventh-day extraembryonic fragments can produce giant cells in ectopic sites (Grobstein, 1950; Gardner and Papaioannou, 1975; Gardner and Rossant, 1976; Diwan and Stevens, 1976). It is not known how long this potential is retained by diploid trophoblast cells.

It appears that the diploid trophoblast derived from the polar trophectoderm may continuously produce giant cells. Thus, the original distinction drawn between mural and polar trophectoderm based solely on cell position may not represent a distinct determinative event. Control over giant cell formation may stem from the ICM. Some workers have suggested that cells in contact with the ICM (or its later derivatives) continue to divide while cells removed from the ICM transform into giant cells (Gardner, 1971, 1972; Barlow and Sherman, 1972; Gardner et al., 1973; Ansell and Snow, 1974).

1.4 Prospects and Conclusions

Considerable advances have been made in establishing the fate map of
the early mouse embryo and in elucidating the early determinative
events in embryogenesis. However, many gaps in our knowledge are
apparent, and more basic experiments are required. What are the future lines of research into early embryogenesis likely to be? Extension
of studies of cell fate in the postimplantation embryo is an obvious
step. Implantation, however, renders the embryo inaccessible to many
manipulative procedures. Only limited success has been achieved with
the injection of cells into postimplantation embryos in utero (Weiss-
man, Papaioannou, and Gardner, in preparation). Ideally, experimental
manipulation of postimplantation stages requires the development of
methods of culturing such embryos that are as reliable and reproducible
as those available for culturing preimplantation stages (Whittingham,
1975). As a result of advances in this field, it is now possible to culture
rat embryos from the ninth to the eleventh days and for one to two
days between the tenth and fifteenth days (New, 1966, 1967, 1973;
Cockcroft, 1973, 1976; Steele, 1975). Some experiments have been performed with this system, particularly in teratogenesis (Morriss and
Steele, 1974; Steele et al., 1974), but cell fate experiments would be
difficult to interpret because development is rarely completely normal.
Embryos can also be cultured over the critical implantation period from
the blastocyst to the early egg cylinder (Jenkinson and Wilson, 1970;
Hsu, 1971, 1972, 1973; Spindle and Pedersen, 1973; Sherman, 1975b;
McLaren and Hensleigh, 1975) and even from 2-cell to early somite
stages (Hsu et al., 1974), but as yet development in this system is not
reproducible enough to give reliable information on cell fate.

Apart from studying the normal embryo, two other approaches to the
understanding of embryogenesis are being made. One is the attempt to
elucidate the effect of wild type genes in normal development by studying mutants that affect early mouse embryogenesis (McLaren, 1973; see
chapters 3 and 4). However, the role of these genes in controlling
specific developmental events is a matter of controversy.

Teratocarcinomas (see chapter 7) eventually might provide material
for studying and manipulating differentiation in vitro without the problems of low cell number and low viability of intact embryos in culture.
This is an interesting approach for the future, but the degree of corre-

spondence between teratocarcinoma cells and embryonic cells is not yet well defined. Biochemical studies of development (see chapter 2) will also be important in defining the cellular events of embryogenesis in molecular terms. Information from all these approaches will need to be combined to produce a clear picture of cell fate, potency, and determination in early embryogenesis.

References

ADAMS, C. E. (1965).
The influence of maternal environment on preimplantation stages of pregnancy in the rabbit. *In* Preimplantation Stages of Pregnancy, G. E. W. Wolstenholme and M. O'Connor, eds. (London: Churchill), pp. 345–376.

ANDREEVA, L. F. (1964).
Investigation of the synthesis of desoxynucleic acid and of the kinetics of cell populations in giant cells and in cell trophoblast of the placenta. *In* Research into Cell Cycles and Metabolism of Nucleic Acid during Cell Differentiation, L. N. Zhinkin and A. A. Zavarzin, eds. (Moscow-Leningrad: Nauka), pp. 136–147. (In Russian).

ANSELL, J. D., and SNOW, M. H. L. (1975).
The development of trophoblast *in vitro* from blastocysts containing varying amounts of inner cell mass. J. Embryol. Exp. Morph. *33*, 177–185.

ASSHETON, R. (1894).
A re-investigation into the early stages of the development of the rabbit. Quart. J. Micro. Sci. *37*, 113–164.

ASSHETON, R. (1898).
The segmentation of the ovum of the sheep, with observations on the hypothesis of a hypoblastic origin for the trophoblast. Quart. J. Micro. Sci. *41*, 205–262.

ASSHETON, R. (1908).
The blastocyst of *Capra*. Guy's Hospital Reports *62*, 209–239.

AUSTIN, C. R. (1961).
The Mammalian Egg. Oxford: Blackwells.

BARLOW, P. W., OWEN, D. A. T., and GRAHAM, C. F. (1972).
DNA synthesis in preimplantation mouse embryos. J. Embryol. Exp. Morph. *27*, 431–445.

BARLOW, P. W., and SHERMAN, M. I. (1972).
The biochemistry of differentiation of mouse trophoblast: studies on polyploidy. J. Embryol. Exp. Morph. *27*, 447–465.

BILLINGTON, W. D. (1965).
The invasiveness of mouse trophoblast and the influence of immunological factors. J. Reprod. Fertil. *10*, 343–352.

BONNEVIE, K. (1950).
New facts on mesoderm formation and proamnion derivatives in the normal mouse embryo. J. Morph. *86*, 495–545.

CALARCO, P. G., and BROWN, E. H. (1969).
An ultrastructural and cytological study of preimplantation development of the mouse. J. Exp. Zool. *171*, 253–284.

CALARCO, P. G., and EPSTEIN, C. J. (1973).
Cell surface changes during preimplantation development in the mouse. Dev. Biol. *32*, 208–213.

CERISOLA, H., and IZQUIERDO, L. (1969).
Mice embryogenesis and RNA distribution. Arch. Biol. Med. Exp. *6*, 10–16. (In Spanish).

CHAPMAN, V. M., WHITTEN, W. K., and RUDDLE, F. H. (1971).
Expression of paternal glucose phosphate isomerase-1 (Gpi-1) in pre-

8 2 0 9 3

LIBRARY
College of St. Francis
JOLIET, ILL.

implantation stages of mouse embryos. Dev. Biol. 26, 153–158.

CLARKE, A. G. (1969).
Factors affecting the growth of trophoblast transplanted to the testis. J. Reprod. Fertil. 18, 539–541.

COCKCROFT, D. L. (1973).
Development in culture of rat foetuses explanted at 12.5 and 13.5 days of gestation. J. Embryol. Exp. Morph. 29, 473–483.

COCKCROFT, D. L. (1976).
Comparison of in vitro and in vivo development of rat foetuses. Dev. Biol. 48, 163–172.

CONDAMINE, H., CUSTER, R. P., and MINTZ, B. (1971).
Pure-strain and genetically mosaic liver tumors histochemically identified with the β-glucuronidase marker in allophenic mice. Proc. Nat. Acad. Sci. USA 68, 2032–2036.

CONKLIN, E. G. (1905).
The organisation and cell lineage of the ascidian egg. J. Acad. Nat. Sci. (Philadelphia) 13, 5–119.

CRISTIANI, H. (1892).
L'inversion des feuillets blastodermiques chez le rat albinos. Arch. Phys. Norm. et Path. 5th Series. 4, 1–11.

DALCQ, A. (1957).
Introduction to General Embryology. Oxford: Oxford University Press.

DALCQ, A., and JONES-SEATON, A. (1949).
La répartition des éléments basophiles dans l'oeuf du rat et du lapin et son intérêt pour la morphologie. Bull. Clin. Sci. Acad. Roy. Belg. 35, 500–511.

DAVIDSON, E. H. (1968).
Gene Activity in Early Development. New York: Academic Press.

DICKSON, A. D. (1963).
Trophoblastic giant cell transformation of mouse blastocysts. J. Reprod. Fertil. 6, 465–466.

DICKSON, A. D. (1966).
The form of the mouse blastocyst. J. Anat. 100, 335–348.

DIWAN, S. B., and STEVENS, L. C. (1976).
Development of teratomas from the ectoderm of mouse egg cylinders. J. Nat. Cancer Inst. 57, 937–942.

DUCIBELLA, T., ALBERTINI, D. F., ANDERSON, E., and BIGGERS, J. D. (1975).
The preimplantation mammalian embryo: characterization of intercellular junctions and their appearance during development. Dev. Biol. 45, 231–250

DUCIBELLA, T., and ANDERSON, E. (1975).
Cell shape and membrane changes in the eight-cell mouse embryo: prerequisites for morphogenesis of the blastocyst. Dev. Biol. 47, 45–58.

DUVAL, M. (1891).
Le placenta des rongeurs. J. Anat. Physiol. Paris 27, 279–476.

DUVAL, M. (1895).
Etudes sur l'embryologie des Cheiropteres. J. Anat. Physiol. Paris 31 and 32.

ENDERS, A. C. (1971).
The fine structure of the blastocyst. In Biology of the Blastocyst, R. J. Blandau, ed. (Chicago: University of Chicago Press), pp. 71–94.

ENDERS, A. C., and SCHALFKE, S. J. (1965).
The fine structure of the blastocyst: some comparative studies. *In* Pre-implantation Stages of Pregnancy, G. E. W. Wolstenholme and M. O'Connor, eds. (London: Churchill), pp. 29–54.

ENDERS, A. C., and SCHALFKE, S. J. (1967).
A morphological analysis of the early implantation stages in the rat. Am. J. Anat. *120*, 185–226.

FORD, C. E., EVANS, E. P., and GARDNER, R. L. (1975).
Marker chromosome analysis of two mouse chimaeras. J. Embryol. Exp. Morph. *33*, 447–457.

FRASER, A. (1883).
On the inversion of the blastodermic layers in the rat and mouse. Proc. Roy. Soc. B *34*.

GARDNER, R. L. (1968).
Mouse chimaeras obtained by the injection of cells into the blastocyst. Nature *220*, 596–597.

GARDNER, R. L. (1971).
Manipulations on the blastocyst. Advan. Biosciences *6*, 279–296.

GARDNER, R. L. (1972).
An investigation of inner cell mass and trophoblast tissue following their isolation from the mouse blastocyst. J. Embryol. Exp. Morph. *28*, 279–312.

GARDNER, R. L. (1974).
Microsurgical approaches to the study of early mammalian development. *In* Birth Defects and Fetal Development: Endocrine and Metabolic Factors, K. S. Moghissi, ed. (Springfield, Ill.: Thomas), pp. 212–233.

GARDNER, R. L. (1975).
Analysis of determination and differentiation in the early mammalian embryo using intra- and inter-specific chimaeras. *In* The Developmental Biology of Reproduction, C. L. Markert and J. Papaconstantinou, eds. (New York: Academic Press), pp. 207–238.

GARDNER, R. L., and JOHNSON, M. H. (1973).
Investigation of early mammalian development using interspecific chimaeras between rat and mouse. Nature New Biol. *246*, 86–89.

GARDNER, R. L., and JOHNSON, M. H. (1975).
Investigation of cellular interaction and deployment in the early mammalian embryo using interspecific chimaeras between the rat and mouse. *In* Cell Patterning, R. Porter and J. Rivers, eds. (Amsterdam: Associated Scientific Publishers), pp. 183–200.

GARDNER, R. L., and LYON, M. F. (1971).
X-Chromosome inactivation studied by injection of a single cell into the mouse blastocyst. Nature *231*, 385–386.

GARDNER, R. L., and MUNRO, A. J. (1973).
Successful construction of chimaeric rabbit. Nature *250*, 146–147.

GARDNER, R. L., and PAPAIOANNOU, V. E. (1975).
Differentiation in the trophectoderm and inner cell mass. *In* The Early Development of Mammals, M. Balls and A. E. Wild, eds. (London: Cambridge University Press), pp. 107–132.

GARDNER, R. L., PAPAIOANNOU, V. E., and BARTON, S. C. (1973). Origin of the ectoplacental cone and secondary giant cells in mouse blastocysts reconstituted from isolated trophoblast and inner cell mass. J. Embryol. Exp. Morph. *30*, 561–572.

GARDNER, R. L., and ROSSANT, J. (1976). Determination during embryogenesis. *In* Embryogenesis in Mammals, K. Elliott and M. O'Connor, eds. (Amsterdam: Associated Scientific Publishers), pp. 5–25.

GARNER, W., and MCLAREN, A. (1974). Cell distribution in chimaeric mouse embryos, before implantation. J. Embryol. Exp. Morph. *32*, 495–503.

GEHRING, W. (1972). The stability of the determined state in cultures of imaginal disks in *Drosophila*. *In* The Biology of Imaginal Disks, H. Ursprung and R. Nöthiger, eds. (Berlin: Springer-Verlag), pp. 35–58.

GRAHAM, C. F. (1971). The design of the mouse blastocyst. *In* Control Mechanisms of Growth and Differentiation, D. D. Davies and M. Balls, eds. (London: Cambridge University Press), pp. 371–378.

GROBSTEIN, C. (1950). Production of intra-ocular hemorrhage by mouse trophoblast. J. Exp. Zool. *114*, 359–373.

GROBSTEIN, C. (1952). Intra-ocular growth and differentiation of clusters of mouse embryonic shields cultured with and without primitive endoderm and in the presence of possible inductors. J. Exp. Zool. *119*, 355–380.

HADORN, E. (1965). Problems of determination and transdetermination. Brookhaven Symp. *18*, 148–161.

HASTINGS, R. A., and ENDERS, A. C. (1975). Junctional complexes in the preimplantation rabbit embryo. Anat. Rec. *181*, 17–34.

HEAPE, W. (1886). The development of the mole (*Talpa europea*), the ovarian ovum, and segmentation of the ovum. Quart. J. Micro. Sci. *26*, 157–176.

HERBERT, M. C., and GRAHAM, C. F. (1974). Cell determination and biochemical differentiation of the early mammalian embryo. Current Topics Dev. Biol. *8*, 151–178.

HEUSER, C. H., and STREETER, G. L. (1928). Early stages in the development of pig embryos, from the period of initial cleavage to the time of the appearance of limb-buds. Contrib. Embryology *20*, 3–30.

HILLMAN, N., SHERMAN, M. I., and GRAHAM, C. F. (1972). The effect of spatial arrangement on cell determination during mouse development. J. Embryol. Exp. Morph. *28*, 263–278.

HOLTZER, H., WEINTRAUB, H., MAYNE, R., and MOCHAN, B. (1972). The cell cycle, cell lineages, and cell differentiation. Current Topics Dev. Biol. *7*, 229–256.

HSU, Y. C. (1971). Post-blastocyst differentiation *in vitro*. Nature *231*, 100–102.

HSU, Y. C. (1972).
Differentiation *in vitro* of mouse embryos beyond the implantation stage. Nature *239*, 200–202.

HSU, Y. C. (1973).
Differentiation of mouse embryos to the stage of early somite. Dev. Biol. *33*, 403–411.

HSU, Y. C., BASKAR, J., STEVENS, L. C., and RASH, J. E. (1974).
Development *in vitro* of mouse embryos from the two-cell egg stage to the early somite stage. J. Embryol. Exp. Morph. *31*, 235–245.

HUBER, G. C. (1915).
The development of the albino rat, *Mus norvegicus albinus*. I. From the pronuclear stage to the stage of mesoderm anlage; end of the first to the end of the ninth day. J. Morph. *26*, 247–358.

HUBRECHT, A. A. W. (1902).
Furchung und Keimblattbildung bei *Tarsius spectrum*. Verh. Kon. Akad. v. Wetensch., vol. 8. Amsterdam.

HUNT, C. V., and AVERY, G. B. (1971).
Increased levels of DNA during trophoblast giant-cell formation in mice. J. Reprod. Fertil. *25*, 85–91.

HUNT, C. V., and AVERY, G. B. (1976).
The development and proliferation of the trophoblast from ectopic mouse embryo allografts of increasing gestational age. J. Reprod. Fertil. *46*, 305–311.

HUNT, T. H. (1937a).
The development of gut and its derivatives from the mesectoderm and mesentoderm of early chick blastoderms. Anat. Rec. *68*, 349–369.

HUNT, T. H. (1937b).
The origin of entodermal cells from the primitive streak of the chick embryo. Anat. Rec. *68*, 449–459.

HUXLEY, J. S., and DE BEER, G. R. (1934).
Elements of Experimental Embryology. London: Cambridge University Press.

IZQUIERDO, L., and ORTIZ, M. E. (1975).
Differentiation in the mouse morulae. Wilhelm Roux' Archiv. *177*, 67–74.

JENKINSON, E. J., and BILLINGTON, W. D. (1974).
Differential susceptibility of mouse trophoblast and embryonic cells to immune cell lysis. Transplantation *18*, 286–289.

JENKINSON, E. J., and WILSON, I. B. (1970).
In vitro support system for study of blastocyst differentiation in the mouse. Nature *228*, 776–778.

JENKINSON, J. W. (1900).
A reinvestigation of the early stages of the development of the mouse. Quart. J. Micro. Sci. *43*, 61–82.

JENKINSON, J. W. (1902).
Observations on the histology and physiology of the placenta of the mouse. Tijdschr. d. Ned. Dierk. Veren. *7*, 124–198.

JENKINSON, J. W. (1913).
Vertebrate Embryology. Oxford: Clarendon Press.

JOLLY, J., and FÉRESTER-TADIÉ, M. (1936).
Recherches sur l'oeuf du rat et de la

souris. Arch. Anat. Micro. *32*, 323–390.

JONES-SEATON, A. (1950).
A study of cytoplasmic basophily in the egg of the rat and some other mammals. Ann. Soc. Roy. Zool. Belg. *80*, 76–86.

KAUFMAN, M. H., and GARDNER, R. L. (1974).
Diploid and haploid mouse partheno-genetic development following *in vitro* activation and embryo transfer. J. Embryol. Exp. Morph. *31*, 635–642.

KELLY, S. J. (1975).
Studies of the potency of early cleav-age blastomeres of the mouse. *In* The Early Development of Mammals, M. Balls and A. E. Wild, eds. (London: Cambridge University Press), pp. 97–106.

LEVAK-ŠVAJGER, B., and ŠVAJGER, A. (1971).
Differentiation of endodermal tissues in homografts of primitive ectoderm from two-layered rat embryonic shields. Experientia 27, 683–684.

LEVAK-ŠVAJGER, B., and ŠVAJGER, A. (1974).
Investigation on the origin of the de-finitive endoderm in the rat embryo. J. Embryol. Exp. Morph. *32*, 445–459.

LEVAK-ŠVAJGER, B., ŠVAJGER, A., and ŠKREB, N. (1969).
Separation of germ layers in pre-somite rat embryos. Experientia 25, 1311–1312.

LEWIS, W. H., and WRIGHT, E. S. (1935).
On the early development of the mouse egg. Contrib. Embryology 24, 133–144.

MCLAREN, A. (1973).
Embryogenesis. *In* Physiology and Genetics of Reproduction, Part B, E. M. Coutinho and F. Fuchs, eds. (New York: Plenum Press), pp. 297–316.

MCLAREN, A., and BOWMAN, P. (1969).
Mouse chimaeras derived from fusion of embryos differing by nine genetic factors. Nature 224, 236–240.

MCLAREN, A., and HENSLEIGH, H. C. (1975).
Culture of mammalian embryos over the implantation period. *In* The Early Development of Mammals, M. Balls and A. E. Wild, eds. (London: Cambridge University Press), pp. 45–60.

MCLAREN, A., and MICHIE, D. (1956).
Studies on the transfer of fertilised mouse eggs to uterine foster mothers. I. Factors affecting the implantation and survival of native and transferred eggs. J. Exp. Biol. 33, 394–416.

MCMAHON, D. (1974).
Chemical messengers in development: a hypothesis. Science *185*, 1012–1021.

MAYER, J. F., and FRITZ, H. I. (1974).
The culture of preimplantation rat em-bryos and the production of allo-phenic rats. J. Reprod. Fertil. *39*, 1–9.

MINTZ, B. (1962).
Formation of genotypically mosaic mouse embryos. Am. Zoologist *2*, 432.

MINTZ, B. (1964a).
Synthetic processes and early devel-opment in the mammalian egg. J. Exp. Zool. *157*, 85–100.

MINTZ, B. (1964b).
Formation of genetically mosaic
mouse embryos, and early development of "lethal (t^{12}/t^{12})-normal" mosaics. J. Exp. Zool. *157*, 273–292.

MINTZ, B. (1965).
Experimental genetic mosaicism in
the mouse. *In* Preimplantation Stages
of Pregnancy, G. E. W. Wolstenholme
and M. O'Connor, eds. (London:
Churchill), pp. 194–207.

MINTZ, B. (1967).
Gene control of mammalian pigmen-
tary differentiation. 1. Clonal origin of
melanocytes. Proc. Nat. Acad. Sci.
USA *58*, 344–351.

MINTZ, B. (1971).
Allophenic mice of multi-embryo ori-
gin. *In* Methods in Mammalian
Embryology, J. C. Daniel, ed. (San
Francisco: W. H. Freeman),
pp. 186–214.

MINTZ, B. (1974).
Gene control of mammalian differentiation. Ann. Rev. Genet. *8*, 411–470.

MINTZ, B., and BAKER, W. W.
(1967).
Normal mammalian muscle differentiation and gene control of isocitrate
dehydrogenase synthesis. Proc. Nat.
Acad. Sci. USA *58*, 592–598.

MORRISS, G. M., and STEELE,
C. E. (1974).
The effect of excess vitamin A on the
development of rat embryos in culture. J. Embryol. Exp. Morph. *32*,
505–514.

MOSSMAN, H. W. (1937).
Comparative morphogenesis of the fetal membranes and accessory uterine
structures. Contrib. Embryology *26*,
133–247.

MULNARD, J. (1955).
Contribution à la connaissance des
enzymes dans l'ontogenèse. Les
Phosphomonestérases acide et alcaline dans le développement du rat et
de la souris. Arch. Biol. *66*, 527–685.

MULNARD, J. (1961).
Problèmes de structure et d'organisation morphogénétique de l'oeuf mam-
mifère. *In* Symposium on the Germ
Cells and Earliest Stages of Development, S. Ranzi, ed. (Milan: Fonda-
zione A. Baselli, Instituto Lombardo),
pp. 639–688.

MULNARD, J. (1965).
Studies of regulation of mouse ova *in
vitro*. *In* Preimplantation Stages of
Pregnancy, G. E. W. Wolfstenholme
and M. O'Connor, eds. (London:
Churchill), pp. 123–138.

MULNARD, J. (1973).
Formation de blastocystes chimé-
riques par fusion d'embryons de rat et
de souris au stade VIII. C. R. Acad.
Sci. Paris *276*, 379.

MULNARD, J. (1974).
Les localisations de la phosphatase
alcaline dans les phases précoces du
développement embryonnaires du rat
et de la souris. Bull. Acad. Méd.
Belg. *129*, 677–697.

MYSTKOWSKA, E. T. (1975).
Development of mouse-bank vole
interspecific chimaeric embryos. J.
Embryol. Exp. Morph. *33*, 731–744.

MYSTKOWSKA, E. T., and
TARKOWSKI, A. K. (1968).
Observations on CBA-p/CBA-T6T6
mouse chimaeras. J. Embryol. Exp.
Morph. *20*, 33–52.

NADIJCKA, M., and HILLMAN, N. (1974).
Ultrastructural studies of the mouse blastocyst substages. J. Embryol. Exp. Morph. *32*, 675–695.

NAGL, W. (1972).
Giant sex chromatin in endopolyploid trophoblast nuclei of the rat. Experientia *28*, 217–218.

NEEDHAM, J. (1942).
Biochemistry and Morphogenesis. London: Cambridge University Press.

NESBITT, M. N. (1974).
Chimeras versus X-inactivation mosaics: significance of differences in pigment distributions. Dev. Biol. *38*, 202–207.

NESBITT, M. N., and GARTLER, S. M. (1971).
The applications of genetic mosaicism to developmental problems. Ann. Rev. Genet. *5*, 143–162.

NEW, D. A. T. (1966).
Development of rat embryos cultured in blood sera. J. Reprod. Fertil. *12*, 509–524.

NEW, D. A. T. (1967).
Development of explanted rat embryos in circulating medium. J. Embryol. Exp. Morph. *17*, 513–525.

NEW, D. A. T. (1973).
Studies on mammalian foetuses *in vitro* during the period of organogenesis. *In* The Mammalian Fetus *In Vitro*, C. R. Austin, ed. (London: Chapman and Hall), pp. 15–65.

NICHOLAS, J. S., and HALL, B. V. (1940).
Experiments on developing rats. II. The development of isolated blastomeres and fused eggs. J. Exp. Zool. *90*, 441–457.

NICOLET, G. (1967).
La chronologie d'invagination chez le poulet: étude à l'aide de la thymidine tritiée. Experientia *23*, 576–577.

NICOLET, G. (1970).
Analyse autoradiographique de la localisation des différentes ébauches presomptives dans la ligne primitive de l'embryon de poulet. J. Embryol. Exp. Morph. *23*, 79–108.

NÖTHIGER, R. (1972).
The larval development of imaginal disks. *In* The Biology of Imaginal Disks, H. Ursprung and R. Nöthiger, eds. (Berlin: Springer-Verlag), pp. 1–34.

PIGHILLS, E., HANCOCK, J. L., and HALL, J. G. (1968).
Attempted induction of chimaerism in sheep. J. Reprod. Fertil. *17*, 543–547.

PORTER, R., and RIVERS, J., eds. (1975).
Cell Patterning. Amsterdam: Associated Scientific Publishers.

POTTS, D. M., and WILSON, I. B. (1967).
The preimplantation conceptus of the mouse at 90 hours *post coitum*. J. Anat. *102*, 1–11.

RAUBER, H. (1875).
Die Entwickelung des Kaninchens. Sitzungsb. der Naturforsch. Gesellsch. zu Leipzig.

REINIUS, S. (1965).
Morphology of the mouse embryo from the time of implantation to mesoderm formation. Z. Zellforsch. *68*, 711–723.

ROBINSON, A. (1892).
Observations upon the development of the segmentation cavity, the archenteron, the germinal layers, and

the amnion in mammals. Quart. J.
Micro. Sci. *33*, 369–455.

ROBINSON, A. (1904).
Lectures on the early stages in the development of mammalian ova and on
the formation of the placenta in different groups of mammals. I. J. Anat.
Physiol. *38*, 1–19.

RODÉ, B., DAMJANOV, I., and
ŠKREB, N. (1968).
Distribution of acid and alkaline phosphatase activity in early stages of rat
embryos. Bulletin Scientifique, Con-
seil des Académies de la RPF, Yugo-
slavia, A, *13*, 304.

ROSSANT, J. (1975a).
Investigation of the determinative
state of the mouse inner cell mass. I.
Aggregation of isolated inner cell
masses with morulae. J. Embryol.
Exp. Morph. *33*, 979–990.

ROSSANT, J. (1975b).
Investigation of the determinative
state of the mouse inner cell mass. II.
The fate of isolated inner cell masses
transferred to the oviduct. J. Embryol. Exp. Morph. *33*, 991–1001.

ROSSANT, J. (1976a).
Investigation of inner cell mass determination by aggregation of isolated rat
inner cell masses with mouse morulae. J. Embryol. Exp. Morph. *36*,
163–174.

ROSSANT, J. (1976b).
Postimplantation development of blastomeres isolated from 4- and 8-cell
mouse eggs. J. Embryol. Exp. Morph.
36, 283–290.

RUGH, R. (1968).
The Mouse: Its Reproduction and
Development. Minneapolis: Burgess.

SCHLAFKE, S. J., and ENDERS,
A. C. (1967).
Cytological changes during cleavage
and blastocyst formation in the rat. J.
Anat. *102*, 13–32.

SELENKA, E. (1883).
Keimblatter und Primitivorgane der
Maus. Weisbaden.

SELENKA, E. (1884).
Die Blätter umkehrung im Ei der Na-
gathiere. Weisbaden.

SHERMAN, M. I. (1972).
Biochemistry of differentiation of
mouse trophoblast: esterase. Exp.
Cell. Res. *75*, 449–459.

SHERMAN, M. I. (1975a).
The role of cell-cell interaction during
early mouse embryogenesis. *In* The
Early Development of Mammals, M.
Balls and A. E. Wild, eds. (London:
Cambridge University Press),
pp. 145–166.

SHERMAN, M. I. (1975b).
Long term culture of cells derived
from mouse blastocysts. Differentiation *3*, 51–67.

SHERMAN, M. I., MCLAREN, A.,
and WALKER, P. M. B. (1972).
Mechanism of accumulation of DNA
in giant cells of mouse trophoblast.
Nature New Biol. *238*, 175–176.

SIMMONS, R. L., and RUSSELL,
P. S. (1962).
The antigenicity of mouse tropho-
blast. Ann. N. Y. Acad. Sci. *99*, 717.

SIMMONS, R. L., and
WEINTRAUB, J. (1965).
Transplantation experiments on pla-
cental aging. Nature *208*, 82–83.

SNELL, G. D., and STEVENS, L. C. (1966).
Early embryology. *In* Biology of the Laboratory Mouse, E. L. Green, ed. (New York: McGraw-Hill), 2nd ed., pp. 205–245.

SNOW, M. H. L. (1973a).
The differential effect of [^{3}H]-thymidine upon two populations of cells in pre-implantation mouse embryos. *In* The Cell Cycle in Development and Differentiation, M. Balls and F. S. Billett, eds. (London: Cambridge University Press), pp. 311–324.

SNOW, M. H. L. (1973b).
Abnormal development of pre-implantation mouse embryos grown in vitro with [^{3}H]thymidine. J. Embryol. Exp. Morph. *29*, 601–615.

SNOW, M. H. L. (1976).
The immediate postimplantation development of tetraploid mouse blastocysts. J. Embryol. Exp. Morph. *35*, 81–86.

SOBOTTA, J. (1903).
Die Entwichlung des Eies der Maus vom Schluss der Furchungsperiode bis zum Auftreten der Amniosfalten. Arch. Mikroskop. Anat. *61*, 274–330.

SOBOTTA, J. (1911).
Die Entwicklung des Eies der Maus vom ersten Auftreten des Mesoderms an bis zur Ausbildung der Embryoanlage und dem Auftreten der Allantois. Arch. Mikroskop. Anat. *78*, 271–352.

SOLTER, D., DAMJANOV, I., and ŠKREB, N. (1973).
Distribution of hydrolytic enzymes in early rat and mouse embryos—a reappraisal. Z. Anat. Entwicklungsgesch. *139*, 119–126.

SOLTER, D., and KNOWLES, B. B. (1975).
Immunosurgery of mouse blastocyst. Proc. Nat. Acad. Sci. USA *72* 5099–5102.

SOTELO, J. R., and PORTER, K. R. (1959).
An electron microscope study of the rat ovum. J. Biophys. Biochem. Cytol. *5*, 327–342.

SPINDLE, A. I., and PEDERSEN, R. A. (1973).
Hatching, attachment and outgrowth of mouse blastocysts *in vitro*: fixed nitrogen requirements. J. Exp. Zool. *186*, 305–318.

STEELE, C. E. (1975).
The culture of postimplantation mammalian embryos. *In* The Early Development of Mammals, M. Balls and A. E. Wild, eds. (London: Cambridge University Press), pp. 61–80.

STEELE, C. E., JEFFREY, E. H., and DIPLOCK, A. J. (1974).
The effect of vitamin E and synthetic antioxidants on the growth *in vitro* of explanted rat embryos. J. Reprod. Fertil. *38*, 115–123.

STERN, M. S. (1973).
Chimaeras obtained by aggregation of mouse eggs with rat eggs. Nature *243*, 472–473.

STERN, M. S., and WILSON, I. B. (1972).
Experimental studies on the organisation of the preimplantation mouse embryo. I. Fusion of asynchronously cleaving eggs. J. Embryol. Exp. Morph. *28*, 247–254.

TARKOWSKI, A. K. (1961).
Mouse chimaeras from fused eggs. Nature *190*, 857–860.

TARKOWSKI, A. K. (1964).
Patterns of pigmentation in experimentally produced mouse chimaerae.
J. Embryol. Exp. Morph. *12*, 575–585.

TARKOWSKI, A. K. (1965).
Embryonic and post-natal development of mouse chimaeras. *In* Preimplantation Stages of Pregnancy, G. E. W. Wolstenholme and M. O'Connor, eds. (London: Churchill), pp. 183–193.

TARKOWSKI, A. K., and WROBLEWSKA, J. (1967).
Development of blastomeres of mouse eggs isolated at the four- and eight-cell stage. J. Embryol. Exp. Morph. *18*, 155–180.

THEILER, K. (1972).
The House Mouse. Berlin: Springer-Verlag.

VAKAET, L. (1962).
Some new data concerning the formation of the definitive endoblast in the chick embryo. J. Embryol. Exp. Morph. *10*, 38–57.

VAN BENENDEN, E. (1875).
De la maturation d'oeuf, de la fécondation, et des premiers phénomènes embryonnaires chez les Mammiferes, d'apres les observations faites chez le lapin. Bull. Acad. Roy. Sci. Belg.

VAN BENENDEN, E. (1880).
Recherches sur l'embryologie des Mammiferes: la formation des feuillets chez le lapin. Arch. Biol. *1*.

VAN BLERKOM, J., BARTON, S. C., and JOHNSON, M. H. (1976).
Molecular differentiation in the preimplantation mouse embryo. Nature *259*, 319–321.

VOGT, W. (1925).
Gestaltungsanalyse am Amphienkeim mit örlicher Vitalfärbung. Vorwort über Wege and Ziele. I. Teil. Methodik and Wirkungweise der örlicher Vitalfärbung mit Agar als Farbträger. Wilhelm Roux Arch. Entw. Mech. Org. *106*, 542–610.

WEISS, P. (1939).
Principles of Development. New York: Holt.

WESTON, J. A. (1967).
Cell marking. *In* Methods in Developmental Biology, F. H. Wilt and N. K. Wessells, eds. (New York: T. Y. Crowell), pp. 723–736.

WHITTINGHAM, D. G. (1975).
Fertilization, early development and storage of mammalian ova *in vitro*. *In* The Early Development of Mammals, M. Balls and A. E. Wild, eds. (London: Cambridge University Press), pp. 1–24.

WILSON, E. (1925).
The Cell in Development and Heredity. 3rd ed. New York: Macmillan.

WILSON, I. B., BOLTON, E., and CUTTLER, R. H. (1972).
Preimplantation differentiation in the mouse egg as revealed by microinjection of vital markers. J. Embryol. Exp. Morph. *27*, 467–479.

WILSON, I. B., and STERN, M. S. (1975).
Organisation in the preimplantation embryo. *In* The Early Development of Mammals, M. Balls and A. E. Wild, eds. (London: Cambridge University Press), pp. 81–96.

WIMBER, D. E. (1964).
Effects of intracellular radiation with

tritium. Advan. Radiation Biol. *1*, 85–115.

ZEILMAKER, G. H. (1973).
Fusion of rat and mouse morulae and formation of chimaeric blastocysts. Nature *242*, 115–116.

ZYBINA, E. V. (1961).
Endomitosis and polyteny of trophoblast giant cells. Dokl. Akad. Nauk. SSSR *140*, 1177–1180. (In Russian).

ZYBINA, E. V. (1963a).
Cytophotometric determination of DNA content in nuclei of trophoblast giant cells. Dokl. Akad. Nauk. SSSR *153*, 1428–1431. (In Russian).

ZYBINA, E. V. (1963b).
Autoradiographic and cytochemical investigations of nucleic acids (DNA and RNA) in the endomitotic cycle of trophoblast giant cells. *In* Cell Morphology and Cytochemistry (Moscow: Nauka), pp. 35–44. (In Russian).

ZYBINA, E. V. (1970).
Anomalies of polyploidization of the cells of the trophoblast. Tsitologiya *12*, 1081–1091. (In Russian).

ZYBINA, E. V., and MOS'YAN, I. A. (1967).
Sex chromatin bodies during endomitotic polyploidization of trophoblast cells. Tsitologiya *9*, 265–272. (In Russian).

ZYBINA, E. V., and TIKHOMIROVA, M. M. (1963).
The question of endomitotic polyploidization in giant cells of the trophoblast. *In* Cell Morphology and Cytochemistry (Moscow: Nauka), pp. 53–63. (In Russian).

2 THE MOLECULAR BIOLOGY OF THE PREIMPLANTATION EMBRYO

Jonathan Van Blerkom and Cole Manes

2.1 Introduction

It is a tribute to the research enterprise currently directed toward the molecular biology of the early mammalian embryo that new chapters are being added to the unfolding story almost every year. We are aware of the numerous reviews of this subject in the recent literature (e.g., Biggers and Stern, 1973; Graham, 1973; Epstein, 1975; Manes, 1975; Schultz and Church, 1975), and these will be cited frequently. It is our primary purpose here to call attention to some quite recent developments in the field, to attempt to fit these developments into the framework of previously established concepts, and to identify what appear to be some important gaps in our present information. Although we have endeavored to include all the relevant information available as of the time of writing, it is probable that some reports have escaped us and will not receive the notice they deserve.

We also intend to devote what may seem to be a disproportionately large section to describing recent findings in the area of protein synthetic patterns in preimplantation embryos. We ask to be excused this bias. Many of the findings come from work done in our laboratory, and we hope the reader will share our excitement and enthusiasm for this subject.

It is becoming conventional in reviews of this sort to caution that the vast majority of our knowledge comes from the in vitro study of pre-implantation stages of mouse and rabbit embryos. Generalizations from these studies that claim to speak for all mammals, especially these residing in vivo, must therefore be presented tentatively. As will be discussed below, the emerging picture of embryo-uterine interactions casts the reproductive tract in a permissive role, one that *allows* the endogenous embryonic program of development to unfold, rather than one that in some important sense *directs* embryonic development by way of special "maternal factors." By ultrastructural and electrophoretic criteria, the initial stages of embryogenesis in vivo are closely paralleled in vitro (see section 2.3.4). It is also encouraging to note that as preliminary information begins to come in from studies on other species, the generalizations prompted by the mouse and rabbit work

Work by the authors reported in this paper was supported by a research grant from the National Institute of Child Health and Human Development, U. S. Public Health Service. This chapter summarizes the literature to June 1976.

have tended to gain support (Brinster, 1974). For these reasons, and without throwing caution entirely to the winds, we feel it is permissible now to generalize more boldly than conventional wisdom would have allowed only a few years back.

2.2 Nucleic Acid Metabolism

2.2.1 Deoxyribonucleic Acid

The consequences of fertilization
The transition of the ovulated egg from a "resting" cell, whose last round of DNA replication may have occurred some twenty to thirty years previously, to an "activated" cell that will resume DNA synthesis within a few hours is normally brought about by sperm penetration and is accompanied by a reversal in membrane potential (Cross et al., 1973). The role of this change in membrane potential in triggering the metabolic activities that lead to DNA synthesis is probably central; similar activation of the egg can be accomplished by a number of physical agents, e.g., temperature and osmotic and electric shock (Van Blerkom and Runner, 1976). The membrane potential of the mammalian egg, as with other eggs, appears to be closely related to the activity of various membrane transport systems (Epel, 1972).

Labeling of the pronuclei with ^{3}H-thymidine begins about three hours after sperm penetration (Mintz, 1964; Graham, 1973). It is not yet known whether there is an obligatory step involving RNA or protein synthesis prior to the resumption of DNA synthesis. Noronha et al. (1972) have demonstrated the appearance of ribonucleotide reductase activity in the sea-urchin egg within one hour after fertilization. Puromycin blocks this enzymatic activity, but Actinomycin D does not— facts suggesting that the peptide is synthesized off a maternally inherited but previously untranslated RNA template. By analogy, it may be expected that some of the initial enzymes active in the mammalian zygote will be derived from RNA templates synthesized before fertilization (see section 2.2.2).

Purine and pyrimidine biosynthesis
There is as yet no direct evidence that the mammalian zygote possesses the host of metabolic pathways for purine and pyrimidine synthesis, riboside synthesis and reduction, nucleoside phosphorylation,

and trinucleotide polymerization. The fertilized eggs of both mouse and rabbit will undergo multiple rounds of cell division in vitro in the absence of an exogenous source of nucleic acid bases (Whitten, 1971). Unless the egg stores these precursors in excess prior to fertilization, this fact alone constitutes presumptive evidence for the existence of these pathways. By recent accounts, there is no "excess" of DNA polymer in the mature mouse egg (Pikó and Matsumoto, 1976). Its DNA content can be accounted for totally by the diploid complement of nuclear DNA—about 5–6 pg—as expected in eggs prior to the second meiotic division and whose first polar bodies have been removed, plus about 2–3 pg of mitochondrial DNA per egg. Nor does there appear to be a large precursor pool in the fertilized egg; as Mintz (1964) pointed out, the ready incorporation of exogenous thymidine by the zygote and early embryo is at least indirect evidence against such a pool.

Graves and Biggers (1970) demonstrated furthermore that 1-cell mouse embryos exposed to ^{14}C-bicarbonate will incorporate label into material that appears to be nucleic acid. It was not determined in that study whether the ^{14}C became incorporated into the nucleic acid bases themselves or into the sugar moiety. Nonetheless, the collective findings indicate that the mammalian zygote and early embryo rely upon an environmental supply of only simple substrates from which to construct their macromolecules.

DNA polymerases
Although considerable attention has been devoted to different molecular forms of RNA polymerase in recent years (see section 2.2.2), a similar delineation of functional roles for the DNA polymerases has not been forthcoming. It is evident that in some cells DNA synthesis is associated not only with replication of the genome but also with DNA repair and with gene amplification. It is thus a possibility that these functions will ultimately be found to reside with molecularly distinct enzymes.

Hoage and Cameron (1975) recently raised the possibility of gene amplification during mammalian oogenesis. By autoradiography they have detected ^{3}H-thymidine incorporation in the nuclei of maturing mouse oocytes. This incorporation appears to be primarily juxtanucleolar during the period of most rapid growth in the oocyte, and Hoage and Cameron suggest that it might reflect amplification of ribosomal RNA genes. To our knowledge, there is as yet no indisputable evidence for gene amplification in any cell except the oocyte, although it has been invoked as a potential mechanism of differentiation in somatic cells.

Pedersen and Cleaver (1975) recently demonstrated the "unscheduled" synthesis of DNA associated with repair activity in the mouse morula and blastocyst. The repair activity is inherited in Mendelian fashion in humans (Cleaver, 1968), and it does not appear to be essential for early embryogenesis, because children homozygous for the deficiency are known clinically. It is not known whether a higher-than-average level of embryonic and fetal wastage is associated with the homozygous deficient condition. The standard method for demonstrating the enzyme activity in cells is to expose them to ultraviolet light to induce chromosomal breakage, then to assay for thymidine incorporation into otherwise nonreplicating DNA. As Pedersen and Cleaver (1975) point out, the mouse embryo in vivo is not normally exposed to ultraviolet light, and the presence of the repair enzyme in these early stage embryos suggests that other forms of DNA perturbation might be a potential problem. In this light, the recent finding by Mukai et al. (1976) that malonaldehyde is mutagenic in bacteria might be relevant. Malonaldehyde is a commonly formed compound in cells utilizing molecular oxygen as an electron receptor in biological oxidations. The lipid peroxidation that occurs under these conditions and the malonaldehyde formation (Demopoulos, 1973) that follows might well pose a threat to the DNA of the mammalian blastocyst, given its high rate of oxygen utilization (Brinster, 1974).

Nothing yet is known regarding the activities or subcellular distribution of the α and β DNA-dependent DNA polymerases in the pre-implantation mammalian embryo. Sherman and Kang (1973) assayed these activities as well as an RNA-dependent DNA polymerase in the eleventh-day mouse embryo, trophoblast, and uterine decidua. Although the specific activity of all enzymes was considerably higher in embryonic tissue than in trophoblast or decidua, they found no evidence of unique enzymes in any of these tissues or any significant differences in the proportion of the various enzymes or in their subcellular distribution. They suggest that the various DNA polymerase activities are coordinately controlled and are highest in rapidly replicating and dividing cells.

Of considerable interest is the recent report that mammalian sperm heads contain an RNA-dependent DNA polymerase as well as an endogenous RNA primer (Witkin et al., 1975). The reaction requirements of this enzyme and its stimulation by the synthetic template $dT_{12-18} \cdot$ poly(rA) are remarkably similar to those of the enzyme obtained from intracisternal A-type viruslike particles in mouse plasmacytoma and from oncornaviruses (Yang and Wivel, 1974). Although morphologically characteristic A-type viruslike particles appear and disappear at specific developmental stages in preimplantation embryos (Calarco and Szollosi, 1973; Chase and Pikó, 1973; figure 2.1), the embryonic particles have not been characterized biochemically (see chapter 6). That these particles do not depend upon sperm penetration of the egg is evident from the fact that they also appear in parthenogenetically activated eggs (Biczysko et al., 1974; Van Blerkom and Runner, 1976). Nonetheless, it is important to examine potential roles for the sperm enzyme in normal early embryogenesis, particularly in view of the recent report that "fertilization" of somatic cells by heterologous sperm elicits the appearance of fetal antigens in these cells (Higgins et al., 1975).

Modifications of DNA
At the time of this writing, there is no information regarding "modifications" of embryonic DNA during the early development of mammals. In the late cleavage stage of the sea urchin—a stage that precedes the functional, or gene-directed, cell differentiation at gastrulation by several cell divisions—Scarano et al. (1965) have found 5-methyl cytosine in the embryonic DNA. In view of recent proposals that methylation and other modifications of DNA could result in functional "restrictions" of portions of the genome (Holloday and Pugh, 1975; Sager and Kitchin, 1975), a similar examination of early mammalian DNA might be fruitful. It should be noted that such functional "restriction" of one X chromosome in female mammals occurs either during the preimplantation period (Issa et al., 1969) or about the time of implantation (Epstein, 1972). In marsupials this restriction does not appear to be random and involves only the paternally inherited X chromosome. This characteristic strongly suggests that some form of prior DNA modification must be occurring (Sager and Kitchin, 1975; see chapter 3).

It is worth noting also that resistance to gamma radiation in bacteria is associated with an absence of methylated bases in the bacterial DNA (Schein et al., 1972). There are variations in gamma radiation sensitiv-

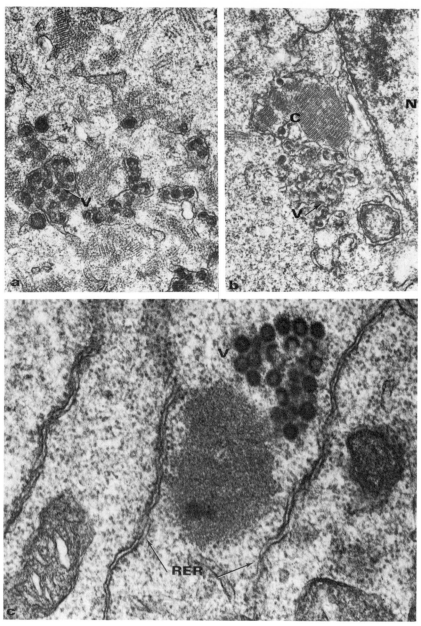

2.1 Intracisternal A-type particles clustered within elongated and distended cisternae of the endoplasmic reticulum (V) of a 2-cell mouse embryo. × 34,500. (b) Cluster of A-type particles (V) adjacent to the nuclear membrane (N) and a crystalloid body (C) in a 4-cell mouse embryo. × 33,500. (c) Group of intercisternal A-type particles (V) located between cisternae of the rough-surfaced endoplasmic reticulum (RER) in a trophectoderm cell of a 108-hr rabbit embryo. × 60,000.

ity during the early embryonic development of the mouse (Goldstein et al., 1975) that might reflect differences in DNA modification.

Mitochondrial DNA and maternal inheritance
The ancestry of mitochondrial DNA in the mammalian embryo has been a matter of some dispute because no clear examples of maternal inheritance are known in mammals, as would be expected if embryonic mitochondria are virtually all from the oocyte. Even a small contribution of mitochondria from the single sperm that enters the egg cytoplasm at fertilization (Szollosi, 1965) would be numerically overwhelmed by the vast numbers of mitochondria present in the egg (Pikó and Matsumoto, 1976), unless sperm mitochondria enjoyed some replicative advantage. The issue now appears to be settled, at least in one case: Hutchinson et al. (1974) have examined the mitochondrial DNA in the offspring of reciprocal crosses between the horse and donkey. The characteristic fragments produced by restriction endonuclease digestion of the mitochondrial DNA clearly show that it is predominantly, if not entirely, of maternal origin. Thus, maternal or "cytoplasmic" inheritance in mammals appears to be a distinct possibility.

2.2.2 Ribonucleic Acid

General background
The questions regarding the rate and extent of genetic transcription in the early mammalian embryo, the identification of the products of this transcription, and the relationship of this transcriptive activity to the morphogenetic and differentiative events of embryogenesis have been discussed at length in several recent reviews (Epstein, 1975; Manes, 1975; Schultz and Church, 1975) and will not be taken up again here. Unlike DNA synthesis, RNA synthesis has been detected only after the first cleavage. Its detection in the zygote might be a problem of precursor uptake, because all studies to date rely upon the incorporation of exogenous radiolabeled nucleoside into an acid-insoluble product. Although the extent of the endogenous nucleotide pool is unknown and although there are changes in uptake of exogenous precursor during development, it seems to be agreed that there is at least a modest increase in the rate of RNA synthesis at about the time of nucleolar activation (Tasca and Hillman, 1970; Epstein and Daentl, 1971; Karp et al., 1973). Following nucleolar activation, the embryo synthesizes all

the major classes of RNA. Full activation of the nucleolus occurs no later than the 4-cell stage in the mouse and by the 64-cell stage in the rabbit (figure 2.2). Low levels of ribosomal gene transcription, perhaps with a high rate of ribosomal precursor RNA degradation, apparently occur at earlier stages (Schultz, 1973; Clegg and Pikó, 1975).

The accumulated evidence points to the transcriptional activation of the embryonic genome within the first few cleavage divisions. However, the exact role of these early gene products in determining the observed patterns of protein synthesis and ultrastructural morphology is not yet clear (see section 2.3.4).

Inherited RNA
Although the mammalian egg at fertilization is relatively small in comparison to eggs of lower vertebrates, it is still many times larger than the average somatic cell and provides the embryo with its total protoplasmic mass during the cleavage divisions that ensue over the next two to three days. The unfertilized egg contains all the requisite RNA species for protein synthesis, and active protein synthesis is in fact detectable in the egg prior to fertilization (see section 2.3.2). Burkholder et al. (1971) presented evidence for a crystalline lattice of ribosomes in the mouse egg, which might be a form of storage for "excess" ribosomes made during oocyte maturation because these structures disappear shortly after fertilization. Bachvarova (1975) found that 20 percent of the ribosomes in the ovulated mouse ovum sediment in a sucrose gradient as structures heavier than 80S monosomes and remain undissociated in 0.5 M KCl, salt conditions that normally cause ribosomes to dissociate into subunits. The mechanism by which these ribosomes become available for postfertilization protein synthesis is not known. Young et al. (1973) confirmed the presence of ribosomal RNA in the unfertilized mouse egg and, in addition, identified presumptive transfer RNA. Still, these two species do not account for the total RNA content of the egg. Schultz (1975), using the technique of hybridization to ^{3}H-polyuridylic acid, identified and quantitated RNA containing poly(A) sequences in the unfertilized rabbit egg. The amount of this presumptive messenger RNA does not change within the first ten hours after fertilization; almost half of this RNA is unattached to ribosomes before fertilization, and within ten hours, this free fraction drops to about one-third. Such studies have not yet been carried out on mouse eggs, but it is probable that the results will be similar.

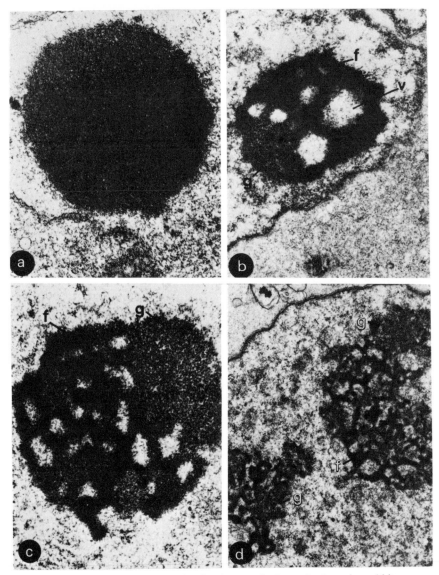

2.2 Four successive stages of nucleolar development in the preimplantation rabbit embryo. (a) The nucleoli of a 12-hr embryo are electron-dense, spheroidal structures composed of a highly compacted, fibrillar material. × 48,000. (b) Characteristic of nucleoli in 24–36-hr embryo is the presence of vacuoles (v) within the fibrillar matrix (f) and the appearance, for the first time during the preimplantation period, of an eccentrically located granular component (g). × 31,500. (c) By 48 hr, nucleoli have a reticulated appearance and contain distinct granular (g) and fibrillar (f) elements. × 38,000. (d) At 60 hr, nucleoli are extremely hypertrophic and extensively reticulated structures in which granular (g) and fibrillar (f) material is distributed along anastomosing networks of nucleolonemas. × 13,500. From Van Blerkom et al. (1973); reprinted with permission from *Developmental Biology*.

Modifications of embryonic RNA

Because it is becoming clear that the functional role of genetic transcripts is in large part determined by posttranscriptional modifications
of the RNA molecule, evidence for such modifications during early
mammalian embryogenesis might be essential to explain gene expression during this period. One such modification is the addition of methyl
groups to either the purine or pyrimidine bases or to the 2'-OH position
of the ribose molecule. Manes and Sharma (1973) have analyzed the
methylation of purine and pyrimidine bases in the tRNA of the preimplantation rabbit embryo by means of thin-layer chromatography. In
the late morula–early blastocyst (the earliest stage studied), they detected all of the commonly formed methylated bases, except 6-methyl-
adenine. This base was, however, detectable at low levels in the later
blastocyst. Thus, a number of highly specific tRNA methyltransferases
are active in the embryo at this stage.

The total in vivo methylation of tRNA in the preimplantation rabbit
embryo decreases following blastocyst formation (Manes and Sharma,
1973). It is not known whether this decrease is due to the accumulation
in blastocysts of a competitive substrate for the methyltransferases
(Kerr and Borek, 1972) and whether it also occurs in embryos reared in
vitro. Recently, Clandinin and Schultz (1975) have found that methionyl-tRNA becomes increasingly methylated following blastocyst
formation in rabbit embryos; an increased affinity for ribosomes accompanies its increased methylation. Presumably, these two reports do
not conflict, because the former assayed total tRNA and the latter the
specific methionyl-tRNA.

The recent interest in modifications of messenger RNA via guanine
methylation at the 5' terminus (Both et al., 1975) has prompted speculation that this mechanism might be involved in posttranscriptional regulation of protein synthesis. It will be of interest to see whether this
mechanism is involved in the changing patterns of protein synthesis
seen during early cleavage in mammalian embryos (see section 2.3.2),
during the period of relative ribosome scarcity. The answer to this
question might require the use of embryonic messenger RNA in cell-
free translation systems.

The synthesis and modification of messenger RNA at the 3' terminus
by the addition of a polyadenylic acid sequence occurs in preimplantation rabbit embryos as early as the 16-cell stage (Schultz et al., 1973)
and in the mouse embryo as early as the 2-cell stage (Clegg and Pikó,

1975). These RNAs rapidly become associated with polysomes in the cleaving rabbit embryo (Schultz, 1973) and in the rabbit blastocyst have half-lives as long as 18 hr (Schultz, 1974). Thus, templates for protein synthesis appear to be supplied by the embryonic genome very early during embryogenesis.

RNA processing
This topic is approached here primarily to point out that virtually nothing is known regarding the processing of nuclear RNA during early mammalian development. Gillespie and Gallo (1975) have presented evidence supporting the hypothesis that the "usual" mechanisms for processing high-molecular-weight nuclear RNA into relatively low-molecular-weight cytoplasmic RNA might not be completely operative in embryonic and tumor cells. These authors term this aberrant mode "paraprocessing" and suggest that the large 60–70S cytoplasmic viral RNAs found in many tumor cells might arise in this manner from nuclear transcripts. Clegg and Pikó (1975) found a newly synthesized 55S RNA that is polyadenylated in the preimplantation mouse embryo and becomes especially prominent at the 4- to 16-cell stage when intracisternal A-type particles are visible. They did not determine whether this 55S RNA is cytoplasmic or nuclear. Slater et al. (1973) found that transcripts of this size, which become polyadenylated in the cytoplasm shortly after fertilization of the sea-urchin egg, are synthesized before, rather than after, fertilization. The relationship of the 55S RNA in the mouse embryo to A-type particles is not known. Genetic transcripts of this size used for postfertilization protein synthesis presumably would be polycistronic unless they contained long nongenetic sequences.

The appearance of A-type particles in the mouse oocyte, their disappearance as oocytes mature, and reappearance at the 2-cell stage of embryogenesis (Calarco and Szollosi, 1973; figure 2.1) are all curiously correlated with nucleolar activity. This correlation could be purely fortuitous, but it also might contain a clue to the significance of paraprocessing, perhaps in association with extraordinarily rapid transport of genetic transcripts out of the nucleus.

RNA polymerases
Versteegh et al. (1975) have identified the three major classes of DNA-dependent RNA polymerases in preimplantation mouse embryos. These authors employed an in vitro incorporation assay for RNA polymerase

activity under varying concentrations of ammonium sulfate and α-amanitin. Although the 2-cell mouse embryo incorporates uridine into acid-insoluble product, no RNA polymerase activity was detectable in vitro at that stage. At all later stages, activity was detectable and became increasingly insensitive to inhibition by α-amanitin at the morula and blastocyst stages, presumably reflecting ribosomal RNA synthesis by RNA polymerase I. In view of the RNA synthetic capacities previously determined in the preimplantation mouse embryo (Ellem and Gwatkin, 1968; Woodland and Graham, 1969; Pikó, 1970), these results confirm the roles assigned to the various RNA polymerases (Roeder and Rutter, 1970a).

Nonetheless, it is difficult to relate the various RNA polymerase activities observed in vitro to those occurring in the intact cell. Roeder and Rutter (1970b) found only a broad correspondence between RNA polymerase activities in isolated nuclei from sea-urchin embryos and the transcriptive activity known to be occurring in intact embryos. The binding of the polymerase molecule to its chromatin template in vivo could be regulated by factors that are not apparent under in vitro assay conditions (Tocchini-Valentini and Crippa, 1970; Chesterton et al., 1975). Another problem in detecting RNA synthesis in preimplantation embryos by the incorporation of uridine into acid-insoluble product stems from the ready conversion of this base to cytidine. In this laboratory, we find that the preimplantation rabbit embryo at all stages can efficiently aminate uridine to cytidine and incorporate the latter base into DNA. Although Woodland and Graham (1969) were unable to detect this conversion in the mouse embryo, it seems likely to be a problem there also, because the only biosynthetic route known for cytidine synthesis is by way of uridine amination. Thus, unless special precautions are taken, such as including deoxycytidine in the labeling medium, the acid-insoluble incorporation of radiolabeled uridine by preimplantation embryos cannot be taken as de facto evidence of RNA synthesis.

Bernstein and Mukherjee (1972), using cell-fusion techniques and autoradiography to investigate the low level of RNA synthesis in the 2-cell mouse embryo, have demonstrated apparently inhibitory properties in the cytoplasm of these embryos. Moore et al. (1975) have found that follicular fluid (albeit bovine) inhibits DNA-dependent RNA synthesis in ascites tumor cells. It is tempting to speculate that this inhibitory material is passively acquired by the egg and transmitted to the early stage embryo, to be either metabolized or diluted during early cleavage.

2.3 Protein Synthesis

As the ultimate expression of genetic information, proteins are molecular determinants of both cell structure and function. A detailed knowledge of the complexity and diversity of translation during early mammalian embryogenesis, as well as the identification of stage- and tissue-specific gene products, is central to an understanding of how differential gene expression enables the embryo to develop progressively. Recent advances in the in vitro culture of preimplantation embryos, as well as the development of high-resolution electrophoretic techniques for the analysis of the products of translation have contributed to an enormous increase in our knowledge of protein synthetic activity during early development. In the past, most of the studies of embryonic translation have addressed questions of metabolism and changes in enzyme activity during development. Because this particular topic has been discussed in detail in several reviews, it will not be presented here. For information relevant to embryonic metabolism and enzyme activities, the reader's attention is directed to the excellent and comprehensive reviews of Biggers and Stern (1973) or of Brinster (1973, 1974). In this section we will consider primarily quantitative and qualitative changes in the pattern of protein synthesis both in the mature oocyte and during the preimplantation stages of development of the rabbit and mouse embryo.

Most of the data concerning embryonic translation are derived from the study of embryos reared in vitro. Thus, it becomes rather fundamental to question whether embryos cultured in vitro develop in a fashion comparable to their in vivo counterparts. Evidence will be presented which demonstrates that in vivo and in vitro development are strictly comparable at both the molecular and cellular levels. Because the preimplantation development of the mammalian embryo occurs in an "open-system," i.e., the oviduct and uterus, it is important to consider whether the physical and chemical environment supplied by the reproductive tract functions in the regulation or control of embryonic-protein synthesis. In examining this question, two systems will be discussed: (1) protein synthetic activity following the termination of interrupted pregnancy (facultative delayed implantation), as it occurs in the mouse, and (2) protein synthetic activity during normal or uninterrupted pregnancy, as it occurs in the rabbit. Much of the available information concerning these questions is still incomplete and, as will become evident, requires continued research. However, the study of protein synthesis during the preimplantation period has revealed important insights into the nature of differential gene expression during early development and has indicated endogenous and exogenous mechanisms by which it might be regulated.

2.3.1 Quantitative Aspects of Protein Synthesis during the Preimplantation Period

In 1964, Mintz reported that preimplantation embryos are engaged in the synthesis of protein. Monesi and Salfi (1967) were the first to apply direct radioactive counting techniques to the quantitation of embryonic protein synthesis. However, as Epstein and Smith (1973) noted, the rate of embryonic protein synthesis is a complex function into which such parameters as rates of uptake, internal pool size, and differential amino acid transport must be taken into account. Simply measuring incorporation of isotope into protein cannot serve as an independent measure of synthesis—uptake and incorporation are not necessarily proportional. Most studies of the rate of embryonic protein synthesis are based on the assumption that the endogenous amino acid pool is very small relative to the amount of amino acid being taken up from the medium (Epstein and Smith, 1973). The only comprehensive, direct measurement of total endogenous amino acid pools is the work of Petzoldt et al. (1973). Their observations of preimplantation rabbit embryos demonstrate quantitative changes in total amino acids, as well as variation in the proportion of several amino acids, one to another, during the preimplantation period. Because the rate of protein synthesis is a function of uptake as well as internal pool size, Epstein and Smith (1973) attempted to distinguish between an energy-dependent, active, amino acid transport system and a non-energy-requiring, facilitated process. Their evidence suggests that during the preimplantation development of the mouse, the uptake of at least leucine and lysine involves a facilitated process such that inhibitors of energy metabolism do not affect uptake. Although Epstein and Smith (1973) were unable to demonstrate a significant effect on uptake of lowered exogenous concentrations of sodium, Borland and Tasca (1974) have recently shown the existence of a sodium-dependent amino acid transport system in the mouse. Their results indicate that whereas leucine and methionine transport are sodium independent during early cleavage and methionine transport by the morula is partially sodium dependent, the transport of both amino acids at the blastocyst stage is completely sodium dependent. Thus, it becomes apparent from these studies that changes in the availability of exogenous ions, such as sodium, might be potential factors in the quantitative regulation of protein synthesis at each stage of development. As Borland and Tasca (1974) have suggested, the activation of a sodium-dependent amino acid transport system in mouse embryos could be part of the overall preparation during late cleavage for the increased protein synthesis and growth that are initiated at the blastocyst stage. The use of electron-probe microanalysis to carefully and precisely determine both endogenous and exogenous pools of ions during development (Borland et al., 1976) should provide information relevant to possible mechanisms involved in the control of translation.

A second potential factor in the regulation of embryonic translation is the availability and metabolism of ATP. Ginsberg and Hillman (1975a) demonstrated three major shifts in ATP synthesis during the preimplantation stages of mouse development: (1) between the 2-cell and late 4-cell, (2) between the 8-cell and late morula, and (3) between the late morula and late blastocyst stages. This observation is especially important in the light of evidence indicating altered rates of ATP metabolism in mouse embryos carrying two of the recessive lethal T-complex alleles (Ginsberg and Hillman, 1975b; Hillman, 1975; see chapter 4). The studies of Ginsberg and Hillman indicate that the lethality that occurs in early t^{w32}/t^{w32} morulae and in late t^{12}/t^{12} morulae is accompanied by excessive rates of ATP synthesis up to the cleavage stage immediately preceding the stage of lethality, with a subsequent drop in ATP synthesis to levels below that observed in controls (wild type). It has yet to be determined whether alterations in the synthesis of ATP by these mutants affect rates of amino acid uptake and/or incorporation. However, these findings do point out the potential role of ATP synthesis in the regulation of amino acid uptake and incorporation.

Although the parameters that ultimately determine the absolute rate of embryonic protein synthesis require additional study, the results of numerous observations offer a general insight into quantitative changes in protein synthetic activity during early development. For example, studies of protein synthesis during mouse development indicate a relatively low and constant rate of incorporation (of several different amino acids in a variety of culture media) during early cleavage and a marked increase at the blastocyst stage (Monesi and Salfi, 1967; Tasca and Hillman, 1970; Brinster, 1971, 1973; Epstein and Smith, 1973; Van Blerkom and Brockway, 1975a). Similar findings have been reported for the rabbit (Manes and Daniel, 1969; Karp et al., 1974). At the subcellular level, increases in the rate of protein synthesis occur concurrently with the maturation of the nucleolus (figure 2.2) and the appearance of a dense population of cytoplasmic ribosomes and polysomes (Hillman and Tasca, 1969; Van Blerkom et al., 1973; Van Blerkom and Runner, 1976; see also figure 2.11a). In addition, Manes (1971) has shown that a "burst" of rRNA synthesis occurs at the initiation of blastocyst formation in the rabbit. Collectively, these observations indicate that the availability of rRNA and ribosomes could have a fundamental role in the timing and regulation of the onset of elevated rates of protein synthesis in preimplantation mammalian embryos. In this regard, the work of Bachvarova (1976) is extremely interesting because she has shown that most ribosomes in mouse oocytes prior to ovulation are inactive in protein synthesis and apparently are stored for use during embryonic development. In addition, it was reported that a large fraction of egg ribosomes are present in a dissociated state (40S and 60S particles) suggesting a distinct difference from ribosomes in other eukaryotic cells. The elucidation of mechanisms that govern the timing of nucleolar maturation (i.e., appearance of a granular element), rRNA transport from the nucleus, and the assembly of ribosomes active in protein synthesis should provide specific clues to the quantitative regulation of protein synthesis in the embryo.

2.3.2 Qualitative Aspects of Protein Synthesis during the Preimplantation Period

In the past, the analysis of qualitative patterns of protein synthesis has been severely restricted by the relatively enormous numbers of embryos required and the comparatively poor electrophoretic separation of proteins on polyacrylamide gels. Manes and Daniel (1969) were the first to study qualitative aspects of embryonic protein synthesis. Their evidence indicated that (1) proteins synthesized by late blastocyst stage rabbit embryos were not identical to proteins synthesized by cleavage

stage embryos and (2) the pattern of protein synthesis of 84-hr blastocysts[1] more closely resembled the pattern of cleavage stage embryos
than it did that of 132-hr blastocysts. Recent advances in the detection
of proteins by polyacrylamide gel electrophoresis have resulted in an
increase of approximately two orders of magnitude in the resolving
ability of gels compared to procedures previously available. As a consequence of this increase in resolution, many of the interpretations of
earlier results (Manes and Daniel, 1969; Petzoldt, 1972; Petzoldt et al.,
1972; Petzoldt, 1975) have to be reevaluated. The information obtained
from these analyses has led to an enormous increase in the ability to
view the complexity and diversity of translation, as well as in the
definition of "stage- and tissue-specific" proteins. Although much of
this work is, at present, a "mapping-function," it is nevertheless an
important prerequisite in the understanding of differential gene expression during early embryogenesis. At the outset, it must be noted that in
the discussion of qualitative changes, we are concerned with radioactive bands or spots that are detectable at one stage of development and
not at another. Whether these differences are genuine qualitative
changes or merely represent peptides present in one case in amounts
below the resolving ability of the system is occasionally difficult to
determine. This caveat must be kept in mind in the interpretation of
complex electrophoretic patterns of protein synthesis.

One-dimensional gel electrophoresis
The application of sodium dodecyl sulfate (SDS), polyacrylamide gel
electrophoresis (PAGE), followed by autoradiography, has greatly
facilitated the visualization of the products of translation. In our laboratory, we have employed an SDS gradient, slab gel system which permits the side-by-side comparison of proteins synthesized at different
developmental stages (Van Blerkom and Manes, 1974; Van Blerkom
and Brockway, 1975a). Under the denaturing conditions of an SDS-
PAGE system, most proteins have a uniform negative charge and consequently migrate to the anode at a rate that is a function of their

---

[1]Gestation age for mouse embryos is defined here as it is in the other chapters of this
monograph, i.e., day of observation of the sperm plug is considered the first day of
pregnancy. Since there is a considerable delay between the time of mating and the time
of ovulation and subsequent fertilization of rabbit embryos (c. 12 hr), the terminology
used in the case of rabbit embryo gestation age is *hours following the estimated time of
fertilization.*

molecular weight. The presence of 2-mercaptoethanol in the SDS system results in the dissociation of proteins into component polypeptides, which are resolved on the gels as bands. The casting of the gel into a gradient of acrylamide (heavy at the bottom, light at the top) ensures the sharp resolution of relatively low-molecular-weight polypeptides which, in a nongradient system, tend to form diffuse bands at the lower regions of the gel [see Van Blerkom and Manes (1974) for a complete description of the system and its construction]. Because the visualization of proteins by chemical staining requires a relatively large amount of protein (and therefore embryos) and our primary interest has concentrated on newly synthesized proteins, the majority of our work has employed the radiolabeling of protein by exposure of embryos in vitro to ^{14}C-labeled amino acids or to high specific activity ^{35}S-methionine (usually with a specific activity greater than 250–300 Ci/mmole). Following electrophoresis, the slab gels are dried under vacuum and exposed to x-ray film for varying lengths of time (Van Blerkom and Manes, 1974). The one-dimensional gels shown in this section are autoradiographic patterns of newly synthesized proteins.

The autoradiograph presented in figure 2.3 illustrates qualitative changes in the pattern of protein synthesis during the first 132 hr following fertilization of the rabbit egg. Three results were most evident from these patterns: (1) the majority of qualitative changes occur between fertilization and approximately 60 hr of development; (2) newly fertilized eggs synthesize proteins across a broad spectrum of molecular weights, from approximately 8000 to greater than 200,000 daltons; and (3) the duration of the synthesis of some of the polypeptides composing the bands is extremely brief (Van Blerkom and Manes, 1974). An identical analysis of protein synthesis in preimplantation mouse embryos demonstrated very similar results (Van Blerkom and Brockway, 1975a). In the mouse, numerous qualitative changes take place during the first few cell divisions following fertilization; the pattern "stabilizes" between the 4- and 8-cell stage (figure 2.4). As with the rabbit, major qualitative changes occur during very brief intervals of time. Epstein and Smith (1974) reported very similar findings in their study of protein synthesis during the preimplantation development of the mouse. An examination of the protein synthetic patterns of preimplantation rat (J. Van Blerkom and M. A. H. Surani, unpublished) and hamster embryos (J. Van Blerkom and M. H. Johnson, unpub-

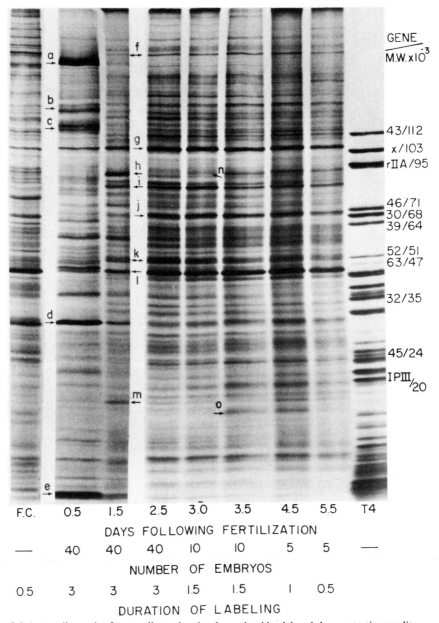

GENE

$\overline{\text{M.W.} \times 10^{-3}}$

43/112
x/103
rⅡA/95
46/71
30/68
39/64
52/51
63/47
32/35
45/24
IPⅢ/20

F.C.	0.5	1.5	2.5	3.0	3.5	4.5	5.5	T4

DAYS FOLLOWING FERTILIZATION

—	40	40	40	10	10	5	5	—

NUMBER OF EMBRYOS

0.5	3	3	3	1.5	1.5	1	0.5

DURATION OF LABELING

2.3 Autoradiograph of a one-dimensional polyacrylamide slab gel demonstrating qualitative changes in the pattern of protein synthesis during cleavage (day 0.5 to day 2.5, i.e., 12–60 hr postfertilization) and blastocyst development (day 3 [72 hr] onward) in preimplantation rabbit embryos grown in vivo or in vitro. The protein patterns of follicle cells (F.C.) are presented in the first column and should be compared with protein patterns of early-cleavage-stage embryos. Molecular weights can be estimated by comparing embryonic bands with prereplicative bacteriophage T4 proteins shown in the last column adjacent to which are approximate molecular weight values ($\times 10^{-3}$). From Van Blerkom and Manes (1974); reprinted with permission from *Developmental Biology*.

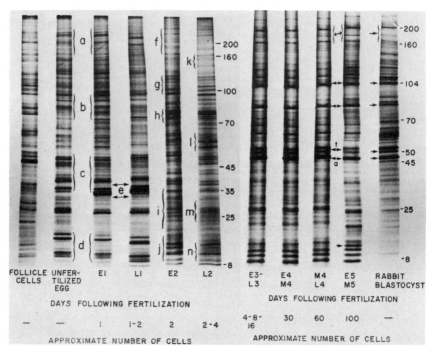

2.4 Autoradiograph of a one-dimensional polyacrylamide slab gel demonstrating qualitative changes in the pattern of protein synthesis in mouse follicle cells, unfertilized eggs, cleavage-stage embryos, and blastocysts. Embryos were exposed to culture medium containing^{35}S-1-methionine at three intervals during development—early (E), mid (M), and late (L) of each day of the preimplantation period. The protein patterns of a rabbit blastocyst (far right) illustrate major areas of similarity in the protein synthetic patterns of the two species. Approximate molecular weights are given on the right hand side ($\times$ 10^{-3}). From Van Blerkom and Brockway (1975a); reprinted with permission from *Developmental Biology*.

lished) indicates that the sequence of rapid changes during early cleavage, followed by the stabilization of protein patterns, is not restricted to the rabbit and mouse. In both rat and hamster, protein synthetic patterns stabilized subsequent to the 4- to 8-cell stage.

Several problems in interpretation are inherent in one-dimensional analyses of protein synthesis. The most critical problem is that the complexity of individual electrophoretic bands is unknown. Any particular band could be composed of numerous polypeptides that have similar molecular weights, and consequently, qualitative changes within that population might not result in an obvious change in the net intensity of the band. A second problem concerns the fact that although the resolution of proteins is usually quite good, only those proteins that are present, either collectively or individually, in high concentrations can be visualized by electrophoresis and autoradiography. Proteins present in the embryo in amounts insufficient to be detected individually without further separation and concentration probably produce the relatively high "gray" background between the autoradiographic bands. A third major limitation in the electrophoretic analysis of embryonic protein synthesis is that in most studies, intact embryos are analyzed. If differential protein synthesis were occurring between cells of, for example, the inner cell mass (ICM) and trophectoderm, differences would not be detected. Clearly, this is a great concern, for although an intact embryo might not show qualitative changes in protein synthesis, differential protein synthesis might take place either in different cells or in different regions of a preimplantation embryo.

The development of high-resolution, two-dimensional electrophoretic techniques (O'Farrell, 1975) and the availability of procedures that amplify extremely low levels of incorporated radioactivity (Bonner and Laskey, 1974; Laskey and Mills, 1975) have removed many of the drawbacks inherent in one-dimensional analysis. The following section presents evidence relevant to the complexity of translation in preimplantation mouse and rabbit embryos, as well as results demonstrating differential protein synthesis in the mouse blastocyst.

Two-dimensional gel electrophoresis
The separation of proteins by two-dimensional PAGE as developed by Patrick O'Farrell (1975) and used in our laboratory relies on two properties of protein: isoelectric point and molecular weight. In this two-dimensional procedure, polypeptides are initially separated in cylindrical, isoelectric focusing gels as a function of their isoelectric point. In a typical experiment, polypeptides whose isoelectric point falls between approximately pH 4.5 and pH 7.5 are separated. Following isoelectric focusing, the cylindrical gels are placed upon SDS linear or exponential gradient polyacrylamide slab gels where the polypeptides migrate as a function of molecular weight. Whereas with one-dimensional PAGE a series of bands is obtained, in the two-dimensional PAGE system, individual polypeptides are resolved into spots. The combination of scintillation autoradiography (*fluorography*) and high-resolution, two-dimensional gel electrophoresis permits the detection and visualization of individual polypeptides from embryonic sam-

ples in which the level of incorporated radioactivity is extremely low. The fluorographic amplification of radioactivity contained within the gel is accomplished as follows: after electrophoresis, slab gels are fixed in trichloroacetic acid, washed in 7 percent acetic acid, dehydrated in dimethyl sulphoxide, infiltrated with the scintillant 2,5-diphenyloxazole (PPO), dried under vacuum, and finally exposed to x-ray film at $-70°C$ (Bonner and Laskey, 1974). Beta particles from either ^3H or $^{14}C/^{35}S$ interact with the scintillant. The result is an emission of light, which in turn causes a local blackening on the x-ray film. An approximate hundredfold amplification occurs with ^3H, whereas a tenfold increase occurs with $^{14}C/^{35}S$ (Bonner and Laskey, 1974). One serious drawback in applying this technique to the complex electrophoretic patterns obtained by two-dimensional electrophoresis is a nonlinear film response, i.e., film images produced might not be true representations of the actual distribution of radioactivity in the gel (Laskey and Mills, 1975). However, preexposure of the x-ray film to a single flash of light hypersensitizes the film and greatly increases the efficiency of the fluorographic process, and after preexposure, the density of the fluorographic image is corrected to linearity such that the density of an electrophoretic spot is proportional to the amount of radioactivity that produced it. In our laboratory, the combination of these two procedures has permitted the detection of several hundred polypeptides from a *single embryonic cell* when the level of incorporated radioactivity was on the order of a few hundred counts per minute.

The *autoradiographs* shown in figures 2.5 and 2.6 demonstrate qualitative changes in protein synthesis in both mature rabbit oocytes and preimplantation rabbit embryos. Oocyte samples were derived from eggs ovulated at 12 hr following the administration of luteinizing hormone and then exposed to ^{35}S-methionine for one hour. All oocytes had a single polar body and chromosomal analysis of representative eggs indicated meiotic arrest at metaphase II. Embryos examined at 12 hr postfertilization were all at the 2-cell stage. What is most apparent from a comparison of these two autoradiographs is that not only do several new and major polypeptide spots appear after fertilization, but also several major polypeptides synthesized by oocytes cease to be synthesized by newly fertilized eggs. On the other hand, it is evident that many of the polypeptides synthesized by the 2-cell embryo are the same as those produced by the mature oocyte. Between 12 and 36 hr postfertilization (figures 2.5b and 2.6a), the autoradiographs reveal numerous qualitative changes as the synthesis of several polypep-

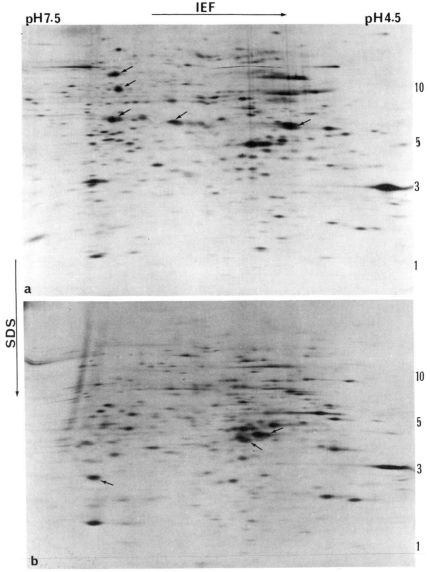

2.5 Autoradiograph of a high-resolution, two-dimensional polyacrylamide slab gel demonstrating qualitative changes in the pattern of protein synthesis in the rabbit egg immediately before and immediately following fertilization. Numerous polypeptides synthesized by (a) a newly ovulated oocyte are no longer being produced by (b) the fertilized egg—only the major polypeptides in this category are indicated by arrows in (a). Several "new" polypeptides are produced by the newly fertilized egg, and the major spots are indicated by arrows in (b). A detailed comparison of these two autoradiographs indicates other polypeptides whose synthesis is either "shut off" or "turned on" after fertilization. The first dimension is isoelectric focusing (IEF) in the range of pH 4.5 and 7.5. The second dimension is a sodium dodecyl sulfate (SDS) polyacrylamide linear (5–15% acrylamide) gradient slab gel. Approximate molecular weights ($\times$ 10^{-4}) are given on the right.

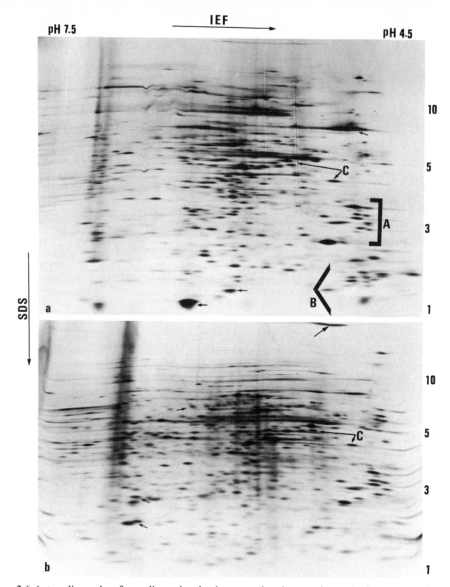

2.6 Autoradiographs of two-dimensional gels comparing the protein synthetic patterns of 36-hr (a) and 60-hr (b) postfertilization rabbit embryos. Some of the major proteins whose synthesis was detected at 36 hr but not at 60 hr (or later) are indicated either by arrows in (a) or are contained within the region denoted by bracket B. Major groups of proteins synthesized both at 36 and 60 hr of development are contained within bracket A or indicated by arrow C. Of the numerous proteins whose synthesis was initially detected at 60 hr, several are marked by arrows in (b). Other examples of proteins that either appear or disappear from the protein synthetic patterns between 36 and 60 hr (unmarked on the autoradiograms) may be observed upon close examination of (a) and (b) (using proteins indicated by arrow C and bracket A as reference points). Approximate molecular weights ($\times$ 10^{-4}) are given on the far right.

tides is "turned on" or "turned off." This type of translational activity continues until approximately 48 to 60 hr, at which time the pattern of protein synthesis stabilizes. Although the autoradiograph shown in figure 2.6b was obtained from a 60-hr embryo, it is nevertheless representative of the polypeptides synthesized between 60 and 120 hr postfertilization. The results obtained with a two-dimensional gel analysis of early mouse development were similar: the protein pattern stabilizes between the 4- and 8-cell stages (J. Van Blerkom, unpublished). Collectively, the evidence supports observations obtained from one-dimensional gels in that the *initial* stages of preimplantation development are accompanied by complex and rapid qualitative changes in protein synthesis, in contrast to the late morula and blastocyst stages, which are characterized by relatively minor qualitative changes.

As previously mentioned, the study of intact embryos does not indicate whether or not differential protein synthesis occurs in different regions of the embryo. In order to determine whether such activity was taking place, protein synthetic patterns of microdissected ICM and trophectoderm of fifth-day mouse blastocysts were analyzed (Van Blerkom et al., 1976). The results of these studies are shown in figures 2.7 and 2.8. A comparison of these *fluorographs* reveals that there are several polypeptides whose synthesis is restricted to the ICM and that a number of other polypeptides are produced only by the trophectoderm. It is clear, therefore, that at least in the fifth-day blastocyst, differential protein synthesis takes place. Combining such very recent techniques as the microdissection of preimplantation embryos (Gardner, 1971), immunosurgery (Solter and Knowles, 1975), and autoradiography or fluorography, with high-resolution, two-dimensional gel electrophoresis should make possible direct testing of the hypothesis that cell position determines selective gene expression and, thus, differentiation in the preimplantation embyro (Tarkowski and Wroblewski, 1967; Hillman et al., 1972; Herbert and Graham, 1974). Such experiments will involve the analysis of protein synthesis in individual blastomeres of cleavage-stage embryos, cells of the ICM and trophectoderm, and cells of postimplantation tissues. The initial findings presented here indicate "stage- and tissue-specific" markers of early development and, in the one example to date, the spatial distribution of the cells producing them.

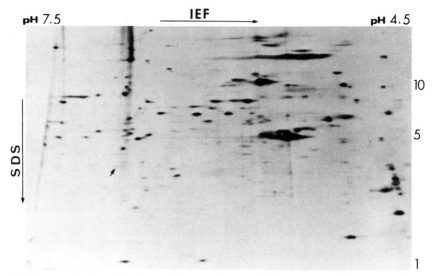

2.7 Fluorograph of proteins derived from an intact mid-5th-day mouse blastocyst and separated by high-resolution, two-dimensional polyacrylamide gel electrophoresis. Approximate molecular weights ($\times$ 10^{-4}) are given to the right of the fluorograph. From Van Blerkom et al. (1976); reprinted with permission from *Nature*.

It was initially intriguing that major changes in the qualitative pattern of protein synthesis occur during cleavage rather than during the blastocyst stage. It is during the blastocyst stage that the first observable event of differentiation in the mammalian embryo takes place, namely, the formation of the ICM and trophectoderm. However, from an examination of fine structural changes during preimplantation development, it is apparent that the major organelle systems undergo fundamental reorganizations during cleavage rather than during the blastocyst stage (Van Blerkom et al., 1973). It is tempting to speculate that the comparatively rapid turning on and off of the synthesis of numerous polypeptides is related to sequential changes in, for example, nucleolar and mitochondrial fine structure (see section 2.3.4). At present, the accumulated evidence neither supports nor disproves this hypothesis.

What the accumulated evidence does support is the fact that many of the polypeptides synthesized during the initial stages of postfertilization development are the same polypeptides synthesized by the mature oocyte. One interpretation of this observation is that some portion of

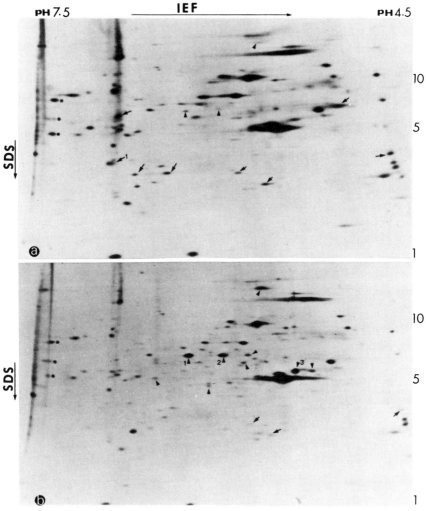

2.8 Fluorograph of proteins derived from separated (a) ICM and (b) trophectoderm of microdissected mid-5th-day mouse blastocysts. The arrows in (a) point to polypeptide spots that are absent in the trophectoderm preparations and relatively intense in the ICM samples. Polypeptides limited to the trophectoderm are marked with solid triangles in (b); major polypeptides are numbered. Faint traces of ICM spots in trophectoderm preparations are marked with arrows in (b). The arrow in figure 2.7 indicates a polypeptide that is a comparatively minor component of the *intact* embryo but becomes a major component of microdissected ICM (arrow 1, figure 2.8a) and is not observed in the trophectoderm. Approximate molecular weights are given to the right of the fluorograph ($\times$ 10^{-4}). From Van Blerkom et al. (1976); reprinted with permission from *Nature*.

protein synthesis associated with early cleavage is simply the "winding down" of translational activity carried over from the mature oocyte into the newly fertilized egg. Studies of the poly(A)-containing RNA of unfertilized and fertilized rabbit eggs have shown similar amounts of the putative mRNA in both oocytes and newly fertilized eggs (Schultz, 1975). Furthermore, the demonstration that the rate of protein synthesis in fertilized eggs is insensitive to α-amanitin at concentrations that inhibit RNA synthesis indicates that maternal mRNA makes an important contribution to protein synthesis in the early stages of cleavage (Manes, 1973). From both fine structural (Van Blerkom et al., 1973) and biochemical studies (Manes, 1971; Karp et al., 1974), it is known that, relative to their numbers at the blastocyst stage, ribosomes are comparatively scarce in the cytoplasm of the cleaving embryo. It seems plausible to suggest that the availability of ribosomes during cleavage has a fundamental role in the control of quantitative and possibly qualitative aspects of protein synthesis. In the rabbit embryo, the stage during which the protein synthetic patterns stabilize is the same stage in which (1) the nucleolus contains an appreciable granular element (see section 2.3.4); (2) the synthesis of significant amounts of rRNA is first detected (Manes, 1971); (3) an increase in the density of cytoplasmic ribosomes is evident; and (4) an increase in the relative rate of protein synthesis is observed (Karp et al., 1974; J. Van Blerkom, unpublished). The first 24 hr of preimplantation development therefore might involve the translation of primarily maternal mRNA on primarily maternal ribosomes with little actual embryonic contribution to either component of protein synthesis. Because the majority of polypeptide spots characteristic of the oocytes are no longer detectable in the stabilized patterns of midmorula and older embryos, it would follow that with the gradual production of embryonic ribosomes (with a large burst of rRNA at the blastocyst stage), and presumably template RNA (Schultz, 1973), maternal mRNA carried over from the oocyte either is not translated with the same efficiency as previously or is no longer present in amounts comparable to the initial stages of the postfertilization period. The information required to support this hypothesis is knowledge of the extent to which the embryonic genome participates in protein synthesis, the half-lives of maternal mRNA in the early embryo, and the efficiency of translation of maternal and embryonic template RNA. Further complicating the interpretation of the autoradiographic polypeptide patterns is that not only are the chemical identities of the majority of the electrophoretic spots unknown, but it remains to be determined also whether these polypeptides are the products of maternal (preformed) templates differentially activated throughout the preimplantation period. That the time of synthesis of a particular polypeptide might not be temporally correlated with the stage of development at which it is functional is another possibility.

A comparison of oocyte and embryonic protein patterns clearly shows that some of the polypeptides produced by mature oocytes are no longer synthesized by fertilized eggs. One highly speculative interpretation of this observation is that after fertilization, a set of maternal mRNA is selectively "inactivated" and, therefore, is unavailable for translation. The purpose of such activity would be to remove template RNA that codes for proteins required for maturation and/or fertilization but not for embryonic development. In the rabbit, the period of time between egg maturation and fertilization is about 12 hr, whereas the duration of the synthesis of some major polypeptide spots is approximately 6 hr or less (Van Blerkom and McGaughey, 1977). No mechanism is known by which the embryo could selectively identify and inactivate a class of mRNA. Alternatively, some oocyte mRNAs might be extremely short-lived and, after fertilization, the disappearance of these templates might coincide with the termination of the synthesis of numerous polypeptides.

Clearly, the ultimate validity of all of these interpretations and speculations requires the accumulation of very difficult and demanding experimental evidence. In the course of further analyses, alternate interpretations will become evident and a more coherent picture of the molecular basis of preimplantation development undoubtedly will emerge. In terms of our present understanding of the available information, the molecular biology of early mammalian embryogenesis is itself in its infancy.

2.3.3 Protein Synthesis Following the Termination of Facultative Delayed Implantation

Perhaps the single most definitive example of the maternal regulation of embryonic development is the phenomenon of delayed implantation. Although delayed implantation is a normal feature of the reproductive physiology of a wide variety of mammals (Enders, 1963; Daniel, 1970), in the mouse and rat delay either occurs naturally, as during lactational delay (McLaren, 1971), or can be induced experimentally if a bred animal is ovariectomized during the first few days of the preimplantation period—facultative delayed implantation (McLaren, 1971; Surani, 1975; Van Blerkom and Brockway, 1975b). During the period of delayed implantation, the embryo exists in the uterus as an unattached blastocyst and remains in this condition for varying lengths of time until delay is terminated. The state of arrested embryonic development is characterized by a cessation of mitosis and DNA synthesis (McLaren, 1968) and dramatic depressions in the levels of RNA synthesis (Prasad et al., 1968), protein synthesis (Weitlauf, 1969), and glucose metabolism (Menke and McLaren, 1970). The delay can be terminated either by the administration of estrogen to the mother (Prasad et al., 1968), in which case implantation usually follows within 24 hr, or by the removal of the embryos from the uterus followed by culture in vitro (Gulyas and Daniel, 1969; Weitlauf, 1969; Van Blerkom and Brockway, 1975b). The release of embryos from delay and their attachment to the endometrium occurs in two phases: the resumption of RNA and protein synthesis within 1 hr after estrogen administration (Dass et al., 1969), followed by the attachment of the embryo 24 hr later (Psychoyos, 1969).

The mechanism(s) responsible for the onset, maintenance, and termination of delayed implantation is unclear. Both uterine secretion activators and inhibitors of development have been postulated (see McLaren, 1973). What is clear is that the capacity of the embryo for reactivation after its development has been arrested for a prolonged period of time offers the investigator an ideal opportunity to study embryonic protein synthesis in relation to preimplantation development. In our laboratory, considerable attention has focused on two questions:

1. When embryos are released from delay, do they synthesize a class of 'new' proteins which may be directly related to the ability of the blastocyst to resume development and implant?
2. Are protein patterns of embryos released from delay in vivo (with estrogen) different from the patterns of embryos released from delay in vitro (in culture)?

The design of some of the principal experiments is as follows: embryos that had been in implantation delay for as many as 17 days (i.e., 17 days past the time at which they would normally have implanted on

the fifth day) either were removed from the uterus and cultured in vitro
or were removed from the uterus at timed intervals from 30 min to 24
hr after injection with 17β-estradiol. Following removal from the
uterus, embryos were immediately exposed to culture medium containing ^{35}S-methionine or were placed in unlabeled culture medium for as
many as 15 hr and then exposed to label [see Van Blerkom and Brockway (1975b) for additional details]. In a second series of experiments,
embryos treated as described above were exposed to culture medium
containing α-amanitin at a dose known to inhibit RNA synthesis in the
mouse—1μg/ml (Golbus et al., 1975).

The results of these studies, shown in figure 2.9, demonstrate that no
qualitative or quantitative differences in the pattern of protein synthesis
are detectable when embryos are released from delay (1) in vivo or in
vitro, (2) in the presence or absence of α-amanitin, (3) during prolonged
initial culture in unlabeled medium, and (4) after as many as 24 hr of
exposure to a uterine environment stimulated by 17β-estradiol (Van
Blerkom and Brockway, 1975b). The protein synthetic patterns obtained from delayed embryos were identical to the patterns observed in
normal blastocysts just prior to implantation (figure 2.9). In addition,
these studies revealed that following the release from delay, the pattern
of protein synthesis appears very rapidly and *in its entirety* with the
same relative band intensities as are observed in predelay blastocysts—
there is no time lag before certain autoradiographic bands appear. Thus
the embryo requires no "new" proteins to reactivate its development,
at least at the limit of detection of the analysis.

The interpretation of these results is based on one-dimensional gels
and, until repeated with the two-dimensional system, must be considered tentative. It is equally important to determine whether exposure of
embryos to α-amanitin does indeed inhibit RNA polymerase II activity
at the concentration used (see section 2.2.2). However, the evidence
does suggest the possibility that the resumption of protein synthesis
following the termination of the delay does not require the immediate
resumption of mRNA synthesis. Perhaps the onset of the developmental arrest involves a storage or inactivation of mRNAs that can be
rapidly activated when conditions favorable for continued development
appear in the embryonic environment. This situation would be analogous to the activation of preformed maternal mRNA following fertilization of sea-urchin eggs (Spirin, 1969). What mechanisms are responsi-

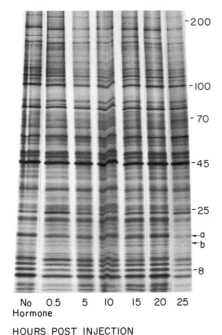

No   0.5   5   10   15   20   25
Hormone

HOURS POST INJECTION

2.9 Autoradiograph demonstrating the qualitative pattern of protein synthesis during the first 25 hr following release of

mouse embryos from delayed implantation in vivo (i.e., following injection of 17β-estradiol and progesterone). The pattern of protein synthesis of embryos released from delay in vitro by removal from the uterus with subsequent culture is shown in the first column (no hormone). Protein patterns derived from embryos recovered from the uterus 5 and 10 min post injection were identical to the patterns shown in 0.5 hr. In addition, delayed blastocysts that had been removed from the uterus and placed in culture medium containing α-amanitin (0.1 to 10 μg/ml) did synthesize proteins, and upon one-dimensional electrophoresis, gave the same spectrum of autoradiographic bands as is shown in this figure. Arrows a and b mark bands that can be somewhat variable in different preparations. Arrow a points to what appears to be a single band but is actually two separate bands. Arrow b indicates two light bands that occasionally migrate as a single band. Approximate molecular weights are given on the far right ($\times 10^{-3}$). From Van Blerkom and Brockway (1975b); reprinted with permission from *Developmental Biology*.

ble for such a process in mammalian embryos is unknown. However, many viral and cellular mRNAs are activated either by methylation of the 5'-terminal nucleotide or by the addition of 5'-terminal methylguanosine (Griffin, 1975). The removal of the 5'-terminal methylguanosine from globin mRNA, for example, prevents its translation (Muthukrishnan et al., 1975). In other systems, adenylation is an important factor in the activation of mRNA (Slater et al., 1973). Recently, Hickey et al. (1976) demonstrated that neither polyadenylation nor methylation of mRNA is responsible for the activation of preformed mRNA in fertilized sea-urchin eggs. Whether any of these mechanisms could operate in the mammalian embryo during delayed implantation remains to be determined.

If the mRNAs that code for the proteins resolved on the gels are unusually stable during the prolonged period of delay, then the release from this state could be associated simply with the termination of some other limiting condition(s), such as (1) the resumption or acceleration of the rate of rRNA synthesis or ribosome production or (2) a marked change in the rate of uptake of nucleic acid and/or protein precursors. Support for both hypothesis exists. For example, Wu and Meyer (1974) have shown that shortly after the estrogen-mediated termination of delay in the rat, an increase in the population of cytoplasmic ribosomes and polysomes takes place, Weitlauf (1974) has reported that actinomycin D prevents the complete reactivation of delayed mouse embryos, possibly by inhibiting rRNA synthesis.

In support of the second hypothesis, Aitken (1974) has presented evidence indicating that the termination of delay in the roe deer is accompanied by a sudden and marked increase in the calcium concentration of the uterine secretions. In 1966, Gwatkin reported that the absence of arginine and leucine in culture medium caused mouse blastocysts to enter a reversible state of delayed development although, it was later discovered, both amino acids are normally present in the uterine secretions of a delaying mouse (Gwatkin, 1969). More recently, Borland and Tasca (1975) noted that the chlorpromazine inhibition of methionine uptake and incorporation induces in the mouse blastocyst a condition analogous to delayed implantation. Collectively, these observations suggest that ionic conditions in the embryonic environment could regulate blastocyst growth by mediating the activity of transport systems and thus influencing the intracellular availability of critical small ions, such as amino acids. The evidence also points out the potential role of ions as pleiotypic modulators of blastocyst development in which the initial control of, for example, amino acid transport ultimately results in the regulation of processes responsible for metabolism, macromolecular synthesis, cell division, and perhaps, mRNA stability.

What is required for a more incisive and therefore less speculative analysis of delayed implantation is a comprehensive body of experimental data. The following are some of the questions that need attention:

1. What is the fate of mRNA during delay?
2. Are embryonic mRNAs chemically modified?
3. Are there significant changes in the ionic composition of the uterine

secretions during delay and coincident with the activation of the embryo?

4. Are metabolism and macromolecular syntheses truly shut down, or is, for example, the rate of mRNA turnover simply reduced during delay?

Amassing such information will provide genuine insight into a fascinating developmental phenomenon.

2.3.4 A Comparison of Embryonic Development In Vivo and In Vitro

To reiterate, much of the present knowledge and understanding of the molecular basis of preimplantation mammalian development derives from the study of embryos reared in vitro. Implicit in the interpretation of experimental results is the assumption that development in vitro parallels development in vivo. To show that this is indeed the situation is fundamental to the validity of the studies that have been discussed. In the following section, we provide evidence that embryos reared in vitro undergo the same sequential, developmental events as are observed in embryos grown entirely in vivo.

Fine structural observations of embryos developed in vivo and in vitro
At the level of the light microscope, cultured mouse and rabbit embryos undergo cleavage and blastocyst formation in an apparently normal manner. However, an examination of embryonic development at the fine structural level yields several fundamental criteria by which to gauge the normality of embryogenesis in vitro. In both mouse and rabbit embryos, major submicroscopic changes take place during the preimplantation period (Calarco and Brown, 1969; Hillman and Tasca, 1969; Van Blerkom et al., 1973). The major submicroscopic events include nucleolar maturation, mitochondrial development, the appearance of cytoplasmic ribosomes, polysomes, the rough-surfaced endoplasmic reticulum, crystalline inclusions, and the formation of intercellular junctional complexes. Our studies of preimplantation mouse and rabbit embryos have involved the culture of embryos from the 1-cell to the expanding blastocyst stage in several different culture media, all of which were devoid of maternal macromolecules normally present in the female reproductive tract during the preimplantation period [see Van

Blerkom et al. (1973) for references to maternal genital tract
macromolecules].

One of the most obvious fine structural changes to occur during the
preimplantation period is the maturation of the nucleolus; stages of
nucleolar maturation in the rabbit embryo are shown in figure 2.2.
Shortly after fertilization, the nucleolus appears as a spherical,
electron-dense structure composed of a highly compacted, fibrillar material (figure 2.2a). At approximately the 4- to 8-cell stage, vacuoles, or
"holes," appear in the fibrillar matrix and a small granular component
appears for the first time (figure 2.2b). By the midmorula stage (about
16–32 cells, 48 hr postfertilization), a larger granular component exists
and nucleoli are more reticulated than previously (figure 2.2c). The
final observed change in nucleolar morphology coincides with the initiation of blastocyst formation between 60 and 72 hr. Nucleoli at this
stage of embryogenesis are extremely hypertrophic and reticulated
structures in which granular and fibrillar material is distributed along
highly anastomosing networks of nucleolonemas (figures 2.2d and
2.11b). The same ultrastructural events and temporal schedule that
lead to fully matured nucleoli in vivo occur in embryos cultured from
the 1-cell stage in vitro.

A second major fine structural event to take place during the preimplantation period in the rabbit is the differentiation of mitochondria.
Prior to 48 hr postfertilization, mitochondria are comparatively small
spherical electron-dense organelles. These mitonchondria contain few
cristae, and those that are present are oriented in a circular fashion
about the periphery of a highly electron-dense mitochondrial matrix
(figure 2.10a). At 48 hr, many mitochondria are somewhat elongated
and contain cristae that partially penetrate the electron-dense matrix
(figure 2.10b). Coincident with the formation of the blastocyst, mitochondria undergo a marked change in fine structure. Most mitochondria in blastocyst-stage embryos are elongated, contain a matrix of
low-to-moderate electron density, and possess numerous cristae (figure
2.11a). The cristae have a lamellar structure and are arranged in a
stacklike manner, perpendicular to the long axis of the organelle (figure
2.11a). A second morphological form of mitochondria encountered in
some cells of the rabbit blastocyst contains a matrix of moderate-to-
high electron density and distended cristae surrounding electron-
translucent spaces (figures 2.11b,c and 2.12). Mitochondrial develop-

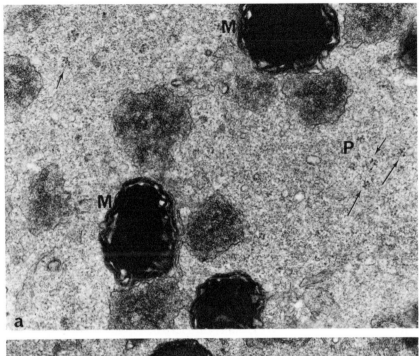

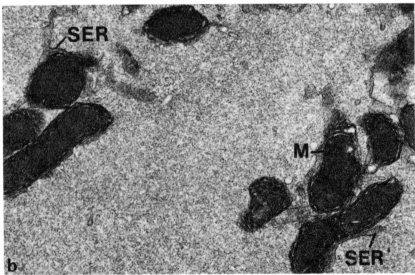

2.10 Section at high magnification of the cytoplasm of (a) 12-hr and (b) 36-hr rabbit embryos grown in vivo. Micrograph (a) illustrates the relative scarcity of ribosomes and polysomes in the cytoplasm (P) as well as the appearance of mitochondria (M) (× 33,000). Mitochondria at this stage are relatively small structures containing few cristae and possessing an electron-dense matrix throughout which the cristae do not penetrate. At 36 hr of development (b), mitochondria (M) still have an electron-dense matrix, but the organelle is somewhat elongated and cristae do penetrate the matrix to varying degrees (× 22,000). Note the association of mitochondria and vesicles of the smooth-surfaced endoplasmic reticulum (SER). From Van Blerkom et al. (1973); reprinted with permission from *Developmental Biology*.

ment in cultured embryos is identical to that observed in embryos grown entirely in vivo.

In contrast to later stages, cleaving embryos contain comparatively few cytoplasmic ribosomes and polysomes (figure 2.10). At the initiation of blastocyst formation, ribosomes and polysomes densely populate the cytoplasm (figures 2.11 and 2.12). The increase in ribosomal density takes place in embryos cultured from the 1-cell stage just as it does in embryos grown entirely in vivo. During cleavage, cisternae of the rough-surfaced endoplasmic reticulum are relatively scarce, whereas profiles of the smooth-surfaced endoplasmic reticulum are quite abundant (figure 2.10b). The rough-surfaced endoplasmic reticulum is initially detected as a major cytoplasmic component at the early blastocyst stage and becomes increasingly prominent with continued blastocyst expansion and growth (figures 2.11 and 2.12). The time course of appearance of the rough-surfaced endoplasmic reticulum is the same in embryos cultured in vitro.

Perhaps the most important factor governing the formation and maintenance of the blastocyst is the presence of junctional complexes between the cells of the trophectoderm (Van Blerkom et al., 1973). Cleavage-stage embryos rarely contain specialized intercellular junctional complexes, and in the rabbit, membranous associations between cells generally involve microvilli. Coincident with the onset of blastocoel formation is the appearance of tight junctions with fused membranes along the lateral surfaces of adjacent trophectoderm cells (figure 2.12). At approximately 84 hr, early or forming desmosomes containing filaments oriented parallel to the plasma membrane appear at the apical borders of trophectoderm cells (figure 2.13a). By 108 hr, elaborate desmosomes composed of filaments that originate from deep within the cytoplasm and converge at the cell membrane are evident at the lateral margins of trophectoderm cells (figure 2.13b). Junctional complexes composed of tight junctions, intermediate junctions, and desmosomes are characteristic of trophectoderm cells but are rare between cells of the ICM, or in areas of contact between the ICM and trophectoderm. There are no significant variations in the formation, organization or spatial distribution of junctional complexes in embryos grown in vivo or in vitro.

The only major difference encountered between rabbit embryos grown in vivo and in vitro is the absence of crystalline inclusions in the expanding blastocyst stage in cultured embryos (figure 2.14). Although

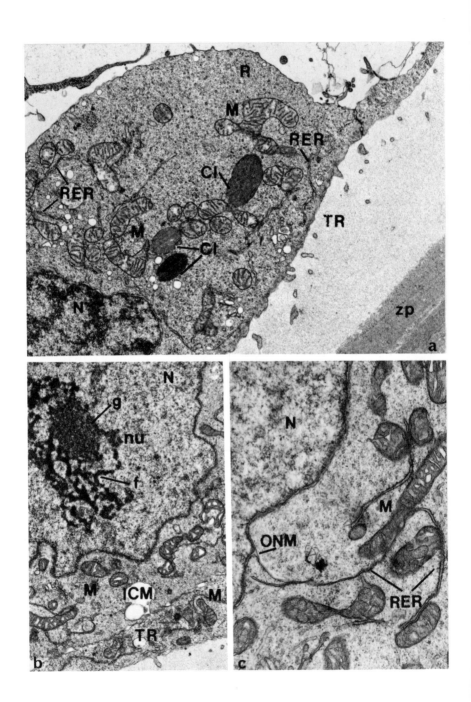

2.11 (a) Section through a trophectoderm cell (TR) of a 108-hr rabbit embryo that had been developing entirely in vivo. Mitochondria (M) in these cells have a matrix of low-to-moderate electron density and possess a larger complement of cristae than in earlier embryos. Cristae in these mitochondria have a lamellar structure and are generally oriented perpendicular to the long axis of the organelle. Of particular interest in this cell are the cisternae of the rough-surfaced endoplasmic reticulum (RER), the crystalline inclusions (CI), and the obvious abundance of cytoplasmic ribosomes and polysomes (R). N = nucleus; Zp = zona pellucida. × 13,000. (b) Section through the trophectoderm (TR) and inner cell mass (ICM) of a 108-hr rabbit embryo that had developed in vivo. This electron micrograph illustrates both the appearance of a fully differentiated nucleolus (nu) with distinct fibrillar (f) and granular (g) regions and the presence of two apparent morphological forms of mitochondria (M). Mitochondria in the trophectoderm cell (TR) are elongated, contain a matrix of moderate electron density and possess cristae with a lamellar structure. Mitochondria in the inner cell mass (ICM) are typically elongated, contain a matrix of moderate-to-high electron density, and possess distended cristae surrounding electron-translucent spaces. Mitochondria in ICM cells usually have a vacuolated appearance. × 11,500. (c) Continuity between the outer nuclear membrane (ONM) and cisternae of the rough-surfaced endoplasmic reticulum (RER) of a 108-hr rabbit embryo developing in vivo. Note that this particular element of the RER is in contact with several mitochondria (M). This section is through an ICM cell. × 30,000. From Van Blerkom et al. (1973); reproduced with permission from *Developmental Biology*.

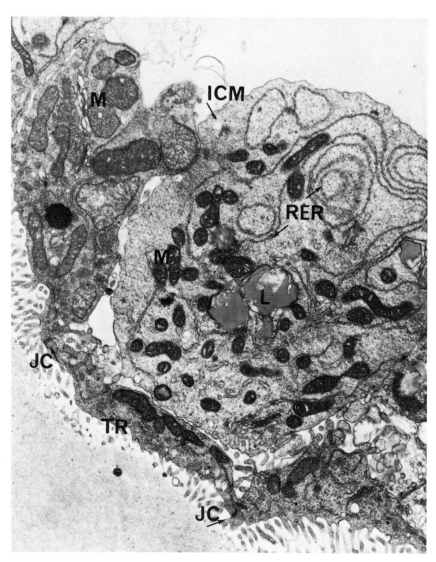

2.12 Section through the trophectoderm (TR) and ICM of a 132-hr rabbit embryo cultured entirely in vitro from the 1-cell to expanding blastocyst stage. Cisternae of the rough-surfaced endoplasmic reticulum (RER) are prominent in the cytoplasm as are ribosomes and polysomes. Mitochondria (M) are fully developed and junctional complexes (JC) are in evidence between trophectoderm cells. Lipid droplets (L) are also present. × 15,000. From Van Blerkom et al. (1973); reprinted with permission from *Developmental Biology*.

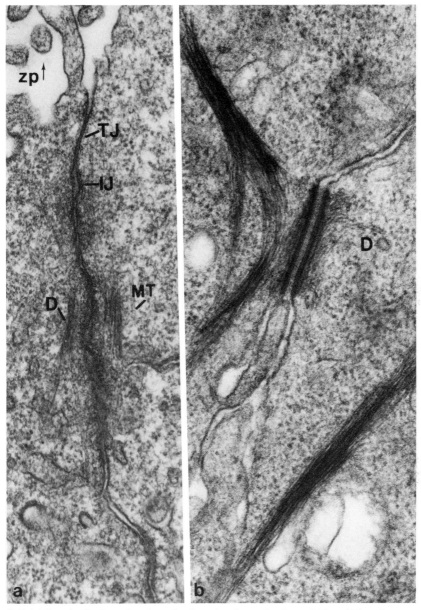

2.13 (a) Longitudinal section through a junctional complex in an 84-hr rabbit embryo cultured for 60 hr in vitro (placed in culture at the 1-cell stage). Note the tight junction (TJ), intermediate junction (IJ), and the early, or forming, desmosome (D) with filaments oriented parallel to the plasma membrane. A microtubule (MT) also appears in this electron micrograph of a trophectoderm cell. The arrow points toward the zona pellucida (zp). × 49,000. (b) Longitudinal section through a junctional complex in a 108-hr rabbit embryo grown in vivo. Only the desmosomal portion (D) of the junctional complex is shown. Note the filaments that originate deep within the cytoplasm and converge at the desmosome and the filaments that are oriented parallel to the plasma membrane at the desmosome. × 46,000. From Van Blerkom et al. (1973); reproduced with permission from *Developmental Biology*.

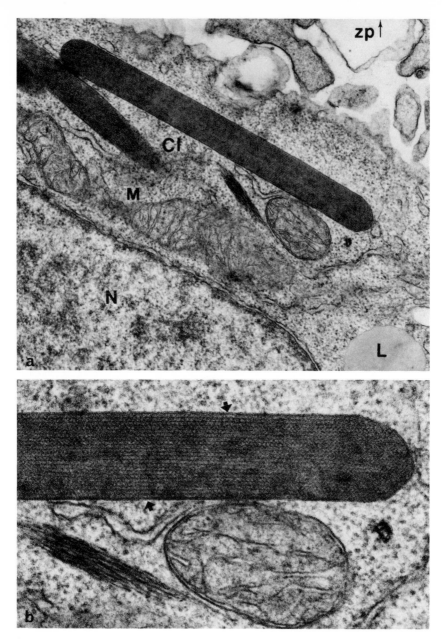

2.14 Section through a trophectoderm cell of a 108-hour postfertilization rabbit embryo grown entirely in vivo. Of particular interest in (a) is the large crystalline inclusion (CI). These crystalline bodies appear for the first time during the preimplantation period at about 108 hr and are never observed in embryos cultured in vitro from the 1-cell stage. Also illustrated in this micrograph are mitochondria (M) with a matrix of low electron density and lamellar cristae. A large lipid droplet (L) is located next to the nucleus (N). × 30,500. At higher magnification (b), it can be observed that the crystalline inclusion is membrane bound (arrows) and has a discernible regularity. × 83,500. From Van Blerkom et al. (1973); reproduced with permission from *Developmental Biology*.

the developmental significance and function of the inclusions is unknown, their absence does not seem to have an adverse effect on the
ability of the embryo to develop normally during the preimplantation
stages in vitro.

With the exception of membrane-bound crystalline inclusions, all the
fine structural changes described for the rabbit in the preceeding para-
graphs occur also during the early development of the mouse embryo—
albeit the time scale during which these processes occurs is shorter in
the mouse and some minor differences in organization exist (Calarco
and Brown, 1969; Hillman and Tasca, 1969; Van Blerkom and Runner,
1976). For example, nucleolar maturation in the mouse begins with the
reticulation and elaboration of a peripheral network of granular material, which surrounds large solid fibrillar cores (Calarco and Brown,
1969). This is in contrast to nucleolar maturation in the rabbit, where
reticulation involves the entire fibrillar matrix. Also of interest is the
fact that the fine structural development of some parthenogenetic
mouse embryos derived from eggs activated and cultured to the expanding blastocyst stage in vitro parallels the changes that occur in
normally fertilized eggs grown entirely in vivo (Van Blerkom and
Runner, 1976). Collectively, the evidence indicates that the submicroscopic development of the preimplantation mammalian embryo is regulated by an endogenous program that requires only a suitable physical
and chemical environment, which, up to a point, can be duplicated in
vitro. Additional observations of mouse parthenotes suggest that the
immediate participation or even the presence of the paternal genome
might not be necessary for the expression of such a program.

Comparison of protein synthetic patterns in preimplantation embryos grown in vivo and in vitro

A second very useful set of criteria by which to monitor the normality
of embryonic development in vitro is an analysis of protein synthetic
patterns. To determine whether embryos grown entirely in vivo or in
vitro undergo the same qualitative changes in the pattern of protein
synthesis, rabbit embryos were exposed for brief periods of time to
radioactive amino acids in vitro either immediately following removal
from the reproductive tract or upon transfer from unlabeled culture
media, into which they had been placed shortly after fertilization (Van
Blerkom and Manes, 1974). At all stages of cleavage and until the
expanding blastocyst stage, qualitative changes in the pattern of protein

synthesis occurred in an identical manner in embryos grown both in vivo and in vitro (figure 2.3). Although this interpretation is based on one-dimensional autoradiographic patterns and must be reexamined by two-dimensional procedures, the findings strongly suggest that at the level of protein synthesis, in vivo and in vitro development are quite comparable.

2.4 Lipid Metabolism

Of the various biochemical aspects of early mammalian embryogenesis, lipid metabolism appears to have suffered relatively more neglect than others. In view of the current realization that the properties of plasma membranes are central in regulating cell growth and division, cell motility, and cell fusion, it seems likely that increasing attention will focus on the lipid composition and metabolic capabilities of embryos in the next few years and, in particular, on membrane lipids and their roles in transport phenomena and implantation.

In 1966, Huff and Eik-Nes demonstrated that the late rabbit blastocyst is capable of incorporating label from ^{14}C-acetate into cholesterol and pregnenolone. This demonstration was the first evidence for the existence in the early embryo of biosynthetic pathways involving mevalonic acid and squalene as intermediates. In 1968, Fridhandler reported that the late rabbit blastocyst also incorporates label from ^{14}C-acetate into material extracting as fatty acids. Finkelstein (1973), studying the fatty acid composition of the rabbit blastocyst by gas-liquid chromatography, found that six—palmitic, palmitoleic, stearic, oleic, linoleic, and arachidonic—constituted over 95 percent of the total. The two essential fatty acids, linoleic and arachidonic, together make up 20 percent of the total in the late blastocyst; because mammalian cells cannot synthesize them, they must come from the maternal diet. In the light of Fridhandler's earlier results, it seems likely that the other fatty acids in the blastocyst are of embryonic origin. Finkelstein further showed that the rabbit blastocyst is able to use fatty acids as energy sources as well as in the synthesis of phospholipids.

Recently, Nadijcka and Hillman (1975) have shown by autoradiographic means that the cleavage stage mouse embryo is capable of neutral lipid synthesis. Extraordinarily large deposits of these lipids, which can be labeled with ^{14}C-pyruvate or ^{3}H-palmitic acid, accumulate in cleavage stages of the t^n lethal homozygotes—t^{12}, t^{w32}, and t^6. These

lipid deposits are also associated with abnormally high ATP levels
(Ginsberg and Hillman, 1975), and this group suggests that the primary
genetic defect of t mutations might be faulty regulation of ATP synthesis and that the lipids accumulate secondarily due to an abnormal diversion of acetyl CoA into lipid synthesis in the presence of excess
ATP (see chapter 4).

The metabolism of arachidonic acid into prostaglandins is a pathway
of potentially great significance in mammalian reproduction, one that is
just beginning to be investigated. Dickmann and Spilman (1975) have
reported the detection of prostaglandins in rabbit blastocysts by radioimmunoassay. The origin of the prostaglandins could not be determined
in their study, although they suggest that prostaglandins are in fact
synthesized by the blastocyst and act as mediators in steroid biosynthesis. A possible role for prostaglandins in early mammalian embryogenesis is suggested by the recent observation by El-Banna et al. (1976)
that indomethacin, an inhibitor of prostaglandin synthesis, interrupts
pregnancy in the rabbit if administered at about the time of implantation. It can be predicted that a considerable amount of attention will be
devoted to the further elucidation of the role of prostaglandins in mammalian embryogenesis and implantation in the near future.

2.5 Hormones

2.5.1 General

It has long been suspected that the mammalian embryo, even before
implantation, is not an entirely passive resident in the reproductive
tract. Recent reports are beginning to justify that suspicion. Emission
of the "signals" that the reproductive tract receives from the embryo
apparently begins during early cleavage. Kent (1975) has identified a
tetrapeptide synthesized, or at least secreted, by 2- to 4-cell hamster
embryos that inhibits further ovulatory cycles in the ovary. The tetrapeptide has been sequenced as threonine-proline-arginine-lysine-NH_2.
The maturation cycle of eggs in the hamster ovary is interrupted by
removal of the oviduct; thus, it is conceivable that the embryonic tetrapeptide counteracts this oviductal influence (Kent, 1975). In the mouse,
the cleaving embryo exerts a locally inhibitory effect on thymidine
incorporation in the oviductal epithelium (Freese et al., 1973). This
inhibitory effect involves not only the immediately adjacent oviduct,

but also the epithelium that lies "downstream" from the embryo. The "upstream" epithelium promptly recovers its ability to incorporate thymidine—and presumably to resume cell division—once the embryo has passed by. It would be of great interest to investigate whether a similar effect occurs in the hamster oviduct and whether ovulation-inhibitory materials are secreted by cleaving embryos of species other than the hamster.

An effect of the preimplantation embryo upon uterine metabolism has also been demonstrated. Renfree (1972) found not only that the embryo provokes more protein secretion in the wallaby uterus, but also that the luminal proteins of the pregnant and nonpregnant uterine horns differ qualitatively. By resorting to embryo transfer, she clearly demonstrated that the uterine changes depended upon the presence of the embryo rather than upon proximity to the corpus luteum. Beier and Kühnel (1973) likewise found that the electrophoretic patterns of uterine-fluid proteins in the rabbit are not the same in pseudopregnancy as in pregnancy. Van Hoorn and Denker (1975) presented evidence that the blastocyst locally inhibits amino acid arylamidase activity in the rabbit uterus at the time of implantation. And Fuchs and Beling (1974) found that circulating progesterone levels are consistently higher during early pregnancy than during pseudopregnancy in the rabbit.

Thus, the reality of embryonic signals is reasonably well established. With the exception of the tetrapeptide isolated from the hamster embryo, however, their chemical nature (or physical nature?) remains to be determined. Prominent candidates are the steroid hormones and gonadotropins.

2.5.2 Steroids

The demonstration by Huff and Eik-Nes (1966) that rabbit blastocysts could metabolize ^{14}C-acetate into sterols and steroids appears to have been the first evidence for the existence of these metabolic pathways in early mammalian embryos. The embryos used in that study, however, were late blastocysts removed from the uterus shortly before the expected time of implantation, then maintained in vitro over the periimplantation period. Thus, the question regarding steroid biosynthesis in embryos at earlier stages was not resolved. In addition, the steroid identified is pregnenolone, rather than progesterone.

In 1973, Perry et al. presented radioimmunoassay evidence that the pig blastocyst at 14 days post coitum possesses enzymatic activities corresponding to aromatase, 17–20 desmolase, and 3-sulfatase—all of which are involved in the synthesis of estrone and 17β-estradiol. These activities are not detectable in 10-day post coitum pig blastocysts, but because definitive implantation does not occur in the pig until about 18 days post coitum, their findings constituted the first evidence for biologically active steroid synthesis in the preimplantation mammalian embryo. Even in the 16-day post coitum blastocyst they detected only a very minor conversion of pregnenolone to progesterone.

Detection of steroid biosynthetic activity in preimplantation embryos other than the pig has been both claimed and disputed. There are reports of histochemical evidence for the presence of Δ^5,3β-hydroxysteroid dehydrogenase activity, the enzyme converting pregnenolone to progesterone in morula-stage embryos of the rat, mouse, hamster, and rabbit (Dickmann and Dey, 1974; Dickmann and Gupta, 1974; Dickmann et al., 1975). The method employed in these histochemical studies relies upon nitroblue tetrazolium reduction, with subsequent formazan deposition in embryonic tissues, in the presence of a suitable substrate. Unfortunately, the substrate chosen was dehydroepiandrosterone rather than pregnenolone, and so there is some doubt regarding the specificity of the reaction observed. Chew and Sherman (1975) used pregnenolone as substrate in a highly sensitive biochemical assay for Δ^5,3β-hydroxysteroid dehydrogenase, but they were unable to detect activity prior to the ninth day of gestation in the mouse embryo—and then primarily in the polyploid giant cells of the trophoblast.

Perhaps it is fair to conclude that, with the exception of the pig blastocyst, there is at this moment no indisputable evidence for steroid biosynthesis in mammalian embryos before implantation. Thus, the status of the steroids as important embryonic signals during the preimplantation period awaits further investigation, perhaps concentrating on the estrogens.

2.5.3 Gonadotropins

The fact that the corpus luteum is maintained during pregnancy but not during pseudopregnancy implies the synthesis and secretion of a luteotropin by the products of conception. An alternative source of the luteotropin might be the pituitary, where its secretion could be prolonged by yet another signal from the conceptus. Elevated levels of gonadotropin can be detected in human plasma six to eight days after egg fertilization, or about the time of expected implantation (Saxena et al., 1974). The possibility that the preimplantation embryo produces such a substance and that it serves as a "signal" to the reproductive tract that fertilization has indeed occurred has stimulated several research efforts to detect a gonadotropin in the preimplantation embryo. Blastocoelic fluid from the late rabbit blastocyst has been assayed for gonadotropic activity by a radioreceptor method (Haour and Saxena, 1974) with positive results and by a bioassay method (Sundaram et al., 1975) with negative results. The radioreceptor assay could not distinguish between a substance similar to human chorionic gonadotropin or to luteinizing hormone, nor could it establish unequivocally the source of the material as embryonic. The bioassay method relied upon testosterone production by isolated mouse testes in response to human chorionic gonadotropin. Rabbit blastocoelic fluid failed to induce testosterone production. Serum samples taken from rabbits at intervals throughout pregnancy likewise failed to induce testosterone production, however. The only positive response reported in this study, in serum taken 1–2 hr after coitus, reflected a postcoital surge of pituitary luteotropin. Thus, by whatever means the rabbit maintains the corpus luteum throughout pregnancy, it is evidently not by the production of material that behaves like human chorionic gonadotropin or luteotropin in this bioassay.

It is possible that a gonadotropin is synthesized by the preimplantation embryo but remains tightly bound to the embryonic surface and does not enter either the blastocoelic fluid or the general circulation. Such cell surface material would scarcely function as a "gonadotropin," but it might have a role in the immunologic protection of the embryo and later placenta (Adcock et al., 1973). In this respect, the investigation by Wiley (1974) might be more relevant. Using indirect immunofluorescence, she found material cross-reacting with human chorionic gonadotropin on the surface of the mouse morula. Unfortunately, with respect to its expected biological role, the amount of this material appears to decline after blastocyst formation and is "undetectable" in the mouse placenta. The significance of these findings awaits clarification.

References

General

BIGGERS, J. D., and STERN, S. (1973).
Metabolism of the preimplantation mammalian embryo. Adv. Reprod. Physiol. 6, 1–59.

EPSTEIN, C. J. (1975).
Gene expression and macromolecular synthesis during preimplantation embryonic development. Biol. Reprod. 12, 81–105.

GRAHAM, C. F. (1973).
Nucleic acid metabolism during early mammalian development. In The Regulation of Mammalian Reproduction, S. J. Segal, ed. (Springfield, Ill.: Charles C Thomas), pp. 286–298.

MANES, C. (1975).
Genetic and biochemical activities in preimplantation embryos. In The Developmental Biology of Reproduction, C. L. Markert and J. Papaconstantinou, eds. (New York: Academic Press), pp. 133–163.

DNA Synthesis

BICZYSKO, W., SOLTER, D., GRAHAM, C., and KOPROWSKI, H. (1974).
Synthesis of endogenous type-A virus particles in parthenogenetically stimulated mouse eggs. J. Nat. Cancer Inst. 52, 483–489.

BRINSTER, R. L. (1974).
Embryo development. J. Animal Sci. 38, 1003–1012.

CALARCO, P. G., and SZOLLOSI, D. (1973).
Intracisternal A particles in ova and preimplantation stages of the mouse. Nature New Biol. 243, 91–93.

CHASE, D. G., and PIKÓ, L. (1973).
Expression of A- and C-type particles in early mouse embryos. J. Nat. Cancer Inst. 51, 1971–1975.

CLEAVER, J. E. (1968).
Defective repair replication of DNA in xeroderma pigmentosum. Nature 218, 652–656.

DEMOPOULOS, H. B. (1973).
The basis of free radical pathology. Fed. Proc. 32, 1859–1861.

EPEL, D. (1972).
Activation of an Na^+-dependent amino acid transport system upon fertilization of sea urchin eggs. Exp. Cell Res. 72, 74–89.

EPSTEIN, C. J. (1972).
Inactivation of the X-chromosome. New England J. Med. 286, 318–319.

GOLDSTEIN, L. S., SPINDLE, A. I., and PEDERSEN, R. A. (1975).
X-ray sensitivity of the preimplantation mouse embryo in vitro. Radiation Res. 62, 276–287.

GRAVES, C. N., and BIGGERS, J. D. (1970).
Carbon dioxide fixation by mouse embryos prior to implantation. Science 167, 1506–1508.

HIGGINS, P. J., BORENFREUND, E., and BENDICH, A. (1975).
Appearance of foetal antigens in somatic cells after interaction with heterologous sperm. Nature 257, 488–489.

HOLLIDAY, R., and PUGH, J. E. (1975).
DNA modification mechanisms and gene activity during development. Science 187, 226–232.

HUTCHINSON, C. A., NEWBOLD, J. E., POTTER, S. S., and EDGELL, M. H. (1974).
Maternal inheritance of mammalian mitochondrial DNA. Nature 251, 536–538.

ISSA, M., BLANK, C. E., and ATHERTON, G. W. (1969).
The temporal appearance of sex chromatin and of the late-replicating X-chromosome in blastocysts of the domestic rabbit. Cytogenetics 8, 219–237.

MINTZ, B. (1964).
Synthetic processes and early development in the mammalian egg. J. Exp. Zool. 157, 85–100.

MUKAI, F. H., and GOLDSTEIN, B. D. (1976).
Mutagenicity of malonaldehyde, a decomposition product of peroxidized polyunsaturated fatty acids. Science 191, 868–869.

NORONHA, J. M., SHEYS, G. H., and BUCHANAN, J. M. (1972).
Induction of a reductive pathway for deoxyribonucleotide synthesis during early embryogenesis of the sea urchin. Proc. Nat. Acad. Sci. USA 69, 2006–2010.

PEDERSEN, R. A., and CLEAVER, J. E. (1975).
Repair of UV damage to DNA of implantation-stage mouse embryos in vitro. Exp. Cell Res. 95, 247–253.

PIKÓ, L., and MATSUMOTO, L. (1976).
Number of mitochondria and some properties of mitochondrial DNA in the mouse egg. Dev. Biol. 49, 1–10.

SAGER, R., and KITCHIN, R. (1975).
Selective silencing of eukaryotic DNA. Science 189, 426–433.

SCARANO, E., IACCARINE, M., GRIPPO, P., and WICKELMANS, D. (1965).
On methylation of DNA during development of the sea urchin embryo. J. Mol. Biol. 14, 603–607.

SCHEIN, A., BERDAHL, B. J., LOW, M., and BOREK, E. (1972).
Deficiency of the DNA of Micrococcus radiodurans in methyladenine and methylcytosine. Biochim. Biophys. Acta 272, 481–485.

SHERMAN, M. I., and KANG, H. S. (1973).
DNA polymerases in midgestation mouse embryo, trophoblast, and decidua. Dev. Biol. 34, 200–210.

SZOLLOSI, D. G. (1965).
The fate of sperm middle-piece mitochondria in the rat egg. J. Exp. Zool. 159, 367–378.

VAN BLERKOM, J., and RUNNER, M. N. (1976).
The fine structural development of preimplantation mouse parthenotes. J. Exp. Zool. 196, 113–123.

WHITTEN, W. K. (1971).
Nutrient requirements for the culture of preimplantation embryos in vitro. Advan. Biosciences 6, 129–139.

WITKIN, S. S., KORNGOLD, G. C., and BENDICH, A. (1975).
Ribonuclease-sensitive DNA-synthesizing complex in human sperm heads and seminal fluid. Proc. Nat. Acad. Sci. USA 72, 3295–3299.

YANG, S. S., and WIVEL, N. A. (1974).
Characterization of an endogenous RNA-dependent DNA polymerase associated with murine intracisternal A particles. J. Virol. *13*, 712–720.

RNA Synthesis

BACHVAROVA, R. (1975).
Mouse egg ribosomes. J. Cell. Biol. *67* (Suppl.), 14a.

BERNSTEIN, R. M., and MUKHERJEE, B. B. (1972).
Control of nuclear RNA synthesis in 2-cell and 4-cell mouse embryos. Nature *238*, 457–459.

BOTH, G. W., BANERJEE, A. K., and SHATKIN, A. J. (1975).
Methylation-dependent translation of viral messenger RNAs *in vitro*. Proc. Nat. Acad. Sci. USA *72*, 1189–1193.

BURKHOLDER, G. D., COMINGS, D. E., and OKADA, T. A. (1971).
A storage form of ribosomes in mouse oocytes. Exp. Cell Res. *69*, 361–371.

CHESTERTON, C. J., COUPAR, B. E. H., BUTTERWORTH, P. H. W., BUSS, J., and GREEN, M. H. (1975).
Studies on the control of ribosomal RNA synthesis in HeLa Cells. Eur. J. Biochem. *57*, 79–83.

CLANDININ, M. Y., and SCHULTZ, G. A. (1975).
Levels and modification of methionyl-transfer RNA in preimplantation rabbit embryos. J. Mol. Biol. *93*, 517–528.

CLEGG, K. B., and PIKÓ, L. (1975).
Patterns of RNA synthesis in early mouse embryos. J. Cell Biol. *67* (Suppl.), 72a.

CROSS, M. H., CROSS, P. C., and BRINSTER, R. L. (1973).
Changes in membrane potential during mouse egg development. Dev. Biol. *33*, 412–416.

ELLEM, K. A. O., and GWATKIN, R. B. L. (1968).
Patterns of nucleic acid synthesis in the early mouse embryo. Dev. Biol. *18*, 311–330.

EPSTEIN, C. J., and DAENTL, D. L. (1971).
Precursor pools and RNA synthesis in preimplantation mouse embryos. Dev. Biol. *26*, 517–524.

GILLESPIE, D., and GALLO, R. C. (1975).
RNA processing and RNA tumor virus origin and evolution. Science *188*, 802–811.

HOAGE, T. R., and CAMERON, I. L. (1975).
DNA synthesis in mature mouse oocytes. J. Cell Biol. *67* (Suppl.), 174a.

KARP, G., MANES, C., and HAHN, W. E. (1973).
RNA synthesis in the preimplantation rabbit embryo: radioautographic analysis. Dev. Biol. *31*, 404–408.

KERR, S. J., and BOREK, E. (1972).
The tRNA methyltransferases. Adv. Enzymology *36*, 1–27.

MANES, C., and SHARMA, O. K. (1973).
Hypermethylated tRNA in cleaving rabbit embryos. Nature *244*, 283–284.

MOORE, G. P. M., LINTERN-MOORE, S., PETERS, H., BYSKOV, A. G., ANDERSEN, M., and FABER, M. (1975).
The inhibition of DNA-dependent RNA synthesis in Yoshida ascites

cells by bovine follicular fluid. J. Cell. Physiol. *86*, 31–36.

PIKÓ, L. (1970).
Synthesis of macromolecules in early mouse embryos cultured *in vitro*: RNA, DNA, and a polysaccharide component. Dev. Biol. *21*, 257–279.

ROEDER, R. G., and RUTTER, W. J. (1970a).
Specific nucleolar and nucleoplasmic RNA polymerases. Proc. Nat. Acad. Sci. USA *65*, 675–682.

ROEDER, R. G., and RUTTER, W. J. (1970b).
Multiple ribonucleic acid polymerases and ribonucleic acid synthesis during sea urchin development. Biochemistry *9*, 2543–2553.

SCHULTZ, G. A. (1973).
Characterization of polyribosomes containing newly-synthesized messenger RNA in preimplantation rabbit embryos. Exp. Cell Res. *82*, 168–174.

SCHULTZ, G. A. (1974).
The stability of messenger RNA containing polyadenylic acid sequences in preimplantation rabbit embryos. Exp. Cell Res. *86*, 190–193.

SCHULTZ, G. A. (1975).
Polyadenylic acid-containing RNA in unfertilized and fertilized eggs of the rabbit. Dev. Biol. *44*, 270–277.

SCHULTZ, G. A., and CHURCH, R. B. (1975).
Transcriptional patterns in early mammalian development. *In* The Biochemistry of Animal Development, R. Weber, ed., vol. 3. (New York: Academic Press), pp. 47–90.

SCHULTZ, G. A., MANES, C., and HAHN, W. E. (1973).
Synthesis of RNA containing poly-adenylic acid sequences in preimplantation rabbit embryos. Dev. Biol. *30*, 418–426.

SLATER, I., GILLESPIE, D., and SLATER, D. W. (1973).
Cytoplasmic adenylation and processing of maternal RNA. Proc. Nat. Acad. Sci. USA *70*, 406–411.

TASCA, R. J., and HILLMAN, N. (1970).
Effects of actinomycin D and cycloheximide on RNA and protein synthesis in cleavage stage mouse embryos. Nature *225*, 1022–1025.

TOCCHINI-VALENTINI, G. P., and CRIPPA, M. (1970).
Ribosomal RNA synthesis and RNA polymerase. Nature *228*, 993–995.

VAN BLERKOM, J., MANES, C., and DANIEL, J. C. Jr. (1973).
Development of preimplantation rabbit embryos *in vivo* and *in vitro*. I. An ultrastructural comparison. Dev. Biol. *35*, 262–282.

VERSTEEGH, L. R., HEARN, T. F., and WARNER, C. M. (1975).
Variations in the amounts of RNA polymerase forms I, II, and III during preimplantation development in the mouse. Dev. Biol. *46*, 430–435.

WOODLAND, H. R., and GRAHAM, C. F. (1969).
RNA synthesis during early development of the mouse. Nature *221*, 327–332.

YOUNG, R. J., STULL, G. B., and BRINSTER, R. L. (1973).
RNA in mouse ovulated oöcytes. J. Cell. Biol. *59*, 372a.

Protein Synthesis

AITKEN, R. J. (1974).
Calcium and zinc in the endometrium
and uterine flushings of the roe deer
(*Capreolus capreolus*) during delayed
implantation. J. Reprod. Fertil. *40*,
333–340.

BACHVAROVA, R. (1975).
Mouse egg ribosomes. J. Cell Biol. *67*
(Suppl.), 14a.

BIGGERS, J. D., and STERN, S.
(1973).
Metabolism of the preimplantation
mammalian embryo. Advan. Reprod.
Physiol. *6*, 1–59.

BONNER, W. M., and LASKEY,
R. A. (1974).
A film detection method for tritium-
labelled proteins and nucleic acids in
polyacrylamide gels. Eur. J. Biochem.
46, 83–88.

BORLAND, R. M., BIGGERS, J. D.,
and LECHENE, C. P. (1976).
Kinetic aspects of rabbit blastocoele
fluid accumulation: an application of
electron probe microanalysis. Dev.
Biol. *50*, 201–211.

BORLAND, R. M., and TASCA, R. J.
(1974).
Activation of a Na^+-dependent amino
acid transport system in preimplantation mouse embryos. Dev. Biol. *36*,
169–182.

BORLAND, R. M., and TASCA, R. J.
(1975).
Inhibition of L-methionine uptake and
incorporation by chlorpromazine in
preimplantation mouse embryos. J.
Reprod. Fertil. *42*, 473–481.

BRINSTER, R. L. (1971).
Uptake and incorporation of amino
acids by the preimplantation mouse

embryo. J. Reprod. Fertil. *27*,
329–338.

BRINSTER, R. L. (1973).
Protein synthesis and enzyme consti-
tution of the preimplantation mamma-
lian embryo. *In* The Regulation of
Mammalian Reproduction, S. Segal,
R. Crozier, P. A. Corfman, and P. G.
Condliffe, eds. (Springfield, Ill.:
Charles C Thomas), pp. 302–334.

BRINSTER, R. L. (1974).
Embryo development. J. Animal Sci.
38, 1003–1012.

CALARCO, P. G., and BROWN,
E. H. (1969).
An ultrastructural and cytological
study of preimplantation development
of the mouse. J. Exp. Zool. *171*,
253–283.

DANIEL, J. C. (1970).
Dormant embryos of mammals.
BioSciences *20*, 411–415.

DASS, C. M. S., MOHLA, S., and
PRASAD, M. R. N. (1969).
Time sequence of action of estrogen
on nucleic acid and protein synthesis
in the uterus and blastocyst during delayed implantation in the rat. Endocri-
nology *85*, 528–536.

ENDERS, A. C., ed. (1963).
Delayed Implantation. Chicago: University of Chicago Press.

EPSTEIN, C. J., and SMITH, S. A.
(1973).
Amino acid uptake and protein synthesis in preimplantation mouse embryos. Dev. Biol. *33*, 171–184.

EPSTEIN, C. J., and SMITH, S. A.
(1974).
Electrophoretic analysis of proteins
synthesized by preimplantation mouse
embryos. Dev. Biol. *40* 233–244.

GARDNER, R. L. (1971).
Manipulations on the blastocyst. Advan. Biosciences 6, 279–296.

GINSBERG, L., and HILLMAN, N. (1975a).
Shifts in ATP synthesis during preimplantation stages of mouse embryos. J. Reprod. Fertil. 43, 83–89.

GINSBERG, L., and HILLMAN, N. (1975b).
ATP metabolism in t^n/t^n mouse embryos. J. Embryol. Exp. Morph. 33, 715–723.

GOLBUS, M. S., CALARCO, P., and EPSTEIN, C. J. (1975).
The effects of inhibitors of RNA synthesis (α amanitin and actinomycin D) on preimplantation mouse embryogenesis. J. Exp. Zool. 186, 207–216.

GRIFFIN, B. (1975).
"Enigma variations" of mammalian messenger RNA. Nature 255, 9.

GULYAS, B. J., and DANIEL, J. C., Jr. (1969).
Incorporation of labelled nucleic acid and protein precursors by diapausing and nondiapausing blastocysts. Biol. Reprod. 1, 11–20.

GWATKIN, R. B. L. (1966).
Amino acid requirement for attachment and outgrowth of mouse blastocysts in vitro. J. Cell. Physiol. 68, 335–343.

GWATKIN, R. B. L. (1969).
Nutritional requirements for post-blastocyst development in the mouse. Int. J. Fertil. 14, 101–105.

HERBERT, M. C., and GRAHAM, C. F. (1974).
Cell determination and biochemical differentiation of the early mammalian embryo. Curr. Top. Dev. Biol. 8, 151–178.

HICKEY, E. D., Weber, L. A., and BAGLIONI, C. (1976).
Translation of RNA from unfertilised sea urchin eggs does not require methylation and is inhibited by 7-methylguanosine-5'-monophosphate. Nature 261, 71–72

HILLMAN, N. (1975).
Studies of the T-locus. In The Early Development of Mammals, M. Balls and A. E. Wild, eds. (London: Cambridge University Press), pp. 189–206.

HILLMAN, N., SHERMAN, M. I., and GRAHAM, C. (1972).
The effect of spatial arrangement on cell determination during mouse development. J. Embryol. Exp. Morph. 28, 263–278.

HILLMAN, N., and TASCA, R. J. (1969).
Ultrastructural and autoradiographic studies of mouse cleavage stages. Am. J. Anat. 126, 151–174.

KARP, G. C., MANES, C., and HAHN, W. E. (1974).
Ribosome production and protein synthesis in the preimplantation rabbit embryo. Differentiation 2, 65–73.

LASKEY, R. A., and MILLS, A. D. (1975).
Quantitative film detection of ^{3}H and ^{14}C in polyacrylamide gels by fluorography. Eur. J. Biochem. 56, 335–341.

MCLAREN, A. (1968).
A study of blastocysts during delay and subsequent implantation in lactating mice. J. Endocrinol. 42 453–464.

MCLAREN, A. (1971).
Blastocysts in the mouse uterus: the effect of ovariectomy, progesterone

and oestrogen. J. Endocrinol. *50*, 515–526.

MCLAREN, A. (1973).
Blastocyst activation. *In* The Regulation of Mammalian Reproduction, S. J. Segal, R. Crozier, P. A. Corfman, and P. G. Condliffe, eds. (Springfield, Ill.: Charles C Thomas), pp. 321–328.

MANES, C. (1971).
Nucleic acid synthesis in preimplantation rabbit embryos. II. Delayed synthesis of ribosomal RNA. J. Exp. Zool. *176*, 87–95.

MANES, C. (1973).
The participation of the embryonic genome during early cleavage in the rabbit. Dev. Biol. *32*, 453–459.

MANES, C., and DANIEL, J. C., Jr. (1969).
Quantitative and qualitative aspects of protein synthesis in the preimplantation rabbit embryo. Exp. Cell Res. *55*, 261–268.

MENKE, T. M., and MCLAREN, A. (1970).
Carbon dioxide production by mouse blastocysts during lactational delay of implantation or after ovarietomy. J. Endocrinol. *47*, 287–294.

MINTZ, B. (1964).
Synthetic processes and early development in the mammalian egg. J. Exp. Zool. *157*, 85–100.

MONESI, V., and SALFI, V. (1967).
Macromolecular synthesis during early development in the mouse embryo. Exp. Cell Res. *46*, 632–635.

MUTHUKRISHNAN, S., BOTH, G. W., FURUICHI, Y., and SHATKIN, A. J. (1975).
5'-Terminal 7-methylguanosine in eukaryotic mRNA is required for translation. Nature *255*, 33–37.

O'FARRELL, P. H. (1975).
High resolution two-dimensional electrophoresis of proteins. J. Biol. Chem. *250*, 4007–4021.

PETZOLDT, U. (1972).
Protein patterns of the rabbit blastocyst tissues. Cytobiologie *6*, 473–475.

PETZOLDT, U. (1975).
Amino acid incorporation into embryonic proteins during rabbit preimplantation development. Cytobiologie *11*, 490–493.

PETZOLDT, U., BRIEL. G., GOTTSCHEWSKI, G. H. M., and NEUHOFF, V. (1973).
Free amino acids in the early cleavage stages of the rabbit egg. Dev. Biol. *31*, 38–46.

PETZOLDT, U., DAMES, W., GOTTSCHEWSKI, G. H. M., and NEUHOFF, V. (1972).
Das Proteinmuster in frühen Entwicklungsstadien des Kaninchens. Cytobiologie *5*, 272–280.

PRASAD, M. N. R., DASS, C. M. S., and MOHLA, S. (1968).
Action of oestrogen on the blastocysts and uterus in delayed implantation: an autoradiographic study. J. Reprod. Fertil. *16*, 97–104.

PSYCHOYOS, A. (1969).
Hormonal requirements for egg implantation. Advan. Biosciences *4*, 275–290.

SCHULTZ, G. A. (1973).
Characterization of polyribosomes containing newly-synthesized messenger RNA in preimplantation rabbit embryos. Exp. Cell Res. *82*, 168–174.

SCHULTZ, G. A. (1975).
Polyadenylic acid-containing RNA in unfertilized and fertilized eggs of the rabbit. Dev. Biol. 44, 270–277.

SLATER, I., GILLESPIE, D., and SLATER, D. W. (1973).
Cytoplasmic adenylation and processing of maternal RNA. Proc. Nat. Acad. Sci. USA 70, 406–411.

SOLTER, D. and KNOWLES, B. B. (1975).
Immunosurgery of mouse blastocyst. Proc. Nat. Acad. Sci. USA 72, 5099–5102.

SPIRIN, A. S. (1969).
Informosomes. Eur. J. Biochem. 10, 20–35.

SURANI, M. A. H. (1975).
Hormonal regulation of proteins in the uterine secretions of ovariectomized rats and implications for implantation and embryonic diapause. J. Reprod. Fertil. 43, 411–417.

TARKOWSKI, A. K., and WROBLEWSKA, J. (1967).
Development of blastomeres of mouse eggs isolated at the 4- and 8-cell stage. J. Embryol. Exp. Morph. 18, 155–180.

VAN BLERKOM, J., BARTON, S. C., JOHNSON, M. H. (1976).
Molecular differentiation in the preimplantation embryo. Nature 259, 319–321.

VAN BLERKOM, J., and BROCKWAY, G. O. (1975a).
Qualitative patterns of protein synthesis in the preimplantation mouse embryo. I. Normal pregnancy. Dev. Biol. 44, 148–157.

VAN BLERKOM, J., and BROCKWAY, G. O. (1975b).
Qualitative patterns of protein synthesis in the preimplantation mouse embryo. II. During release from facultative delayed implantation. Dev. Biol. 46, 446–451.

VAN BLERKOM, J., and MCGAUGHEY, R. (1977).
Molecular differentiation during the in vivo and in vitro maturation, fertilization and preimplantation development of the rabbit ovum. Submitted for publication.

VAN BLERKOM, J., and MANES, C. (1974).
Development of preimplantation rabbit embryos in vivo and in vitro. II. A comparison of qualitative aspects of protein synthesis. Dev. Biol. 40, 40–51.

VAN BLERKOM, J., MANES, C., and DANIEL, J. C., Jr. (1973).
Development of preimplantation rabbit embryos in vivo and in vitro. I. An ultrastructural comparison. Dev. Biol. 35, 262–282.

VAN BLERKOM, J., and RUNNER, M. N. (1976).
The fine structural development of preimplantation mouse parthenotes. J. Exp. Zool. 196, 113–123.

WEITLAUF, H. M. (1969).
Temporal changes in protein synthesis by mouse blastocysts transferred to ovariectomized recipients. J. Exp. Zool. 171, 481–486.

WEITLAUF, H. M. (1974).
Effect of actinomycin D on protein synthesis by delayed implanting mouse embryos in vitro. J. Exp. Zool. 189, 197–202.

WU, J. T., and MEYER, R. K. (1974).
Ultrastructural changes in rat blastocysts induced by estrogen during delayed implantation. Anat. Rec. *179*,
253–272.

Lipids

DICKMANN, Z., and SPILMAN,
C. H. (1975).
Prostaglandins in rabbit blastocysts.
Science *190*, 997–998.

EL-BANNA, A. A., SACHER, B.,
and SCHILLING, E. (1976).
Effect of indomethacin on egg transport and pregnancy in the rabbit. J.
Reprod. Fertil. *46*, 375–378.

FINKELSTEIN, I. N. (1973).
The fatty acids of early pregnancy
and their metabolism in the preimplantation rabbit embryo. Ph.D.
dissertation, University of Colorado.

NADIJCKA, M., and HILLMAN, N.
(1975).
Autoradiographic studies of t^n/t^n
mouse embryos. J. Embryol. Exp.
Morph. *33*, 725–730.

Hormones

ADCOCK, E. W., TEASDALE, F.,
AUGUST, C. S., COX, S.,
MESCHIA, G., BATTAGLIA, F. C.,
and NAUGHTON, M. C. (1973).
Human chorionic gonadotropin: its
possible role in maternal lymphocyte
suppression. Science *181*, 845–847.

BEIER, H. M., and KÜHNEL, W.
(1973).
Pseudopregnancy in the rabbit after
stimulation by human chorionic
gonadotropin. Hormone Res. *4*, 1–27.

CHEW, N. J., and SHERMAN, M. I.
(1975).
Biochemistry of differentiation of
mouse trophoblast: $\Delta^5,3\beta$-hydroxy-
steroid dehydrogenase. Biol. Reprod.
12, 351–359.

DICKMANN, Z., and DEY, S. K.
(1974).
Steroidogenesis in the preimplantation
rat embryo and its possible influence
on morula-blastocyst transformation
and implantation. J. Reprod. Fertil.
37, 91–93.

DICKMANN, Z., DEY, S. K., and
GUPTA, J. S. (1975).
Steroidogenesis in rabbit preimplanta-
tion embryos. Proc. Nat. Acad. Sci.
USA *72*, 298–300.

DICKMANN, Z., and GUPTA, J. S.
(1974).
$\Delta^5,3\beta$-hydroxysteroid dehydrogenase
and 17β-hydroxysteroid dehydrogenase activity in preimplantation hamster embryos. Dev. Biol. *40*, 196–198.

FREESE, U. E., ORMAN, S., and
PAULOS, G. (1973).
An autoradiographic investigation of
epithelium-egg interaction in the
mouse oviduct. Am. J. Obstet.
Gynec. *117*, 364–370.

FRIDHANDLER, L. (1968).
Intermediary metabolic pathways in
preimplantation rabbit blastocysts.
Fertil. Steril. *19*, 424–434.

FUCHS, A. R., and BELING, C.
(1974).
Evidence for early ovarian recognition
of blastocysts in rabbits. Endocrinol-
ogy *95*, 1054–1058.

GINSBERG, L., and HILLMAN, N.
(1975).
ATP metabolism in t^n/t^n mouse embryos. J. Embryol. Exp. Morph. *33*.
715–723.

HAOUR, F., and SAXENA, B. B.
(1974).
Detection of a gonadotropin in rabbit
blastocyst before implantation. Science *185*, 444–445.

HUFF, R. L., and EIK-NES, K. B.
(1966).
Metabolism *in vitro* of acetate and
certain steroids by six-day-old rabbit
blastocysts. J. Reprod. Fertil. *11*,
57–63.

KENT, H. A., Jr. (1975).
The two to four-cell embryos as
source tissue of the tetrapeptide preventing ovulations in the hamster.
Am. J. Anat. *144*, 509–512.

PERRY, J. S., HEAP, R. B., and
AMOROSO, E. C. (1973).
Steroid hormone production by pig
blastocysts. Nature *245*, 45–47.

RENFREE, M. B. (1972).
Influence of the embryo on the marsupial uterus. Nature *240*, 475–477.

SAXENA, B. B., HASAN, S. H.,
HAOUR, F., and SCHMIDT-
GOLLWITZER, K. (1974).
Radioreceptor assay of human chori-
onic gonadotropin: detection of early
pregnancy. Science *184*, 793–795.

SUNDARAM, K., CONNELL,
K. G., and PASSANTINO, T. (1975).
Implication of absence of HCG-like
gonadotrophin in the blastocyst for
control of corpus luteum function in
pregnant rabbit. Nature *256*, 739–741.

VAN HOORN, G., and DENKER,
H. W. (1975).
Effect of the blastocyst on a uterine
amino acid arylamidase in the rabbit.
J. Reprod. Fertil. *45*, 359–362.

WILEY, L. D. (1974).
Presence of a gonadotropin on the
surface of preimplanted mouse embryos. Nature *252*, 715–716.

3 GENETICS OF EARLY MAMMALIAN EMBRYOGENESIS

Verne M. Chapman,
John D. West, and
David A. Adler

3.1 Introduction

In early mammalian development, much attention has focused on the genetic issues concerned with the timing of embryo gene function and the relative roles of the embryo genome and cytoplasmic factors in directing the initial stages of morphogenesis. Results obtained from the commonly studied embryos of amphibians and marine invertebrates have influenced these studies. Oocytes of these organisms possess amplified amounts of genes coding for ribosomal RNA (Brown and Dawid, 1969), masked messenger RNA (Gross, 1967; Tyler, 1967; Skoultchi and Gross, 1973; Gross et al., 1973), and an extensive array of maternally derived enzymes (Wallace, 1961; Wright and Moyer, 1968).

In mammals, however, biochemical and genetic evidence indicates that gene function begins during early cleavage stages, perhaps as early as the two-cell stage in the mouse (Epstein, 1975; Chapman et al., 1976). Genetic variation has been useful in establishing the timing of gene expression in embryos, and the existence of early lethal genes confirms that gene function is in fact required in embryos for their successful preimplantation development.

Because the embryo genome is functional during the preimplantation period, it is possible to study the developmental regulation of structural genes beginning at this stage. The genetic questions addressed here include:

1. Are there sites in the genome that regulate the expression of structural genes in early embryos?
2. If so, what are the linkage relationships between these regulatory sites and the structural genes?
3. How do different classes of regulatory genes, linked and unlinked, affect such regulation?
4. Are there general regulatory sites in the genome that coordinately activate or regulate a set of genes?

Answering these questions would be facilitated by studying the regulation of genes with biochemically characterized products, such as the structural genes of enzymes. Alleles of these genes should not significantly alter the overall development of the embryo. Sensitive assay

Work reported in this paper was supported by grants from the National Institutes of Health and from the Lalor Foundation.

procedures that measure amounts of gene product either in single embryos or in relatively small sample sizes are technically helpful but not
essential.

One system that has many features resembling control of a single
locus in early embryogenesis is the phenomenon of X-chromosome
dosage compensation (Lyon, 1961). Because this phenomenon has been
widely studied in a variety of mammals, the timing of the event and the
possible genetic mechanisms responsible will receive special attention
here.

3.2 Gene Expression in Early Mammalian Embryos

An important feature of mammalian development is the early functioning of the embryonic genome. This has been established by both genetic and biochemical studies.

3.2.1 Paternal Gene Products

One way to test for transcription and translation of embryo genes is to
use genetic variants of a structural gene product that can be qualitatively identified either electrophoretically or by heat denaturation. Embryo gene expression is determined by examining F_1 hybrid embryos
for the appearance of a heterozygous phenotype. In early embryos,
enzyme activity remaining from synthesis during oogenesis might obscure the actual timing of appearance of paternal gene product and
impair the determination of whether maternally and paternally derived
genes are coordinately expressed. However, in the mouse this general
approach has been successful with electrophoretic variants of glucose
phosphate isomerase (GPI: Chapman et al., 1971; Brinster, 1973),
heat denaturation variants of β-glucuronidase (Wudl and Chapman,
1976), and cell surface alloantigen differences between inbred strains
(Muggleton-Harris and Johnson, 1976).

Two electrophoretic alleles exist for GPI in inbred mice (Carter and
Parr, 1967). These alternative forms can be easily separated by starch
gel electrophoresis and identified with a specific enzyme stain. The
enzyme is a dimer with single bands of activity in the homozygous
forms. The heterozygote phenotype is 3-banded and has an intermediate heteropolymeric form.

Activity levels of GPI in early mouse embryos are initially high. Activity per embryo begins to decrease at about the 8-cell stage. The electrophoretic phenotype between the 2-cell and 8-cell stage is predominantly the maternal type. Hence the GPI activity during this period probably persists from synthesis during oogenesis. In initial studies, Chapman et al. (1971) used pooled sample sizes of 15 to 40 embryos to detect the heteropolymer band of activity in blastocysts. Using a sample size of 500 embryos, Brinster (1973) was able to detect paternal gene product by the 8-cell stage.

Employing a similar experimental approach, Wudl and Chapman (1976) used a heat denaturation variant for β-glucuronidase to determine the expression of paternal gene products for another autosomal locus in early mouse embryos. These experiments were facilitated by a hundredfold increase in activity during the preimplantation period. The C3H strain of mice carries the Gus^h structural allele for β-glucuronidase, which is more heat labile than the product of the Gus^b allele found in the C57BL/6 strain. Specifically, less than 10 percent of the Gus^h β-glucuronidase activity remains after heating for 30 min at 70°C, whereas more than 50 percent of the Gus^b β-glucuronidase activity survives the same heat treatment. In $Gus^{h/b}$ heterozygotes, an intermediate amount of activity survives the heat treatment (Ganschow and Paigen, 1968). Activity levels of β-glucuronidase are also lower in adult tissues of C3H than in C57BL/6, and the hybrid has intermediate levels. Antibody titration of β-glucuronidase activity in Gus^h and Gus^b strain shows that equivalent units of enzyme activity are precipitated by the same amount of antibody (Ganschow, 1975; R. Davey, personal communication). Differences in enzyme activity are thus due to differences in the number of enzyme molecules. Recently, Ganschow (1975) has demonstrated that this effect is caused by differences in rates of synthesis.

Using a microfluorometric assay technique, Wudl and Chapman (1976) demonstrated that a hundredfold increase in β-glucuronidase activity occurred between 60 and 84 hr post coitum (8-cell to early blastocyst stage) in C3H, C57BL/6, and (C3H × C57BL/6)F_1 embryos. Heat denaturation studies on pooled embryo samples at 84 hr showed that a higher percentage of β-glucuronidase activity survived the heat treatment in F_1 hybrid ($Gus^{h/b}$) embryos from C3H females than survived in homozygous Gus^h embryos. These data demonstrate that the paternal

Gus^b allele was transcribed prior to or during the hundredfold increase in activity. A comparison of activity changes in F_1 and C3H embryos also showed a greater increase in activity in F_1 embryos between 36 and 57 hr (2- to 8-cell stage). However, there are reports that C3H embryos might be more delayed in their overall development during early cleavage than C57BL (McLaren and Bowman, 1973) so that β-glucuronidase activity differences might also reflect differences in developmental rates.

To circumvent the problem of strain differences in developmental rates (Chapman et al., 1976) early changes in β-glucuronidase activity were examined in a C57BL/6 strain congenic for the Gus^h allele from C3H. Measurements of β-glucuronidase activity levels were made at 30 and 60 hr post artificial insemination in Gus^b, $Gus^{h/b}$ and Gus^h embryos (table 3.1). The heterozygous $Gus^{h/b}$ were from Gus^h females and Gus^b males. These data show that the Gus^h embryos have lower β-glucuronidase activity than $Gus^{h/b}$ embryos at all stages of development. Furthermore, the amount of increase in activity is more than ten times greater in heterozygous embryos than in homozygous Gus^h embryos. These findings indicate that the paternal Gus^b allele is actively transcribed and translated during the 2- to 8-cell stage. The smaller increase in activity of β-glucuronidase in Gus^h embryos is also consistent with the lower rate of synthesis found for the Gus^h genotype compared with the Gus^b genotype in adult tissues.

Additional evidence that mouse embryo genes are active in early cleavage comes from the detection by Muggleton-Harris and Johnson (1976) of paternal alloantigens on the surface of blastomeres by the 6- to 8-cell stage. These workers used antisera to C3H spleen cells generated in C57BL/10 mice and anti-C57BL/10 sera raised in C3H mice to look for alloantigens on blastomeres. They observed non-H-2 alloantigens on the surfaces of blastomeres at all preimplantation stages. In hybrid combinations they were able to detect anti-C57BL/10 alloantigens on the surfaces of (C3H × C57BL)F_1 embryos as early as the 6- to 8-cell stage. The antisera in these studies probably reacts with a set of antigens on the cell surface and involves a number of gene products. These findings are consistent with the timing of embryo gene expression observed for β-glucuronidase and GPI.

Table 3.1. β-Glucuronidase Activity in Heterozygous $Gus^{h/b}$ Embryos from C57BL/6 $Gus^{h/h}$ Females $\times$ C57BL/6 $Gus^{b/b}$ Males and in Homozygous $Gus^{h/h}$ Embryos from C57BL/6 $Gus^{h/h}$ Females $\times$ C57BL/6 $Gus^{h/h}$ Males

	β-Glucuronidase activity (f moles embryo^{-1} hr^{-1}) $\pm$ S.E.		
Embryo genotype	30 hr [a]	51 hr [a]	60 hr [a]
$Gus^{h/b}$	2.2 $\pm$ 0.28 (34) [b]	3.4 $\pm$ 0.27 (28)	9.3 $\pm$ 0.80 (70)
$Gus^{h/h}$	1.5 $\pm$ 0.06 (36)	1.6 $\pm$ 0.09 (65)	2.0 $\pm$ 0.15 (34)

[a] Post artificial insemination.

[b] Number of individual embryos assayed.

3.2.2 Molecular Evidence of Gene Expression in Early Embryos

The biochemistry of early development is the subject of another article in this monograph (see chapter 2). However, because these data directly pertain to the topic of gene expression during early embryogenesis, an outline of the biochemical evidence that embryo genes are transcribed and translated early in cleavage divisions of the mouse belongs here as well. In particular, RNA synthesis during early development, the effects of inhibitors of RNA synthesis, changes in the pattern of protein synthesis, and the presence of RNA polymerase warrant attention.

Early work on RNA synthesis in the mouse embryo demonstrated incorporation of ^{3}H-uridine at all stages of development, and label was observed in the nucleolus by the 4-cell stage (Mintz, 1964). Subsequent work verified this early finding and also established the classes of RNA synthesized at various stages.

Characterization of RNA synthesis in cleaving mouse embryos has established that high-molecular-weight RNA is made by the 2-cell stage and that 4s RNA, rRNA, and DNA-like RNA are all synthesized by the 4-cell stage (Woodland and Graham, 1969; Knowland and Graham, 1971). The involvement of newly transcribed RNA in the metabolism of the early embryo is further supported by the observation that inhibitors of RNA synthesis such as actinomycin D and α-amanitin are capable of blocking cleavage and normal development of the mouse (Mintz, 1964; Golbus et al., 1973).

Further support for the transcriptional capability of the embryo genome comes from observations on RNA polymerase (Siracusa, 1973; Warner and Versteegh, 1974; Moore, 1975). In these studies, α-amanitin–resistant polymerase was detectable at the 4-cell stage and α-amanitin–sensitive polymerase was first detectable at the 2-cell stage. These findings are consistent with the timing of the synthesis of high-molecular-weight RNA and the first appearance of paternal gene expression in early embryos.

Protein synthesis has been detected at all stages of early development by means of incorporation of labeled amino acids into specific proteins (Epstein and Smith, 1974; Van Blerkom and Brockway, 1975). These studies have demonstrated that the electrophoretic pattern of newly synthesized proteins changes between the 1- and 2-cell stages. Few further qualitative changes in the pattern of proteins synthesized were evident between the late 2-cell stage and the first morphogenetic change in the formation of the blastocyst. Thus, although no direct evidence indicates that these proteins are the products of newly transcribed message, as opposed to stored mRNA, the biochemical and genetic evidence clearly indicates that it is possible for these proteins to be embryo gene products.

3.2.3 Genetic Abnormalities in Early Development

Single gene effects
The class of mutants that best indicates that functioning of embryo genes is essential during early development are the early lethals. These are primarily observed in mice and include the Ts (tail short) locus, Os (oligosyndactylism) locus, the A^y allele of the A (agouti) locus and the t^{12} allele of the T (Brachyury) complex (Smith, 1956; Green, 1966; McLaren, 1974). In each case, the homozygous mutant dies before implantation, but the actual stage of the arrest in development varies among the mutants and is somewhat dependent upon genetic background.

The important conclusion here is that the early death of mutant genotypes supports the general model that mammalian genes are expressed in early embryogenesis. Furthermore, these effects also suggest not only that embryo genes are expressed but also that successful early development requires a functional embryo genome.

Chromosomal abnormalities

Certain chromosomal imbalances also behave in a similar way to early lethals and support the view that successful development requires a functional genome. For example, the OY class of embryos from XO mothers probably does not survive more than a few cleavage divisions (Morris, 1968; Burgoyne and Biggers, 1976).

Imbalances within a chromosomal complement (aneuploidy) have been documented for several mammalian species (see Hamerton, 1971b; White et al., 1974, a,b) and occur with a particularly high frequency by nondisjunction during gametogenesis in hybrids of laboratory mice (Mus musculus) and tobacco mice (Mus poschiavinus) (Tettenborn and Gropp, 1970; Ford, 1972). Little is known about the effects of aneuploidy on preimplantation development. Hypoploid mouse embryos appear to die before midterm (Ford, 1972), although hyperploid embryos sometimes develop beyond implantation and in some cases survive postnatally. An imbalance in the number of autosomes is usually incompatible with the completion of normal development, but imbalances in the number of sex chromosomes are more readily tolerated, although, as indicated above, it is likely that OY mouse embryos die before implantation. The findings of Burgoyne and Biggers (1976) also indicate that, even in culture, preimplantation mouse embryos from XO mothers develop more slowly than those from XX mothers. These authors suggest that this maternal effect is a result of the difference in X-chromosome dosage during oogenesis.

Variation in the number of sets of chromosomes, leading either to polyploidy or to haploidy, is also well known. Triploid and tetraploid mammalian embryos tend to die during the postimplantation stages of development (see Beatty, 1957; Carr, 1971; Niemierko, 1975; Snow, 1975, 1976). Generally, triploid embryos survive better than tetraploids (Carr, 1971), and many tetraploid mouse embryos lack a functional ICM (Snow, 1976). However, on rare occasions tetraploid mice (Snow, 1975) and triploid humans (Walker et al., 1973) have survived until the end of gestation, only to die a few hours later. The production of adult triploid rabbits has also been claimed (Häggqvist and Bane, 1950), although subsequent chromosome analysis (Melander, 1950) revealed that, at least by the time of analysis, the rabbits were complex chromosomal mosaics (Beatty, 1957).

Haploid embryos in several species have been produced parthenogenetically [reviewed by Graham (1974)] and from mouse zygotes by means of microsurgical techniques (Modlinsky, 1975; Tarkowski and Rossant, 1976). These procedures might result in the formation of haploid, diploid, or haploid/diploid mosaic embryos, but haploid embryos have been characterized only in the mouse and rat (Graham, 1974). Haploid mouse embryos form blastocysts and are capable of inducing a decidual response and implanting (Kaufman and Gardner, 1974) but they probably die soon after implantation. Kaufman and Sachs (1975) studied parthenogenetic mouse embryos with 19, 20, or 21 chromosomes. Embryos with only 19 chromosomes normally failed to develop beyond the 2-cell stage, but those with 21 chromosomes developed like those with 20 chromosomes, at least to the morula stage. The latter result again suggests that normal gene function is necessary for development beyond the 2-cell stage.

At present, studies on the development of embryos with abnormal karyotypes give us little insight into the genetic regulation of early mammalian development. Abnormal karyotypes are tolerated to varying degrees. Extra copies of certain chromosomes (such as the X chromosome or chromosome 21 in humans) appear to be more readily tolerated than most. This phenomenon perhaps reflects a capacity to regulate the "dose" of genetic information (for example, by the inactivation of additional X chromosomes) or might simply indicate that a "genetic overdose" of these particular chromosomes is not fatal. The apparent death of OY mouse embryos during preimplantation development suggests a critical role for at least one X-linked gene before implantation, and the death of many embryos with abnormal karyotypes soon after implantation is consistent with a need for ordered gene expression at this stage (Graham, 1974).

Further considerations of the effects of chromosomal abnormalities upon early embryonic and teratocarcinoma cell development appear in chapter 7.

3.2.4 General Genetic Effects

Studies of the "tempo of development" (McLaren and Bowman, 1973) exist for only a few species. The embryos of the rabbit (Lewis and Gregory, 1929), the mouse (Lewis and Wright, 1935), and the monkey (Lewis and Hartman, 1931) have been the subjects of time-lapse cinematography. Castle and Gregory (1929) and Gregory and Castle (1931) observed differences in cleavage rate between races of small and large rabbits. The differences, however, are not statistically significant, and thus the evidence for a genetic effect is weak.

The rate and timing of early development in mammals is most likely a product of many interacting elements, and genetic factors might operate at many levels of the process. Any study of the rate of development must have a time reference point. The most logical choice is the time of fertilization. In vitro fertilization has been achieved, but an accurate timing of this occurrence in vivo is not yet feasible. In addition, the interval between fertilization and the first cleavage might not be related to intermitotic times of later cleavages [this possibility is consistent with classification of the stages of early development in the human by Streeter (1948)]. The former might be influenced by the maternal environment and genotype, whereas the latter might depend primarily on the embryonic genome.

Whitten and Dagg (1961) determined the times of cleavages in two strains of mice that differed in developmental maturity at the tenth day of gestation. They found that the first cleavage occurred earlier in BALB/c-Gn embryos, but 129 strain embryos arrived at the second and third cleavages and formed blastocoel cavities earlier. Embryos from a 129♀ × BALB/c♂ cross showed similar developmental timing to 129 embryos. The timing of the first cleavage of embryos from the reciprocal cross was similar to that of BALB/c embryos, although the timing of successive cleavages was consistently earlier in F_1 embryos compared with BALB/c. They inferred from these findings a "spermatozoan influence" on the intermitotic time between the first and second cleavages. In this case a spermatozoan influence could include a genetic effect of the paternal genome.

McLaren and colleagues have investigated cleavage rates in several strains of inbred mice and lines of the Q-strain selected for large and small body size (Bowman and McLaren, 1970a,b; Allen and McLaren, 1971; McLaren and Bowman, 1973). They observed no differences in cleavage rates among strains. However, they found that C3H embryos were consistently behind C57BL in development, and subsequently Nicol and McLaren (1974) determined that this difference in development was due to retarded sperm transport in C3H females.

A striking maternal effect has been reported for crosses of inbred mice involving the Japanese strain DDK (Wakasugi et al., 1967; Wakasugi, 1973, 1974). When DDK females are mated to males of other strains, most of the embryos die around the time of implantation and many of these embryos have defective trophoblast development. That this effect is not seen in reciprocal matings when DDK males are mated to females of other (alien) strains indicates that the reduced viability is maternally inherited. Transplantation of DDK ovaries showed that (DDK × alien)F_1 embryos still die when fertilization and cleavage occurs in an (alien × DDK)F_1 reproductive tract. This finding implies that the maternal effect is mediated by the cytoplasm of the ova rather than by the maternal environment and so this effect is controlled differently from the retarded development of embryos in C3H mothers (McLaren and Bowman, 1973; Nicol and McLaren, 1974). The normal viability of DDK × DDK embryos suggests that the effect involves two elements, one in the egg cytoplasm and another controlled by the paternal genotype or sperm genotype.

Genetic studies involving reciprocal backcrosses and F_2 matings (Wakasugi, 1974) suggest that the early lethal effect is inherited as a single autosomal gene, but the pattern of expression is rather complex and involves interactions between oocytes and male pronuclei or sperm of specific genotypes. Table 3.2 lists the various crosses and their relative levels of embryo lethality in each kind of cross.

Segregation of a single Mendelian gene occurs in the female progeny of two backcrosses (crosses 4 and 5). In both of these crosses all of the progeny are viable, but when the female progeny are mated with alien males in test crosses two classes of fertility are evident. In backcrosses of F_1 females with DDK males (cross 4), 50 percent of the female progeny have the same fertility as DDK females in test crosses, whereas 50 percent are similar to F_1 females. Among the female progeny of alien females and F_1 males (cross 5), 50 percent are the same as F_1 females and 50 percent are fully fertile, i.e., the latter have the alien phenotype. These findings clearly demonstrate that a factor segregates as a single gene and that it can be inherited through either the paternal or maternal genome.

In the other two backcrosses, crosses 6 and 7 (table 3.2), there is a 50 percent reduction in embryo survival. The surviving female progeny of the DDK female × F_1 male cross 7 were predominantly "DDK type" in fertility tests. Surprisingly, however, in the backcross of F_1 females by alien males (cross 6), the female progeny are equally "F_1 type" and "alien type." Test crosses of the surviving male progeny in backcrosses 6 and 7 to determine whether they were DDK, F_1, or alien type in crosses with DDK females were not reported.

Table 3.2. Summary of Crosses Involving the DDK Strain and DDK × Alien[a] Hybrid Combinations[b]

Cross	Female	Male	Percent Survival	Phenotypes of Surviving Progeny
1	DDK	DDK	100	
2	DDK	alien	<20	
3	alien	DDK	100	
4	(alien × DDK)F_1	DDK	100	(alien × DDK)F_1 = DDK[c]
5	alien	(alien × DDK)F_1	100	(alien × DDK)F_1 = alien
6	(alien × DDK)F_1	alien	50	alien = (DDK × alien)F_1
7	DDK	(alien × DDK)F_1	50	DDK >> (DDK × alien)F_1
8	(alien × DDK)F_1	(alien × DDK)F_1	75	

[a] Alien refers to non-DDK strains such as KK, C57BL/6, and NC.
[b] Taken from the data of Wakasugi et al. (1976) and Wakasugi (1973, 1974).
[c] Ratio of (alien × DDK)F_1 to DDK is 1:1.

Wakasugi has explained these unexpected results by assuming that only one of two kinds of cytoplasmic factors present in the ova of F_1 females is expressed. In brief, the model assumes that:

1. One autosomal locus controls a cytoplasmic factor in the ova and a factor in the sperm. The DDK strain carries the *om* allele (ovum mutant) and other strains carry the + allele.
2. The *om* allele produces o cytoplasmic factors in the oocyte and the + allele produces O factors.
3. The cytoplasmic factors in the ovum interact with the gene at the *om* locus in the sperm. An interaction between an o cytoplasmic factor and a + allele in the sperm is the only combination that results in increased embryonic death.
4. Both o and O factors are present in the ova of F_1 females, but only one factor interacts with the sperm gene.

Thus the segregation of two types of offspring from the backcross of F_1 (*om*/+) females × alien (+/+) males (cross 6 in table 3.2) is explained by assuming that the + gene in the sperm of alien males interacts randomly with an o or an O factor in the F_1 ovum. The majority of surviving progeny will result from the interaction with an O factor because an interaction with an o factor produces embryonic mortality. However, because the type of cytoplasmic factor that interacts with the sperm gene is *independent of the egg genotype*, half of the surviving progeny will be of the F_1 type (*om*/+) and half will be of the alien type (+/+). The segregation of a single gene that affects early embryonic viability, mediated by the egg cytoplasm, is of great interest, and Wakasugi's model deserves further experimental investigation and evaluation.

3.2.5 Summary

In summary, genetic evidence from specific enzyme loci, cell surface alloantigens, and the effects of early lethals support the view that genes function during the very early cleavage stages in the mouse embryo. This conclusion agrees with observations of RNA and protein synthesis during this period. Furthermore, the findings suggest that the mouse embryo genome plays a significant role in the regulation of the early developmental process.

The degree to which the mouse model applies to other mammals is a matter of speculation. The rabbit differs from the mouse in that it has a much larger volume of oocyte cytoplasm, a faster rate of early cleavage, and a different pattern of early RNA synthesis (Manes, 1975). Thus, the onset of transcription of the embryonic genome might occur later in the rabbit than in the mouse, and the early stages of rabbit embryogenesis might be more dependent on maternally derived factors stored in the oocyte. The data available on early development in species other than mouse and rabbit are primarily descriptive. The role in early development of stored macromolecules present in the oocyte is not completely resolved.

Oocyte factors could function in several ways. Development might depend on maternally transmitted masked mRNA or repressor/activator type molecules directing transcription during early embryogenesis. The possible existence of oocyte factors with gene regulatory capability is of interest as a potential source among cleaving blastomeres of asymmetry because it could serve as a signal for differential gene expression. Whether the program for gene expression is determined by the embryo genome or whether it derives from the maternal genome as stored macromolecules might eventually be determined by studying genetic variation in the expression of specific gene products in early embryos. It should be possible to establish experimental systems that choose between maternal and biparental inheritance of regulatory factors controlling gene expression during early embryogenesis.

3.3 Early Lethals

Lethal genes have been defined as "Mendelian units which cause the death of an organism prior to the reproductive stage" (Hadorn, 1961). This discussion concentrates on genes associated with embryonic death during early development.

3.3.1 Identification and Characterization

Lethal genes are typically identified first by the absence of an expected genotypic class in a specific mating. Because of the number of progeny that are necessary to establish that a particular class is indeed missing, only in the mouse, among mammals, have early lethals been found. In the mouse, the T complex and the Ts (tail short), Os (oligosyndactylism), and A (agouti) loci have lethal alleles associated with them (Green, 1966; McLaren, 1974). The alleles of the Ts, Os, and A^y loci are characterized as semidominants in which the heterozygotes show pronounced effects upon different aspects of morphogenesis and/or physiology. However, with respect to lethality, these alleles are recessive. Only the homozygote with two doses of the mutant allele is nonviable. The only recessive alleles known to cause arrested development in the homozygous state but no adverse effects in the heterozygous condition are the t series of alleles in the T complex. These alleles were originally recognized because the combination of a recessive t allele with the T allele causes a tailless but viable condition. (See chapter 4 for more on the subject of the t allelic series.)

Except for the specific case of the t alleles, where true recessives are recognizable, the study of alleles is limited to those having an observable effect on the adult phenotype. This circumstance is favorable for genetic analyses because it permits a direct test of allelism or linkage without the necessity of resorting to awkward test crosses to diagnose heterozygotes. The altered phenotypes of semidominant alleles might also provide clues about the underlying molecular lesions that block development. However, one assumption in relating the semidominant phenotype with early lethality is that the same gene is responsible for both conditions. If the mutational event affects more than a single gene product (e.g., a deletion), the molecular basis of the semidominant phenotype might not be the cause of arrested development.

A^y (agouti yellow)

One example in which the adult phenotype might not correspond to the incidence of early lethality is the yellow phenotype at the agouti locus. The A^y allele is associated with early lethality in the homozygote and a yellow-agouti phenotype in the heterozygote. The A^{vy} (viable yellow) allele, which arose spontaneously in the C3H strain, also produces a yellow-agouti phenotype. However, although homozygosity for A^{vy} does have associated physiological problems, such as obesity, it does not cause early lethality (Dickie, 1962). The observation of recombination between "alleles" of this series supports the idea that the agouti locus is actually a complex locus, and that the gene(s) specifying lethality might differ from those causing the yellow phenotype (Russell et al., 1963).

The yellow mouse and the developmental effects of the homozygous condition have been studied extensively. Cuénot [cited by Ibsen and Steigleder (1917)] and Castle and Little (1910) first observed the absence of a Mendelian class in the progeny of yellow × yellow matings and that yellow mice never bred true. Yellow × yellow matings resulted in yellow:nonyellow progeny in the ratio 2:1, whereas yellow × nonyellow matings gave a 1:1 ratio. The A^y/A^y embryo was first identified as a specific type by Kirkham (1917, 1919) and by Ibsen and Steigleder (1917). They observed abnormal blastocysts arising from yellow × yellow matings at the expected frequency. The abnormal blastocysts appeared shrunken, and their cavity size was smaller than in $A^y/+$ embryos. These embryos were resorbed around the time of implantation. Robertson (1942) confirmed the earlier findings and in addition tested the effect of the maternal environment on A^y/A^y embryos. He transplanted ovaries from A^y/a females into ovariectomized a/a females. In this experiment, abnormal embryos (presumed A^y/A^y) developed somewhat further in the a/a uterus than in the A^y/a uterus, but eventually death and resorption occurred. In addition, Eaton and Green (1962) demonstrated that A^y/A^y embryos could implant in females of various genetic backgrounds. The conclusion is that neither the maternal environment nor the implantation process is the major factor in the lethality of A^y/A^y embryos.

Pederson (1974) cultured embryos from yellow matings ($A^y/a \times A^y/a$) in vitro and observed morphological abnormalities as early as the fourth cleavage division. Presumed A^y/A^y embryos did not show proliferation of trophoblast cells or differentiation of inner cell mass (into endoderm and ectoderm). In an ultrastructural study of A^y/A^y embryos, Calarco and Pedersen (1976) observed abnormal morulae. Some blastomeres of the abnormal morulae appeared to arrest at the 4–8-cell stage, although others seemed to develop normally until degenerative changes occurred at the late blastocyst stage.

Os (oligosyndactylism)

Van Valen (1966) studied in vivo development of Os/Os embryos histologically. Abnormal cells were not observable before the 64-cell stage. However, after the fifth cleavage division, "some cells, often in pairs, become pale and the chromatin becomes fragmented and pycnotic" (Van Valen, 1966). The conclusion, based on the frequent occurrence of these paired cells, was that they are either unable to complete division or cannot reenter interphase. More recently, Paterson (1976) has confirmed that many cells of presumed Os/Os blastocysts become blocked in mitosis, but it appears that trophoblast giant cells (which undergo endoreduplication) survive in vitro. The cause of these degenerative changes is unknown. The heterozygote ($Os/+$) adult is characterized by digital fusion and diabetes insipidus; neither symptom has been related to the lethality of the homozygote.

Ts (tail short)

The tail short mutant (Ts) also produces a number of skeletal anomalies and prenatal anemia (Morgan 1950; Deol, 1961; Green, 1966). Detailed studies of the early development of presumed homozygous mutants are not available. However, there is some evidence that Ts/Ts homozygotes arrest at the morula to blastocyst stage (Paterson, 1976).

3.3.2 Summary

In general, the effects of early lethals support the notion that embryo gene function is required for early development, but more specific information about why these mutants arrest has been difficult to obtain.

In fact, it is difficult to determine whether any of these mutants arrest because of an impairment of intermediary metabolism, and are thus of little developmental interest, or whether a specific gene function controlling development is affected. Nor do we have the information to decide whether a protein has been altered in these mutants or a regulatory site changed. Part of the lack of information derives from the fact that techniques for assaying biochemical parameters in single embryos have not been available until recently, and even these techniques are limited. A second operational difficulty in studying semidominant mutants is that there might be some fertility impairment, so that collection of early embryonic material and unequivocal identification of the genetically nonviable class would be difficult.

Given the relatively few mutations that clearly alter the course of early development and the technical difficulty of identifying additional recessive lethals acting during this stage, it might be pertinent to ask what information lethal genes can reasonably be expected to provide about early development. It would be useful to have mutants that failed to elaborate a group of gene products normally or did not undergo the initial morphogenesis of separating ICM and trophectoderm cells. It might then be possible to apply a genetic approach similar to that employed with the O^{nu} mutant in Xenopus (Brown and Gurdon, 1964) to distinguish gene products necessary for a developmental sequence of events. This approach is analogous to the use of mutants for reconstructing metabolic pathways. As the mouse linkage map is expanded, it is probable that additional lethal or semilethal combinations will be found as deficient classes in genetic studies. However, until we can identify the gene product associated with any of these mutants, we cannot determine whether the primary genetic lesion affects a structural gene, causing an alteration of a critical protein, or whether it affects a regulatory gene and disrupts the developmental program.

3.4 X-Chromosome Dosage Compensation

3.4.1 Single-active-X Hypothesis

Mammalian X-chromosome dosage compensation is a special type of genetic regulation whereby a whole chromosome is probably functionally inert. The most widely accepted mechanism suggested for X-chromosome dosage compensation is known as the "single-active-X hypothesis" or the Lyon hypothesis. Lyon (1961) proposed that the variegated phenotypes of female mammals, heterozygous for X-linked genes, together with the normal female phenotype of XO mice are attributable to the inactivation of one X chromosome in each cell of normal diploid females early in embryonic development. This differentiation of the X chromosomes to active and inactive states is presumed to be usually at random, so that, on average, half of the cells of the embryo will express the maternally derived X-linked genes and half of the cells will express the paternally derived alleles. Consequently, the female embryo becomes a functional mosaic for X-linked genes. It is also postulated that this differentiation is stable, so that the same X chromosome remains active in all the mitotic progeny of a cell.

Direct evidence that the active state of the X chromosome is mitotically inherited comes from in vitro cloning experiments, using cells from females heterozygous for X-linked biochemical markers. These clones, established from single cells, uniformly express only gene products from one X chromosome (Davidson et al., 1963; Danes and Bearn, 1967; Migeon et al., 1968; Romeo and Migeon, 1970; Hamerton et al., 1971; Chapman and Shows, 1976).

The most vigorous challenge to this hypothesis came from Grüneberg, who suggested the "complemental-X" hypothesis as an alternative mechanism for X-chromosome dosage compensation (Grüneberg, 1967, 1969). According to this hypothesis, both X chromosomes are active in each cell, but the total X-linked genetic activity per XX cell is equal to that in XY cells. The relative contributions from the two homologous X chromosomes might vary among cells. However, most of the cytological and genetic data support the single-active-X hypothesis. This evidence has been reviewed by Lyon (1968, 1970, 1972, 1974), Eicher (1970), and Hamerton (1971a).

The functional differentiation (either inactivation or activation) of one of two X chromosomes, as proposed by the Lyon hypothesis, is a genetic event occurring early in mammalian development. (The process is a genetic event both in the sense that it affects genetic expression and in the sense that the event itself is probably under genetic control.) Genetic markers make it possible to detect the two populations of cells present in X-chromosome mosaics, and genetic analysis of the mechanism and control of the differentiation process has begun.

3.4.2 Experimental Approach to the Study of X-Chromosome Expression and Differentiation

Within the context of gene regulation during early mammalian development, the primary issue to be considered is whether there are sites in the genome that regulate X-chromosome expression and more specifically whether such sites are autosomal or intrinsic to the X chromosome. Evidence from the differential expression of various X-linked genes strongly suggests that different X chromosomes have different intrinsic probabilities of being expressed (Cattanach and Isaacson, 1967; Ohno et al., 1973; reviewed by Cattanach, 1975). These data are all subject to the common criticism that it is impossible to distinguish between differential inactivation and cell selection.

This difficulty could be circumvented if it were possible to examine the distribution of X-chromosome activities soon after the differentiation process occurred. This examination requires a knowledge of the timing of the event, whether it occurs in all cells at the same time, and whether the mechanism involves an inactivation of a previously active X chromosome or activation of X-chromosome transcription at the time of differentiation.

3.4.3 Biochemical Genetic Resources

The most suitable tools for studying X-chromosome differentiation are X-linked biochemical variants. In man and a number of other mammals, variants are known for the enzymes glucose-6-phosphate dehydrogenase (G6PD), phosphoglycerate kinase (PGK), hypoxanthine guanine phosphoribosyl transferase (HGPRT), and α-galactosidase. Evidence that the structural genes for each of these enzymes are X-linked in all mammals so far tested supports the theory of mammalian X-chromosome homology advanced by Ohno (1967, 1973). The laboratory mouse is one of the most convenient mammals for studying developmental processes, but although there is evidence for the X linkage of the genes responsible for producing G6PD, PGK, HGPRT, and α-galactosidase (Epstein, 1969, 1972; Kozak, et al., 1974; Kozak et al., 1975; Chapman and Shows, 1976), until recently no useful variants of these enzymes had been found among laboratory mice. The first X-linked enzyme variant reported for the mouse was a phosphorylase-b-kinase deficiency (Lyon, 1967), which is expressed only in skeletal muscle and

is of limited use. Recently an electrophoretic variant for PGK
(Chapman and Nielson, 1977) and a thermostability variant for α-galac-
tosidase (Lusis and West, 1976) have been discovered in two subspecies of Mus musculus. Both subspecies are interfertile with laboratory
mice. Electrophoretic differences for G6PD, PGK, and HGPRT exist
between Mus musculus and Mus caroli (Chapman and Shows, 1976),
but no hybrids have survived to breeding age. Electrophoretic variants
of X-linked enzymes are also available in other species, including man
(Boyer et al., 1962; Kirkman and Hendrickson, 1963), two species of
kangaroos, two species of wallabies (Cooper et al., 1971, 1975), and in
various interspecific hybrids, including mules (Trujillo et al., 1965),
hybrids of wild hares (Ohno et al., 1965), and hybrids of various marsupials (Cooper et al., 1971, 1975).

3.4.4 X-Chromosome Inactivation or X-Chromosome Activation?

There is no conclusive experimental evidence to show whether gene
transcription occurs from both X chromosomes in each cell in the early
embryo. The process of X-chromosome differentiation could be one of
inactivation or one of activation. Lyon (1968) cites evidence from
human studies indicating that individuals who have a normal autosome
complement but more than two X chromosomes still have only one
active X chromosome. Cattanach (1975) suggests that it is difficult to
imagine how a cell with such a karyotype would "know" how many
X chromosomes to inactivate, but that no such conceptual difficulties arise if the process is assumed to be one of activation.
However, the little experimental evidence that is available tends to
favor a process of inactivation. This evidence includes indirect cytological observations, such as the absence of facultative heterochromatin
and the absence of a late-replicating X chromosome until the late
blastocyst stage in all mammals so far examined (reviewed by Lyon,
1972, 1974). There is also evidence that an X chromosome is necessary
for the successful completion of preimplantation development. Morris
(1968) and Burgoyne and Biggers (1976) have claimed that mouse OY
embryos, which lack X chromosomes, fail to complete preimplantation
development. However, the diagnosis of OY embryos was not confirmed by karyotype analysis in either study. There is also some evidence for X-linked gene function in preimplantation mouse embryos.
Epstein (1972) has shown that the activity of HGPRT per embryo in-

creases between the 8-cell and the blastocyst stage. Comparisons of HGPRT activity in pooled samples of embryos from XO and XX mothers suggest that this increase is due to the activity of the embryonic genome, rather than the activation of stored, maternally derived mRNA or enzyme precursors.

Measurement of α-galactosidase activity levels in single mouse embryos from the 2-cell to blastocyst stage (J. D. West, D. A. Adler, and V. M. Chapman, unpublished observations) revealed a fiftyfold increase in activity in the 24-hr period between the 2-cell and 8-cell stage and a continuous increase in enzyme activity throughout preimplantation development. The embryos gave a bimodal distribution of activities at the 8-cell and morula stages. (The distribution for 8-cell embryos of the C3H/HeHa strain is shown in figure 3.1.) If the increase in α-galactosidase activity is indeed due to the transcription of embryo genes, as is the case for a similar increase in β-glucuronidase, the bimodality suggests that both X chromosomes are active early in development and that the two activity classes are actually male and female embryos.

Although the evidence is limited, it seems probable that some X-linked genes are active very early in mouse development, probably before the time that the differentiation of X chromosomes occurs. This suggests that X-chromosome differentiation involves the inactivation of a previously active X chromosome.

3.4.5 Timing of X-Chromosome Differentiation

Differentiation of X chromosomes is widely believed to occur around the time of implantation in most eutherian mammals [see Lyon (1972, 1974) for reviews], although for the mouse the estimates range all the way from the 40–50 cell blastocyst stage (fourth day of gestation) to well after implantation (ninth day of gestation). Most of the approaches used to estimate the timing of X-chromosome differentiation have been indirect and entail use of cytological markers or deduction of the timing from statistical analyses of mosaicism patterns in adult heterozygotes.

Gardner and Lyon (1971) have used X-linked genetic markers in a direct experimental effort to determine the time of X-chromosome differentiation. These authors produced chimeric mice by injecting single cells, taken from mid-fourth-day inner cell masses (ICMs), into each of a series of blastocysts. The donor cells from female embryos were

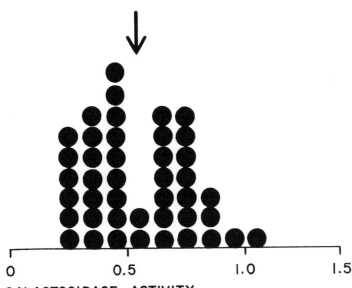

α – GALACTOSIDASE ACTIVITY

(picomoles product embryo^{-1} hr^{-1})

3.1 α-Galactosidase activity in 8-cell C3H/HeHa mouse embryos (54 hr after artificial insemination). Each circle represents a single embryo, and the arrow indicates the mean enzyme activity per embryo. Embryos were assayed using the microassay techniques of Wudl and Chapman (1976). The reaction mixture consisted of 2 mM 4-methyl unbelliferyl-α-D-galactoside, 0.1 M citrate, 0.1% bovine serum albumin and 0.9% NaCl, at pH 4.2. Embryos were incubated for 1 hr at 37°C.

heterozygous for Cattanach's translocation (Cattanach, 1961). If both X chromosomes were active in these female cells, two cell lineages would be produced from a single XX donor cell. Pigment markers made it possible to distinguish these two cell populations from one another and from the recipient cell population in the coat of the adult chimera. Of nine chimeras produced in this way, four showed all three types of pigmentation: therefore, at least some individual cells of the mid-fourth-day ICM are capable of producing daughter cells in which either X chromosome may be expressed. One chimera showed no white (donor) patches, but this cell population might have been present in other tissues. Four chimeras showed no wild-type (donor) color, but this result is predictable if the donor cells were from male embryos. The most reasonable interpretation of these data is that X-chromosome differentiation occurs after the middle of the fourth day of gestation, assuming the experimental manipulation did not reactivate an inactive X chromosome and that the translocation markers are inactivated at the same time as the rest of the X chromosome.

Gardner and Lyon's experiment does not exclude the possibility that X-chromosome differentiation occurs earlier in the cells of the trophectoderm or in some cells of the ICM. Park (1957) suggested that sex chromatin forms earlier in the trophectoderm than in the ICM of human and macaque embryos, and Aitken (1974) proposed that a gradual formation of sex chromatin occurs in the roe-deer blastocyst during delayed implantation. However, sex chromatin is probably an unreliable guide to the time of X-chromosome differentiation, and Plotnick et al. (1971) claim that sex-chromatin formation in the rabbit begins in all regions of the embryo at the same time. Deol and Whitten (1972) have also suggested that X-chromosome differentiation in mice might occur at different times in different tissues, but the assumptions made in their arguments have been questioned (Nesbitt, 1974; West, 1976), and as yet there is no direct genetic evidence on this matter.

3.4.6 Randomness of X-Chromosome Differentiation

Genetic effects
The overall balance between the expression of the two X chromosomes in a mosaic might be altered either by nonrandom X-chromosome differentiation or by cell selection. Rapidly dividing tissues would be likely to show the effects of cell selection most clearly, and Nyhan et al. (1970) have shown that blood cells from human X-inactivation mosaics can comprise a single population when other tissues are demonstrably mosaic. In this case the presumptive selection pressure is against cells in which the active X chromosome produces an HGPRT deficiency. Unbalanced mosaicism has also been reported in some tissues and organs of female mules (Hook and Brustman, 1971), and there is evidence for in vitro cell selection of cultured mule fibroblasts over a period of several months (Hamerton et al., 1971; Rattazzi and Cohen, 1972), resulting in a predominance of cells in which the horse-derived X chromosome is expressed. Earlier, based on the predominance of this cell population in cultured cells, Hamerton et al. (1969) postulated that nonrandom X-chromosome differentiation resulted in preferential activ-

ity of the horse X chromosome. It now seems that in vitro cell selection satisfactorily explains this predominance.

Several lines of mouse X-inactivation mosaics have now been established which are unbalanced for coat color (Cattanach and Isaacson, 1965; Krzanowska and Wabik, 1971; Falconer and Isaacson, 1972). The unbalanced X-inactivation mosaicism seen in some tissues and organs in vivo of both mouse and mule might be due either to cell selection or to nonrandom X-chromosome differentiation. Cell selection seems to be the most likely cause for the imbalance in mules because the imbalance appears to be tissue specific (Hook and Brustman, 1971). It is still unclear how the unbalanced phenotypes arise in mice despite attempts to rule out cell selection (Drews et al., 1974).

Cattanach and Williams (1972) suggested that the imbalance in proportions of the two cell populations in some mouse X-inactivation mosaics is due to genetically determined nonrandom X-chromosome differentiation. The studies of Cattanach and his colleagues (reviewed by Cattanach, 1975) revealed a locus that they called Xce (X-chromosome controlling element), located near the center of the linkage group close to the tabby (Ta) locus. Different alleles of the controlling elements exist, and in females heterozygous at the Xce locus the controlling elements modify the balance of the expression of the two X chromosomes. This imbalance is particularly pronounced in a variant that Ohno and colleagues (Ohno et al., 1973; Drews et al., 1974) have designated O^{hv} (organizer, high variegation); this variant maps in the same region as Xce. This effect would be expected if the unbalanced phenotypes were the result of nonrandom X-chromosome differentiation controlled by the Xce locus. Alternatively, these unbalanced phenotypes could be caused by cell selection if the Xce locus is the site of a deleterious mutation. The X-linked enzyme variants now available in mice should provide a means of discriminating between the two alternative mechanisms by permitting comparison of the proportions of the two cell populations early in development.

Effects of the source of the X chromosome
A rather different type of effect on the balance of the two cell populations is suggested by the cytological data of Takagi and Sasaki (1975). They used heterozygotes in which one of the two X chromosomes carried Cattanach's translocation. In 9th-day embryonic tissues the two cell populations were present in nearly equal proportions, whereas in

some extraembryonic membranes most cells had an inactive X chromosome that was of paternal origin. The result was the same whether the
maternal or paternal chromosome was marked with the translocation.
The preferential activity of maternally derived X chromosomes in the
extraembryonic membranes might result either from selection at the
level of X-chromosome differentiation or from cell selection during
extraembryonic membrane growth.

Female marsupials, like eutherian mammals, have two X chromosomes. Cytological studies of leucocyte cultures (Sharman, 1971) and
electrophoretic analysis of G6PD expression in red blood cells (Richardson et al., 1971) indicated the activity of only the maternally derived
X chromosome in several kangaroo species. This was interpreted to
suggest that X-chromosome dosage compensation in these kangaroos,
and perhaps all marsupials, is a process of paternal X-chromosome
inactivation (or maternal X-chromosome activation). The absence of
paternal X-chromosome expression in red blood cells was confirmed by
using an electrophoretic variant for PGK, but the paternal allele for
PGK was also found to be expressed at a low level in some tissues and
organs (Cooper et al., 1971; VandeBerg et al., 1973; Cooper et al.,
1975). There is no evidence for any expression of the paternal allele for
G6PD in these tissues in a different species, but because no double
heterozygotes are available, it is not clear whether this discrepancy
between PGK and G6PD expression is due to differences between species, differences between loci, or differences in sensitivity between the
two techniques. The paternal allele for both PGK and G6PD is expressed in fibroblast cultures, and there is evidence for the formation
of G6PD heteropolymers, suggesting the possibility of two active X
chromosomes in the same cell in vitro, but this has yet to be confirmed
by cloning experiments (Cooper et al., 1975). Although the paternally
derived X chromosome is not always inactive in marsupials, the tissues
and organs that appear to be mosaic are certainly unbalanced.

Assuming this nonrandom X-chromosome expression is not a result
of cell selection, these findings raise the issue of whether marsupial X-
chromosome dosage compensation is achieved by mechanisms similar
to those acting in eutherian mammals. If both X chromosomes are
initially active in mouse embryos, as is suggested by the expression of
HGPRT and α-galactosidase (see pp. 113–114), the general model that
follows is that both X chromosomes are active during early cleavage

and that a single X chromosome remains active following an inactivation event. In the case of marsupials, the question is whether both X chromosomes are active during early cleavage. If both X chromosomes are active, the evidence suggests that the paternally derived X chromosome has a greater probability of being inactivated. This model implies that the germ cell environment somehow alters a controlling element site on the X chromosome (Chandra and Brown, 1975). Alternatively, if marsupial X chromosomes are inactivated during spermatogenesis and remain inactive in the zygote and subsequent somatic cells, reactivation during oogenesis would eliminate the need for an X-inactivation step in early embryos.

3.4.7 Discussion

The regulation of X-chromosome differentiation might be influenced by autosomal loci as well as by loci intrinsic to the X chromosome itself. Variation for an intrinsic locus controlling the time at which the differentiation event occurs could result in nonrandom X-chromosome differentiation in females heterozygous for such a variant. The additional participation of autosomal genes seems probable from the observed relationship between the number of active X chromosomes and the number of sets of autosomes (Lyon, 1972). If the differentiation process involves autosomal genes, an X-linked locus, such as *Xce,* could be the site at which the initial differentiating stimulus, directed by an autosomal site, is received. Alleles at such a locus could vary in their ability to receive the stimulus, and perhaps in the time required to respond to the stimulus. If, for example, the stimulus is the binding of a molecule to the X chromosome, different alleles of an X-linked regulatory locus might vary in their binding affinity for the molecule.

The search for autosomal and X-linked variants affecting either the timing or randomness of X-chromosome differentiation would be a fruitful approach to the genetic dissection of this control process and might help in the choice among the various molecular models that have been proposed [for reviews see Lyon (1974), Cook (1974), Cattanach (1975), Riggs (1975)]. Despite a large experimental literature on the Lyon hypothesis, there is a paucity of data of the type necessary for making such choices.

3.5 Regulation of Gene Expression

The biochemical and genetic evidence demonstrating embryonic gene function early in preimplantation development, together with the existence of early lethal genes and the block in development resulting from chromosome deficiencies, suggests the importance of specific gene functions in initiating differentiation and maintaining normal morphogenesis. The problem of X-chromosome differentiation concerns the regulation of this process at the level of an entire chromosome. Among the autosomes, this regulation is affected at the level of individual structural genes. Unfortunately, by comparison with X-chromosome inactivation, there is very little information on individual gene regulation during early mammalian development.

During the preimplantation development of the mouse, several changes in enzyme activities have been observed [reviewed by Epstein (1975)], and numerous new proteins are synthesized after the first cleavage (Epstein and Smith, 1974; Van Blerkom and Brockway, 1975; see also chapter 2). Van Blerkom et al. (1976), employing two-dimensional electrophoresis and autoradiography, have observed differences in incorporation of label into proteins between the inner cell mass and trophectoderm. A switch in lactate dehydrogenase activity from the B to the A form at the time of implantation was observed in the mouse (Wolf and Engel, 1972). Unfortunately, structural gene variants that permit a genetic analysis of development have not yet been found for these systems.

As discussed previously, β-glucuronidase expression in preimplantation embryos demonstrates paternal gene function. In addition, it is the one system having genetic variation in the regulation of a single gene in early embryos. Homozygous $Gus^{h/h}$ embryos, from the congenic C57BL/6 $Gus^{h/h}$ strain, have lower β-glucuronidase activity than C57BL/6 $Gus^{b/b}$ embryos at all stages (figure 3.2). Furthermore, the net increase in β-glucuronidase activity is lower in $Gus^{h/h}$ than in $Gus^{b/b}$ embryos. The difference in control apparent between these congenic lines indicates that it is exerted by the region of chromosome 5 that carries the structural gene itself.

More information on the control of individual structural genes during

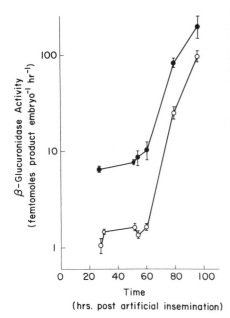

3.2 β-Glucuronidase activities during
preimplantation development of
C57BL/6 $Gus^{h/h}$ (○) and C57BL/6 $Gus^{b/b}$ (●).
Each point represents the mean activity of
30 to 70 individually assayed embryos. The
vertical bars represent the standard error
of the mean. Embryos were assayed using
the β-glucuronidase microassay of Wudl
and Chapman (1976).

development is available from studies of later developmental stages.
By making the reasonable assumption that mechanisms of developmental regulation are fundamentally the same throughout development, one
can ask what implications these later studies have for understanding
early developmental processes. Such systems include β-glucuronidase
in postnatal tissues of mice (Paigen, 1961; Paigen and Ganschow, 1965),
aldehyde oxidase during Drosophila development (Dickinson, 1975), the
postnatal expression of the H-2 complex on erythrocytes (Boubelík et al.,
1975), β-galactosidase in postnatal liver (Chapman et al., 1973; Meisler,
1976; Paigen et al., 1976), esterase in corn (Schwartz, 1962), and
α-amylase in Drosophila (Doane and Abraham, 1976; Abraham and
Doane, 1977). Several of these systems demonstrate that developmental
expression of structural genes can be determined by closely linked
elements acting in a *cis* fashion. In at least one system, developmental
expression is controlled by unlinked genes as well as by closely linked
elements.

3.5.1 β-Glucuronidase in Mice

Some of the features of β-glucuronidase in the mouse, namely the Gus^h and Gus^b allele differences, were described earlier. To reiterate, the Gus^h allele product has different heat denaturation kinetics from Gus^b enzyme and is associated with lower activity levels in nearly all tissues of the adult (Paigen, 1961; Ganschow and Paigen, 1968), although the ratio of β-glucuronidase activity between Gus^h and Gus^b mice is not constant among tissues.

The relative activity per molecule is the same in Gus^h and Gus^b and the differences in activity between Gus^h and Gus^b strains are due to differences in the rate of enzyme synthesis (Ganschow, 1975). The developmental difference between these strains that is especially interesting occurs in liver between 10 and 20 days of age. The specific activity of liver β-glucuronidase decreases markedly in Gus^h mice, whereas the specific activity in Gus^b mice remains relatively constant (Paigen, 1961). In genetic tests, liver activity levels of β-glucuronidase segregated with the heat denaturation characteristics. Thus, the developmental expression of liver β-glucuronidase activity is determined by a site that is closely linked to the glucuronidase structural gene (Paigen, 1961). Recently, K. Smith and R. Ganschow (personal communication) have found that the decrease in β-glucuronidase activity is the result of a decrease in the relative rate of synthesis in Gus^h livers between 10 and 20 days of age. Thus this regulation is exerted at the transcription-translation level and is not due to posttranslational processing or enzyme degradation.

3.5.2 Aldehyde Oxidase in Drosophila

Two developmental patterns of aldehyde oxidase activity have been observed in Drosophila melanogaster (Dickinson, 1975). A doubling of aldehyde oxidase activity occurs in one phenotype at about the time of pupation. Two days later a second increase in activity to adult values takes place. A variation in the developmental pattern occurs in some stocks that show only the second increase in enzyme activity two days after pupation. The latter phenotype has a low pupa:adult activity ratio, whereas the stocks that undergo an increase in aldehyde oxidase in pupae have a high pupa:adult ratio. The F_1 hybrids of these stocks show an intermediate increase in aldehyde oxidase activity at the time of pupation. Aldehyde oxidase activities one day after pupation served as an indicator of the developmental patterns. These activity levels segregated into two classes when F_1 flies were backcrossed with a stock having a low pupa:adult activity ratio, suggesting that a single gene controls the developmental pattern. Linkage analysis with an electrophoretic variant of aldehyde oxidase showed that the developmental phenotype segregates with the structural gene. The developmental factor acts *cis* in its regulation of the linked structural gene because the electrophoretic patterns of F_1 pupae were similar to those seen with pupae homozygous for the allele producing the first rise in activity.

3.5.3 H-2 Complex in Mice

One final system to be considered in detail is the temporal expression of H-2 antigens on the surface of the red blood cell (Boubelík et al., 1975). The model of regulation proposed for this system is similar to the aldehyde oxidase system in Drosophila and β-glucuronidase in mice, in that the H-2 complex is regulated in part by a site at or near H-2 itself. However, the timing of H-2 expression also appears to be regulated by unlinked genes that can override the linked or intrinsic program of the H-2 complex.

In brief, the *H-2^b* allele in strain C57BL/10 is detectable by hemagglutination on red cells at birth, whereas the *H-2^a* allele in strain A is not detectable until three days after birth. The phenotypes are designated *H-2^b* early (bE) and *H-2^a* late (aL). In F_1 hybrid red cells, the H-2 haplotypes are expressed codominantly, that is, *H-2^b* is expressed early and *H-2^a* is expressed late. The occurrence of reversed parental strain timing (aE and bL) in F_2 segregation, in congenic lines (when only the H-2 complex was transferred), and in selected F_2 lines from an aL × bE cross led Boubelík et al. (1975) to propose a model that they call the "remote *cis* effect." Some of the features of this model include:

1. H-2 types have intrinsic developmental timing.
2. A temporal gene unlinked to H-2, called *Tem*, is capable of overriding the intrinsic timing of H-2.
3. The ability of *Tem* alleles to override H-2 is dependent upon a gene linked to *Tem*, called *Rec*, which recognizes alleles of H-2 (that is, *Rec^a* recognizes *H-2^a* and *Rec^b* recognizes *H-2^b*).

4. *Tem* and *Rec* act *cis*. (That is, *Tem^b* causes the normally late *H-2^a* to be expressed early, but only when *Tem^b* and *Rec^a* are in coupling. Conversely, *Tem^a* and *Rec^b* cause *H-2^b* antigen to be produced late.)

Among the various crosses of inbred strains and strains with specific H-2 haplotypes, the data of H-2 expression are consistent with the proposed model. While more work is required to verify the model, this system has been considered here because it introduces a type of genetic regulation of specific genes in mammals involving distant unlinked sites.

If the regulation of X-chromosome differentiation involves both autosomal and X-linked loci, it resembles some of the features of the regulation of H-2 antigens; that is, distant or unlinked regulatory elements have temporal effects that augment the intrinsic program of expression at the H-2 complex. In principle, this model is similar to gene regulation systems described in maize and other organisms by McClintock (1967).

3.5.4 Summary

In summary, the developmental expression of structural genes in mice and Drosophila appears to be programmed, in part, by DNA sequences that are in close proximity to the structural gene. In at least one case, the temporal program specified by a closely linked site is augmented or modified by unlinked genes. Although none of these systems involves early mammalian development, the genetic principles should apply equally to the early embryo. The models they suggest could provide experimental approaches for studying gene regulation during the early initiation of gene expression and during the changes in gene expression that accompany the initial stages of morphogenesis.

3.6 Conclusions

The development of mouse genetic resources, including recombinant inbred lines, congenic lines, new biochemical markers, and the establishment of new inbred lines from diverse unrelated genetic sources, all provide a rich source of material for studying gene regulation in early embryos.

Recombinant inbred lines, derived from strain crosses that are inbred from F_2 sibs, provide an array of inbred lines in which the parental strain genomes have been mixed and distributed into new homozygous combinations. Linkage relationships are established by examining the codistribution among recombinant inbred lines of characters that differ between the parental inbred strains. Because embryos within recombinant inbred lines are uniformly homozygous, it is possible to use pooled samples for biochemical assays. More important, if concern exists about a difference in the developmental pattern of enzyme or protein expression, it is possible to establish the pattern in each of the recombinant inbred lines. Thus, recombinant inbred lines, as well as congenic stocks, provide an alternative to assaying single embryos in many genetic analyses (Bailey, 1971; Taylor et al., 1973, 1975).

The rapid growth of mouse biochemical genetics enhances the probability that genetic variation will be established for some of the enzymes and major proteins that are synthesized during early development. In this case, the recent advances in identifying newly synthesized proteins in embryos will be especially useful (Van Blerkom and Brockway, 1975). These systems, in addition to microassay techniques for monitoring enzyme activity (Galjaard et al., 1974; Hösli et al., 1974; Wudl and Paigen, 1974; Wudl and Chapman, 1976), and the previously discussed report on identifying cell surface antigens in embryos (Muggelton-Harris and Johnson, 1976), will provide methods for studying the regulation of specific genes during the initial stages of mammalian development.

Acknowledgment   The authors are grateful to Kenneth Paigen and Anne McLaren for their helpful criticism and suggestions and to Nancy Horton and Sally Adler for their secretarial assistance. They also thank R. Davey, K. Smith, R. Ganschow, and H. Paterson for access to unpublished experimental results.

References

ABRAHAM, I., and DOANE, W. W.
(1977).
Temporal gene control of mid gut
amylase activity patterns in *Drosophila melanogaster*. Genetics, in
press.

AITKEN, R. J. (1974).
Sex chromatin formation in the blastocyst of the roe deer (*Capreolus
capreolus*) during delayed implantation. J. Reprod. Fertil. *40*, 235-239.

ALLEN, J., and MCLAREN, A.
(1971).
Cleavage rate of mouse eggs from induced and spontaneous ovulation. J.
Reprod. Fertil. *27*, 137-140.

BAILEY, D. W. (1971).
Recombinant inbred strains: an aid to
finding identity, linkage, and function
of histocompatibility and other genes.
Transplantation *2*, 419-422.

BARLOW, P., OWEN, D. A. J., and
GRAHAM, C. F. (1972).
DNA synthesis in the preimplantation
mouse embryo. J. Embryol. Exp.
Morph. *27*, 431-445.

BEATTY, R. A. (1957).
Parthenogenesis and Polyploidy in
Mammalian Development. London:
Cambridge University Press.

BOUBELÍK, M., LENGEROVÁ, A.,
BAILEY, D. W., and MATOUŠEK,
V. (1975).
A model for genetic analysis of programmed gene expression as reflected
in the development of membrane antigens. Dev. Biol. *47*, 206-214.

BOWMAN, P. and MCLAREN, A.
(1970a).
Viability and growth of mouse em-

bryos after *in vitro* culture and fusion.
J. Embryol. Exp. Morph. *23*,
693-704.

BOWMAN, P. and MCLAREN, A.
(1970b).
Cleavage rate of mouse embryos *in
vivo* and *in vitro*. J. Embryol. Exp.
Morph. *24*, 203-207.

BOYER, S. H., PORTER, J. H., and
WEILBACHER, R. G. (1962).
Electrophoretic heterogeneity of glu-
cose-6-phosphate dehydrogenase and
its relationship to enzyme deficiency
in man. Proc. Nat. Acad. Sci. USA
48, 1868-1876.

BRINSTER, R. L. (1973).
Paternal glucose phosphate isomerase
activity in three-day mouse embryos.
Biochem. Genet. *9*, 187-191.

BROWN, D. D., and DAWID, I. B.
(1969).
Developmental genetics. Ann. Rev.
Genet. *3*, 127-154.

BROWN, D. D. and GURDON, J. B.
(1964).
Absence of ribosomal RNA synthesis
in the anucleolate mutant of *Xenopus
laevis*. Proc. Nat. Acad. Sci. USA *51*,
139-146.

BURGOYNE, P. S., and BIGGERS,
J. D. (1976).
The consequence of X-dosage deficiency in the germ line: impaired
development *in vitro* of preimplantation embryos from XO mice. Dev.
Biol. *51*, 109-117.

CALARCO, P. G., and PEDERSEN,
R. A. (1976).
Ultrastructural observations of lethal
Yellow (A^Y/A^Y) mouse embryos. J.
Embryol. Exp. Morph. *35*, 73-80.

CARR, D. H. (1971).
Chromosome studies in selected spontaneous abortions, polyploidy in man. J. Med. Genet. 8, 164–174.

CARTER, N. D., and PARR, C. W. (1967).
Isoenzymes of phosphoglucose isomerase in mice. Nature 216, 511–512.

CASTLE, W. E., and GREGORY, P. W. (1929).
The embryological basis of size inheritance in the rabbit. J. Morph. Physiol. 48, 81–103.

CASTLE, W. E., and LITTLE, C. C. (1910).
On a modified Mendelian ratio among Yellow mice. Science, 32, 868–870.

CATTANACH, B. M. (1961).
A chemically-induced variegated type position effect in the mouse. Z. Vererbungsl. 92, 165–182.

CATTANACH, B. M. (1974).
Position effect variegation in the mouse. Genet. Res. 23, 291–306.;

CATTANACH, B. M. (1975).
Control of chromosome inactivation. Ann. Rev. Genet. 9, 1–18.

CATTANACH, B. M., and ISAACSON, J. H. (1965).
Genetic control over the inactivation of autosomal genes attached to the X chromosome. Z. Vererbungsl. 96, 313–323.

CATTANACH, B. M., and ISAACSON, J. H. (1967).
Controlling elements in the mouse X chromosome. Genetics 57, 331–346.

CATTANACH, B. M., and WILLIAMS, C. E. (1972).
Evidence of nonrandom X chromosome activity in the mouse. Genet. Res. 19, 229–240.

CHANDRA, H. S., and BROWN, S. W. (1975).
Chromosome imprinting and the mammalian X chromosome. Nature 253, 165–168.

CHAPMAN, V. M., ADLER, D. A., LABARCA, C., and WUDL, L. (1976).
Genetic variation of β-glucuronidase expression during early embryogenesis. In Embryogenesis in Mammals, K. Elliott and M. O'Connor, eds. Amsterdam: Associated Scientific Publishers, pp. 115–131.

CHAPMAN, V. M., FELTON, J. S., MEISLER, M. H., and PAIGEN, K. (1973).
A locus determining the developmental pattern of mouse β-galactosidase. Genetics (Suppl.) 74, S44.

CHAPMAN, V. M., and NIELSEN, J. T. (1977).
Manuscript in preparation.

CHAPMAN, V. M., and SHOWS, T. B. (1976).
Somatic cell genetic evidence for the X-chromosome linkage of three enzymes in the mouse. Nature 259, 665–667.

CHAPMAN, V. M., WHITTEN, W. K., and RUDDLE, F. H. (1971).
Expression of paternal glucose phosphate isomerase-1 (Gpi-1) in preimplantation stages of mouse embryos. Dev. Biol. 26 153–158.

COOK, P. R. (1974).
On the inheritance of differentiated traits. Biol. Rev. 49, 51–84.

COOPER, D. W., JOHNSTON, P. G., MURTAGH, C. E., SHARMAN, G. B., VANDEBERG, J. L., and POOLE, W. E. (1975).
Sex-linked isozymes and sex-chromosome evolution and inactivation in kangaroos. In Isozymes, C. L. Markert, ed. New York: Academic Press, vol. 3, pp. 559–573.

COOPER, D. W., VANDEBERG, J. L., SHARMAN, G. B., and POOLE, W. E. (1971).
Phosphoglycerate kinase polymorphism in kangaroos provides further evidence for paternal X-inactivation. Nature New Biol. 230, 155–157.

DANES, B. S., and BEARN, A. G. (1967).
Hurler's syndrome: a genetic study of clones in cell culture with particular reference to the Lyon hypothesis. J. Exp. Med. 126, 509–521.

DAVIDSON, R. G., NITOWSKY, H. M., and CHILDS, B. (1963).
Demonstration of two populations of cells in the human female heterozygous for glucose-6-phosphate dehydrogenase variants. Proc. Nat. Acad. Sci. USA 50, 481–485.

DEOL, M. S. (1961).
Genetical studies on the skeleton of the mouse. XXVIII. Tail-short. Proc. Roy. Soc. Lond. B155, 78–95,.

DEOL, M. S., and WHITTEN, W. K. (1972).
X-chromosome inactivation: Does it occur at the same time in all cells of the embryo? Nature New Biol. 240, 277–279.

DICKIE, M. M. (1962).
A new viable Yellow mutation in the house mouse. J. Heredity 53, 84–86.

DICKINSON, W. J. (1975).
A genetic locus affecting the developmental expression of an enzyme in Drosophila melanogaster. Dev. Biol. 42, 131–140.

DOANE, W. W., and ABRAHAM, I. (1976).
Control gene for mid gut amylase activity pattern in Drosophila. J. Cell Biol. 70, 266a.

DREWS, U., BLECHER, S. R., OWEN, D. A., and OHNO, S. (1974).
Genetically directed preferential X-activation seen in mice. Cell 1, 3–8.

DUNN, G. R. (1972).
Expression of a sex-linked gene in standard fusion chimeric mice. J. Exp. Zool. 181, 1–16.

EATON, C. J., and GREEN, M. M. (1962).
Implantation and lethality of the Yellow mouse. Genetica 33, 106–112.

EICHER, E. M. (1970).
X autosome translocations in the mouse: total inactivation versus partial inactivation of the X chromosome. Advan. Genet. 15, 175–259.

EPSTEIN, C. J. (1969).
Mammalian oocytes: X-chromosome activity. Science 163, 1078–1079.

EPSTEIN, C. J. (1972).
Expression of the mammalian X-chromosome before and after fertilization. Science 175, 1467–1468.

EPSTEIN, C. J. (1975).
Gene expression and macromolecular synthesis during preimplantation embryonic development. Biol. Reprod. 12, 82–105.

EPSTEIN, C. J., and SMITH, S. A. (1974).
Electrophoretic analysis of proteins synthesized by preimplantation mouse embryos. Dev. Biol. *80*, 233–244.

FALCONER, D. S., and ISAACSON, J. H. (1972).
Sex-linked variegation modified by selection in Brindled mice. Genet. Res. *20*, 291–316.

FORD, C. E. (1972).
Gross genome unbalance in mouse spermatozoa: does it influence the capacity to fertilize? *In* The Genetics of the Spermatozoan, R. A. Beatty and S. Gluecksohn-Waelsch, eds. Copenhagen: Boggtrykkiert Forum, pp. 359–369.

GALJAARD, H., VAN HOOGSTRATEN, J. J., DE JOSSELIN DE JONG, J. E., and MULDER, M. P. (1974).
Methodology of quantitative cytochemical analysis of single or small numbers of cultured cells. Histochem. J. *6*, 409–429.

GANSCHOW, R. (1975).
Simultaneous genetic control of the structure and rate of synthesis of murine β-glucuronidase. *In* Isozymes, vol. 4, C. L. Markert, ed. (New York: Academic Press), pp. 633–647.

GANSCHOW, R., and PAIGEN, K. (1968).
Glucuronidase phenotypes of inbred mouse strains. Genetics *59*, 335–349.

GARDNER, R. L., and LYON, M. F. (1971).
X-chromosome inactivation studied by injection of a single cell into the mouse blastocyst. Nature *231*, 385–386.

GOLBUS, M. S., CALARCO, P. G., and EPSTEIN, C. J. (1973).
The effects of inhibitors on RNA synthesis (α-amanitin and Actinomycin D) on preimplantation mouse embryogenesis. J. Exp. Zool. *186*, 207–216.

GRAHAM, C. F. (1974).
The production of parthenogenetic mammalian embryos and their use in biological research. Biol. Rev. *49*, 399–422.

GREEN, M. C. (1966).
Mutant genes and linkages. *In* Biology of the Laboratory Mouse, 2nd ed., E. L. Green, ed. New York: McGraw-Hill, pp. 87–150.

GREGORY, P. W., and CASTLE, W. E. (1931).
Further studies on the embryological basis of size inheritance in the rabbit. J. Exp. Zool. *59*, 199–211.

GROSS, P. R. (1967).
The control of protein synthesis in embryonic development and differentiation. Current Topics Dev. Biol. *2*, 1–46.

GROSS, K. W., JACOBS-LOREN, M., BAGLIONI, C., and GROSS, P. R. (1973).
Cell free translation of maternal messenger RNA from sea urchin eggs. Proc. Nat. Acad. Sci. USA *70*, 2614–2618.

GRÜNEBERG, H. (1967).
Gene action in the mammalian X-chromosome. Genet. Res. *9*, 343–357.

GRÜNEBERG, H. (1969).
Threshold phenomena versus cell heredity in the manifestation of sex-linked genes in mammals. J. Embryol. Exp. Morph. *22*, 145–179.

GRÜNEBERG, H., CATTANACH, B. M., MCLAREN, A., WOLFE, H. G., and BOWMAN, P. (1972). The molars of Tabby chimaeras in the mouse. Proc. Roy. Soc. Lond. *B182*, 183–192.

HADORN, E. (1961). Developmental Genetics and Lethal Factors. London: Methuen.

HÄGGQVIST, G., and BANE, A. (1950). Studies in triploid rabbits produced by Colchicine. Hereditas *36*, 329–334.

HAMERTON, J. L. (1971a). General Cytogenetics. Human Cytogenetics, vol. 1. New York: Academic Press.

HAMERTON, J. L. (1971b). Clinical Cytogenetics. Human Cytogenetics, vol. 2. New York: Academic Press.

HAMERTON, J. L., GIANNELLI, F., COLLINS, F., HALLET, J., FRYER, A., MCGUIRE, V. M., and SHORT, R. V. (1969). Non-random X-inactivation in the female mule. Nature *222*, 1277–1278.

HAMERTON, J. L., RICHARDSON, B. J., GEE, P. A., ALLEN, W. R., and SHORT, R. V. (1971). Non-random X chromosome expression in female mules and hinnies. Nature *232*, 312–315.

HOOK, E. B., and BRUSTMAN, L. D. (1971). Evidence for selective differences between cells with an active horse X chromosome and cells with an active donkey X chromosome in the female mule. Nature *232*, 349–350.

HÖSLI, P., DEBRUYN, C. H. M. M., and OEI, T. L. (1974). Development of a micro HG-PRT activity assay: preliminary complementation studies with Lesch-Nyhan cell strains. *In* Purine Metabolsim in Man, O. Sperling, A. DeVries, and J. B. Wyngaarden, eds. New York: Plenum Press, 811–815.

IBSEN, H. L., and STEIGLEDER, E. (1917). Evidence for the death *in utero* of the homozygous Yellow mouse. Am. Naturalist *51*, 740–752.

KAUFMAN, M. H., and GARDNER, R. L. (1974). Diploid and haploid parthenogentic development following *in vitro* activation. J. Embryol. Exp. Morph. *31*, 635–642.

KAUFMAN, M. H., and SACHS, L. (1975). The early development of haploid and aneuploid parthenogenetic embryos. J. Embryol. Exp. Morph. *34*, 645–655.

KIRKHAM, W. B. (1917). Embryology of the Yellow mouse. Anat. Rec. *11*, 480–481.

KIRKHAM, W. B. (1919). The fate of homozygous Yellow mice. J. Exp. Zool. *28*, 125–135.

KIRKMAN, H. N., and HENDRICKSON, E. M. (1963). Sex-linked electrophorectic difference in glucose-6-phosphate dehydrogenase. Am. J. Hum. Genet. *15*, 241–258.

KNOWLAND, J., and GRAHAM, C. F. (1972). RNA synthesis in the two-cell stage of mouse development. J. Embryol. Exp. Morph. *27*, 167–176.

KOZAK, C., NICHOLS, E., and RUDDLE, F. H. (1975).
Gene linkage analysis by somatic cell hybridization: assignment of adenine phosphoribosyltransferase to chromosome 8 and α-galactosidase to the X-chromosome. Somatic Cell Genet. *1*, 371–382.

KOZAK, L. P., MCLEAN, G. K., and EICHER, E. M. (1974).
X-linkage of phosphoglycerate kinase in the mouse. Biochem. Genet. *11*, 41–47.

KRZANOWSKA, H., and WABIK, B. (1971).
Selection for expression of sex-linked gene *Ms (Mosaic)* in heterozygous mice. Genetica Polonica *12*, 537–544.

LEWIS, W. H., and GREGORY, P. W. (1929).
Cinematographs of living developing rabbit eggs. Science *69*, 226–229.

LEWIS, W. H., and HARTMAN, C. G. (1931).
Three living monkey eggs in cleavage, with motion pictures of one. Anat. Rec. *48*, 53.

LEWIS, W. H., and WRIGHT, E. S. (1935).
On the early development of the mouse egg. Contrib. Embryol. *148*, 113–144.

LUSIS, A. J., and WEST, J. D. (1976).
X-linked inheritance of a structural gene for α-galactosidase in *Mus musculus*. Biochem. Genet. *14*, 849–855.

LYON, J. B., PORTER, J., and ROBERTSON, M. (1967).
Phosphorylase b kinase inheritance in mice. Science *155*, 1550–1551.

LYON, M. F. (1961).
Gene action in the X-chromosome of the mouse *(Mus musculus L)*. Nature *190*, 372–373.

LYON, M. F. (1968).
Chromosomal and subchromosomal inactivation. Ann. Rev. Genet. *2*, 31–52.

LYON, M. F. (1970).
Genetic activity of sex chromosomes in somatic cells of mammals. Phil. Trans. Roy. Soc. Lond. *B259*, 41–52.

LYON, M. F. (1972).
X-chromosome inactivation and developmental patterns in mammals. Biol. Rev. *47*, 1–35.

LYON, M. F. (1974).
Mechanisms and evolutionary origins of variable X-chromosome activity in mammals. Proc. Roy. Soc. Lond. *B187*, 243–268.

MANES, C. (1975).
Genetic and biochemical activities in preimplantation embryos. *In* The Developmental Biology of Reproduction, C. L. Markert and J. Papaconstantinou, eds. New York: Academic Press, pp. 133–163.

MCCLINTOCK, B. (1967).
Genetic systems regulating gene expression during development. Dev. Biol. Suppl. *1*, 84–112.

MCLAREN, A. (1974).
Embryogenesis. *In* Physiology and Genetics of Reproduction, Part B, E. M. Coutinho and F. Fuchs, eds. New York: Plenum Press, pp. 297–316.

MCLAREN, A., and BOWMAN, P. (1973).
Genetic effects on the timing of early development in the mouse.
J. Embryol. Exp. Morph. *30*, 491–497.

MEISLER, M. (1976).
Effects of the *Bgs* locus on mouse β-galactosidase. Biochem. Genet. *14*, 921–932.

MELANDER, Y. (1950).
Chromosomal behavior of the triploid adult rabbit as produced by Häggqvist and Bane after Colchicine treatment. Hereditas *36*, 288–289.

MIGEON, B. R., DER KALOUSTIAN, V. M., NYHAN, W. L., YOUNG, W. J., and CHILDS, B. (1968).
X-linked hypoxanthine-guanine phosphoribosyl transferase deficiency: heterozygote has two clonal populations. Science *160*, 425–427.

MINTZ, B. (1964).
Synthetic processes and early development in the mammalian egg. J. Exp. Zool. *157*, 85–100.

MODLINSKY, J. A. (1975).
Haploid mouse embryos obtained by microsurgical removal of one pronucleus. J. Embryol. Exp. Morph. *33*, 897–905.

MORGAN, W. C. (1950).
A new Tail-short mutation in the mouse. J. Heredity *41*, 208–215.

MORRIS, T. (1968).
The XO and OY chromosome constitutions in the mouse. Genet. Res. *12*, 125–137.

MOORE, G. P. M. (1975).
The RNA polymerase activity of the preimplantation mouse embryo.
J. Embryol. Exp. Morph. *34*, 291–298.

MUGGLETON-HARRIS, A. L., and JOHNSON, M. H. (1976).
The nature and distribution of serologically detectable alloantigens on the preimplantation mouse embryo. J. Embryol. Exp. Morph. *35*, 59–72.

NESBITT, M. N. (1974).
Chimaeras vs. X-inactivation mosaics: significance of differences in pigment distribution. Dev. Biol. *38*, 202–207.

NICOL, A., and MCLAREN, A. (1974).
An effect of the female genotype on sperm transport in mice. J. Reprod. Fertil. *39*, 421–424.

NIEMIERKO, A. (1975).
Induction of triploidy in the mouse by Cytochalasin B. J. Embryol. Exp. Morph. *34*, 279–289.

NYHAN, W. L., BAKAY, B., CONNOR, J. D., MARKS, J. F., and KEEL, D. K. (1970).
Hemizygous expression of glucose-6-phosphate dehydrogenase in erythrocytes of heterozygotes of the Lesch-Nyhan syndrome. Proc. Nat. Acad. Sci. USA *65*, 214–218.

OHNO, S. (1967).
Sex Chromosomes and Sex-Linked Genes. Berlin: Springer-Verlag.

OHNO, S. (1973).
Ancient linkage groups and frozen accidents. Nature *244*, 259–262.

OHNO, S., CHRISTIAN, L., ATTARDI, B. J., and KAN, J. (1973).
Modified expression of the *Testicular Feminization (Tfm)* gene of the mouse

by a "controlling element" gene. Nature New Biol. *245*, 92–93.

OHNO, S., POOLE, J., and GUSTAVSSON, I. (1965). Sex linkage of erythrocyte glucose-6-phosphate dehydrogenase in two species of wild hares. Science *150*, 1737–1738.

PAIGEN, K. (1961). The genetic control of enzyme activity during differentiation. Proc. Nat. Acad. Sci. USA *47*, 1641–1649.

PAIGEN, K., and GANSCHOW, R. (1965). Genetic factors in enzyme realization. Brookhaven Symp. Biol. *18*, 99–115.

PAIGEN, K., MEISLER, M., FELTON, J., and CHAPMAN, V. M. (1976). Genetic determination of the β-galactosidase developmental program in mouse liver. Cell *9*, 533–539.

PARK, W. W. (1957). The occurrence of sex chromatin in early human and macaque embryos. J. Anat. *91*, 369–373.

PATERSON, H. (1976). Private communication. Mouse Newsletter *54*, 47.

PEDERSEN, R. A. (1974). Development of lethal Yellow (A^Y/A^Y) mouse embryos *in vitro*. J. Exp. Zool. *188*, 307–320.

PLOTNICK, F., KLINGER, H. P., and KOSSEFF, A. L. (1971). Sex chromatin formation in pre-implantation rabbit embryos. Cytogenetics *10*, 244–253.

RATTAZZI, M. C., and COHEN, M. M. (1972). Further proof of genetic inactivation

of the X-chromosome on the female mule. Nature *237*, 393–395.

RICHARDSON, B. J., CZUPPON, A. B., and SHARMAN, G. B. (1971). Inheritance of glucose-6-phosphate dehydrogenase variation in kangaroos. Nature New Biol. *230*, 154–155.

RIGGS, A. D. (1975). X-inactivation, differentiation, and DNA methylation. Cytogenet. Cell Genet. *14*, 9–25.

ROBERTSON, G. C. (1942). An analysis of the development of homozygous Yellow mouse embryos. J. Exp. Zool. *89*, 197–231.

ROMEO, G. and MIGEON, B. R. (1970). Genetic inactivation of the α-galactosidase locus in carriers of Fabry's disease. Science *170*, 180–181.

RUSSELL, L. B., MCDANIEL, M. N. C., and WOODIEL, F. N. (1963). Crossing-over within the *A* "locus" of the mouse. Genetics *48*, 907.

SCHWARTZ, D. (1962). Genetic studies on mutant enzymes in maize. III. Control of gene action in the synthesis of pH 7.5 esterase. Genetics *47*, 1609–1615.

SHARMAN, G. B. (1971). Late DNA replication in the paternally derived X chromosome of female kangaroos. Nature *230*, 231–232.

SIRACUSA, G. (1973). RNA polymerase during early development in mouse embryo. Exp. Cell Res. *78*, 460–462.

SKOULTCHI, A., and GROSS, P. R. (1973). Maternal histone messenger RNA detection by molecular hybridization. Proc. Nat. Acad. Sci. USA 70, 2840–2844.

SMITH, L. J. (1956).
A morphological and histochemical investigation of a preimplantation lethal (t^{12}) in the house mouse. J. Exp Zool. 132, 51–84.

SNOW, M. H. L. (1975).
Embryonic development of tetraploid mice during the second half of gestation. J. Embryol. Exp. Morph. 34, 707–721.

SNOW, M. H. L. (1976).
The immediate postimplantation development of tetraploid mouse blastocysts. J. Embryol. Exp. Morph. 35, 81–86.

STREETER, G. (1948).
Developmental horizons in human embryos. Contrib. Embryol. 30, 211–245.

TAKAGI, N., and SASAKI, M. (1975).
Preferential inactivation of the paternally derived X-chromosome in the extraembryonic membranes of the mouse. Nature 256, 640–641.

TARKOWSKI, A. K., and ROSSANT, J. (1976).
Haploid mouse blastocysts developed from bisected zygotes. Nature 259, 663–665.

TAYLOR, B. A., BAILEY, D. W., CHERRY, M., RIBLET, R., and WEIGERT, M. (1975).
Genes for immunoglobin heavy chains and serum prealbumin protein are linked in mouse. Nature 256, 644.

TAYLOR, B. A., HEINIGER, H. J., and MEIER, H. (1973).
Genetic analysis of resistance to

cadmium induced testicular damage in mice. Proc. Soc. Exp. Biol. Med. 143, 629–631.

TETTENBORN, U., and GROPP, A. (1970).
Meiotic nondisjunction in mice and mouse hybrids. Cytogenetics 9, 272–283.

TRUJILLO, J. M., WALDEN, B., O'NEIL, P., and ANSTALL, H. B. (1965).
Sex linkage of glucose-6-phosphate dehydrogenase in the horse and donkey. Science 148, 1603–1604.

TYLER, A. (1967).
Masked messenger RNA and cytoplasmic DNA in relation to protein synthesis and processes of fertilization and determination in embryonic development. Dev. Biol. Suppl. 1, 170–226.

VAN BLERKOM, J., BARTON, S. C., and JOHNSON, M. H. (1976).
Molecular differentiation in the preimplantation mouse embryo. Nature 259, 319–321.

VAN BLERKOM, J., and BROCKWAY, G. O. (1975).
Qualitative patterns of protein synthesis in the preimplantation mouse embryo. Dev. Biol. 44, 148–157.

VANDEBERG, J. L., COOPER, S. D. W., and SHARMAN, G. B. (1973).
Phosphoglycerate kinase. A polymorphism in the wallaby Macropus parryi: activity of both X chromosomes in muscle. Nature New Biol. 243, 47–48.

VAN VALEN, P. (1966).
Oligosyndactylism, an early embryonic lethal in the mouse. J. Embryol. Exp. Morph. 15, 119–124.

WAKASUGI, N. (1973).
Studies on fertility of DDK mice: reciprocal crosses between DDK and C57BL/6J strains and experimental transplantation of the ovary. J. Reprod. Fertil. *33*, 283–291.

WAKASUGI, N. (1974).
A genetically determined incompatibility system between spermatozoan and eggs leading to embryonic death in mice. J. Reprod. Fertil. *41*, 85–96.

WAKASUGI, N., TOMITA, T., and KONDO, K. (1967).
Differences of fertility in reciprocal crosses between inbred strains of mice: DDK, KK and NC. J. Reprod. Fertil. *13*, 41–50.

WALKER, S., ANDREWS, J., GREGSON, N. M., and GAULT, W. (1973).
Three further cases of triploidy in man surviving to birth. J. Med. Genet. *10*, 135–141.

WALLACE, R. A. (1961).
Enzymatic patterns in the developing frog embryo. Dev. Biol. *3*, 486–515.

WARNER, C. M., and VERSTEEGH, L. R. (1974).
In vivo and *in vitro* effect of α-amanitin on preimplantation mouse embryo RNA polymerase. Nature *248*, 678–680.

WEST, J. D. (1976).
Clonal development of the retinal epithelium in mouse chimaeras and X-inactivation mosaics. J. Embryol. Exp. Morph. *35*, 445–461.

WHITE, B. J., TJIO, J. H., VAN DE WATER, L. C., and CRANDALL, C. (1974a).
Trisomy 19 in the laboratory mouse. I. Frequency in different crosses at specific developmental stages and relationship of trisomy to cleft palate. Cytogenet. Cell Genet. *13*, 217–231.

WHITE, B. J., TJIO, J. H., VAN DE WATER, L. C., and CRANDALL, C. (1974b).
Trisomy 19 in the laboratory mouse. II.Intrauterine growth and histological studies of trisomics and their normal littermates. Cytogenet. Cell Genet. *13*, 232–245.

WHITTEN, W. K., and DAGG, C. P. (1961).
Influence of spermatozoa on the cleavage rate of mouse eggs. J. Exp. Zool. *148*, 173–183.

WOLF, U., and ENGEL, W. (1972).
Gene activation during early development of mammals. Humangenetik *15*, 99–118.

WOODLAND, H. R., and GRAHAM, C. F. (1969).
RNA synthesis during early development of the mouse. Nature *221*, 327–332.

WRIGHT, D. A., and MOYER, F. H. (1968).
Inheritance of frog lactate dehydrogenase patterns and the persistance of maternal isozymes during development. J. Exp. Zool. *167*, 197–206.

WUDL, L., and CHAPMAN, V. M. (1976).
The expression of β-glucuronidase during preimplantation development of mouse embryos. Dev. Biol. *48*, 104–109.

WUDL, L., and PAIGEN, K. (1974).
Enzyme measurements on single cells. Science *184*, 992–994.

4 T-COMPLEX MUTATIONS AND THEIR EFFECTS

Michael I. Sherman and Linda R. Wudl

4.1 Introduction

It has been argued that the T complex[1] on chromosome 17 in the mouse specifies gene products essential in development because embryos homozygous for a variety of mutations in this region die at specific stages of gestation. Whether or not T-complex mutants are true "developmental" mutants, they have provided a foundation for the formulation of one of the most popular theories concerning genetic control of differentiation. This chapter is an attempt to describe this genetic system and to evaluate its possible role in the process of mammalian differentiation.

Perhaps the best way to begin to define the T region is to present a brief historical review of the early discoveries that demonstrated the possible importance of this system from the standpoint of research in mammalian development and genetics. A detailed description of early work on the T complex is given by Grüneberg (1952). Dobrovolskaia-Zavadskaia reported the dominant T mutation, called Brachyury, in 1927 and showed that it causes spinal abnormalities and a short-tailed phenotype in heterozygotes. Homozygous T/T mutants died prenatally (Chesley, 1932). When heterozygous $T/+$ mice were outcrossed with other mouse populations, an unexpected tailless phenotype often appeared (Dobrovolskaia-Zavadskaia and Kobozieff, 1932a). Chesley and Dunn (1936) proved these tailless mice were heterozygous for T and a recessive lethal mutation, called t.

Recessive lethal t mutations have been found not only in laboratory stocks but also in very high proportions of mice in wild populations. In fact, they occur with such frequency that they are considered true polymorphisms. In addition to enhancement of the T-mutation effect on

1. In the past, the term "T locus" has been used to describe the region of chromosome 17, just proximal to the centromere, extending from T (Brachyury) to tf (tufted) and characterized by a variety of so-called t (or T) alleles. Since a number of experiments have indicated that more than a single locus is involved, and multiple genetic functions probably comprise this region, we shall adopt here the terms "T region" to denote structural aspects and "T complex" to refer to the functional elements included within the region. When we wish to refer specifically to the Brachyury gene within the T complex, we shall use an italic T.

In view of the complexity of the T region, the term "allele" is, properly speaking, inappropriate. Artzt and Bennett (1975, 1977) introduced an alternative term, "haplotype." We feel the strong immunological implications of this term to be unwarranted on the basis of available evidence at the time of this writing (see section 4.8). Consequently, for want of better terminology, we shall continue to use the term "allele."

tail and spinal structures, t mutants in the homozygous state can be lethal, in which case death occurs prenatally, or semilethal, in which case some t/t embryos survive the gestation period but others do not. A third class of t mutants, observed only in laboratory crosses from preexisting t alleles, are viable. Table 4.1 provides a listing of the T and t mutations discussed in this chapter.

It is not expected that mutations causing lethality in the homozygous state should be so ubiquitous in nature. Two properties of t mutations help to explain this polymorphism. First, many t mutations exhibit "segregation distortion" caused by abnormally high male transmission frequencies. Up to 100 percent of the offspring in a single litter might receive the t allele from the heterozygous male parent. Transmission frequencies of t alleles found in wild populations are usually greater than 90 percent. (See section 4.3 for a detailed discussion of this phenomenon.) The second observation that no doubt has important bearing on the question of the polymorphism of this genetic system is that all t mutations found in the wild suppress crossing over in the T region of chromosome 17. The significance of these two phenotypic effects of t mutations is the subject of section 4.2.

The foregoing description of the features that permit the survival of t-mutant mice in laboratory and feral populations is only part of the complex picture of the genetic system. Homozygous t-mutant embryos may be lethal, semilethal, or viable. If lethal, the affected embryos might die at any of several stages of embryogenesis. A t mutation might interact with T to give a tailless animal, or it might not (see section 4.2). Although some t mutations cause transmission frequency distortion, others do not, and still others actually show a subnormal transmission frequency. Similarly, most t mutations suppress recombination, as indicated above, but some do not. A rare recombinational event in the former class is virtually always accompanied by a change in the t phenotype. Indeed, the T region appears to be an unusually dynamic genetic system.

The aspect of the T complex that is most central to this monograph is the observation that homozygous lethal T and t mutants die at stages in embryogenesis characteristic of their particular "allele." Lethal t mutations can be classified according to the stage at which death occurs. Furthermore, t mutations exhibit at least partial complementation when representatives from two different mutant classes are combined in a compound heterozygote (i.e., t^x/t^y, where t^x and t^y are representa-

Table 4.1. A Survey of Some T and t Alleles and Their Properties

Allelle	Phenotype of Homozygote	Derivation	Complementation Group
T	**Lethal**	**Laboratory stock**	"T"
T^h	Lethal	Laboratory stock	"T"
T^c	Lethal	Irradiation	"T"
T^{hp}	Lethal	Laboratory stock	"T"
T^{Orl}	Lethal	Irradiation	"T"
T^{OR}	Lethal	Irradiation	"T"
t^0	**Lethal**	**Laboratory stock**	t^0
t^{1*}	Lethal	Laboratory stock	$t^{12}(?)$
$t^1\dagger$	Lethal	Laboratory stock	t^0
t^3	Viable	$T/t^{1*} \times T/t^{1*}$	—
t^4	Lethal	$T/t^{1*} \times T/t^{1*}$	t^9
t^6	Lethal	Laboratory stock	t^0
t^8	Viable	$T/t^{1*} \times T/t^{1*}$	—
t^9	**Lethal**	$T/t^{1*} \times T/t^{1*}$	t^9
t^{12}	**Lethal**	$T/t^{1*} \times T/t^{1*}$	t^{12}
t^{h2}	Viable	$T/t^6 \times T/t^6$	—
t^{h3}	Viable	$T/t^6 \times T/t^6$	—
t^{h4}	Viable	$T/t^6 \times T/t^6$	—
t^{h5}	Viable	$T/t^6 \times T/t^6$	—
t^{h6}	Viable	$T/t^6 \times T/t^6$	—
t^{h7}	Lethal	$T/t^6 \times T/t^6$	t^0

Segregation Ratio**	Suppression of Recombination	Key References
N	No	Dobrovolskaia-Zavadskaia (1927), Chesley (1935), Spiegelman (1976)
N	No	Lyon (1959)
N	No	Searle (1966)
N	No	Johnson (1974)
N	No	R. Moutier, cited by Bennett (1975a)
N	No	Bennett et al. (1975)
+ to ++	Yes	Dobrovolskaia-Zavadskaia and Kobozieff (1932), Chesley and Dunn (1936), Dunn and Gluecksohn-Schoenheimer (1940)
++	Yes	Dobrovolskaia-Zavadskaia and Kobozieff (1932), Dunn (1937), Dunn and Gluecksohn-Schoenheimer (1943)
++	Yes	Silagi (1962)
− to N	No	Dunn and Gluecksohn-Waelsch (1953b)
−	No	Dunn and Gluecksohn-Waelsch (1953a), Moser and Gluecksohn-Waelsch (1967)
++	Yes	Carter and Phillips (1950), Dunn and Gluecksohn-Schoenheimer (1950), Nadijcka and Hillman (1975)
N	No	Dunn and Gluecksohn-Waelsch (1953a)
− to N	No	Dunn and Gluecksohn-Waelsch (1953a), Moser and Gluecksohn-Waelsch (1967)
++ to +++	Yes	Dunn and Gluecksohn-Waelsch (1953a), Dunn (1956), Smith (1956), Mintz (1964a), Calarco and Brown (1968), Hillman et al. (1970)
− −	No	Lyon and Meredith (1964a,b)
N	No	Lyon and Meredith (1964a,b)
N	No	Lyon and Meredith (1964a)
− −	Partial	Lyon and Meredith (1964a)
− −	No	Lyon and Meredith (1964a,b)
N	Yes	Lyon and Meredith (1964a,b,c)

Table 4.1 (continued). A Survey of Some T and t Alleles and Their Properties

Allelle	Phenotype of Homozygote	Derivation	Complementation Group
t^{h8}	Viable	$T/t^6 \times T/t^6$	—
t^{h9}	Viable	$T/t^6 \times T/t^6$	—
t^{h10}	Viable	$T/t^6 \times T/t^6$	—
t^{h11}	Viable	$T/t^6 \times T/t^6$	—
t^{h13}	Lethal	$T/t^{h7} \times T/t^{h7}$	t^0
t^{h14}	Viable	$T/t^{h7} \times T/t^{h7}$	—
t^{h15}	Viable	$T/t^{h7} \times T/t^{h7}$	—
t^{h16}	Lethal	$T/t^{h7} \times T/t^{h7}$	t^0
t^{h18}	Lethal	$+/t^6 \times +/t^6$	t^0
t^{w1}	**Lethal**	**Wild (U.S.)**	t^{w1}
t^{w2}	Semilethal	Wild (U.S.)	—
t^{w3}	Lethal	Wild (U.S.)	t^{w1}
t^{w4}	Lethal	Wild (U.S.)	t^0
t^{w5}	**Lethal**	**Wild (U.S.)**	t^{w5}
t^{w6}	Lethal	Wild (U.S.)	t^{w5}
t^{w8}	Semilethal	Wild (U.S.)	—
t^{w10}	Lethal	Wild (U.S.)	t^{w5}
t^{w11}	Lethal	Wild (U.S.)	t^{w5}
t^{w12}	Lethal	Wild (U.S.)	t^{w1}
t^{w13}	Lethal	Wild (U.S.)	t^{w5}
t^{w15}	Lethal	Wild (U.S.)	t^{w5}
t^{w17}	Lethal	Wild (U.S.)	t^{w5}
t^{w18}	Lethal	$+/t^{w11} \times T/+$	t^9
t^{w20}	Lethal	$+/t^{w15} \times T/+$	t^{w1}

Segregation Ratio**	Suppression of Recombination	Key References
– –	No	Lyon and Meredith (1964a)
– –	No	Lyon and Meredith (1964a)
N	No	Lyon and Meredith (1964a)
N	No	Lyon and Meredith (1964a,b)
N	Yes	Lyon and Meredith (1964b)
N	Yes	Lyon and Meredith (1964b)
– –	Yes	Lyon and Meredith (1964b)
+	Partial?	Lyon and Meredith (1964b,c)
N	Yes	Lyon and Meredith, (1964a,b)
+ +	**Yes**	Dunn and Suckling (1956), Bennett et al. (1959a,b)
+ + +	Yes	Dunn and Suckling (1956), Bennett and Dunn (1969)
+ + +	Yes	Dunn and Suckling (1956), Bennett et al. (1959a,b)
+ + +	Yes	Dunn and Suckling (1956)
+ + +	**Yes**	Dunn and Suckling (1956), Bennett and Dunn (1958)
+ + +	Yes	Dunn and Suckling (1956), Bennett and Dunn (1958)
+ +	Yes	Dunn (1957), Bennett and Dunn (1969)
+ + +	Yes	Dunn (1957)
+ + +	Yes	Dunn (1957), Bennett and Dunn (1958)
+ + +	Yes	Dunn (1957), Bennett et al. (1959a,b)
+ + +	Yes	Dunn (1957), Bennett and Dunn (1958)
+ + +	Yes	Dunn (1956, 1957)
+ + +	Yes	Dunn (1957), Bennett and Dunn (1958)
+	Partial?	Bennett and Dunn (1960), Spiegelman and Bennett (1974)
+ + +	Yes	Bennett et al. (1959a,b), Bennett and Dunn (1964)

Table 4.1 (continued). A Survey of Some T and t Alleles and Their Properties

Allelle	Phenotype of Homozygote	Derivation	Complementation Group
t^{w21}	Lethal	$+/t^{w17} \times T/+$	t^{w1}
t^{w29}	Viable	$T/t^{w18} \times T/t^{w18}$	—
t^{w30}	Lethal	$T/t^{w12} \times T/t^{w12}$	t^9
t^{w31}	Viable	$T/t^{w5} \times T/t^{w5}$	—
t^{w32}	Lethal	$T/t^{w10} \times T/t^{w10}$	t^{12}
t^{w34}	Semilethal	$T/t^{w5} \times T/t^{w5}$	—
t^{w36}	Semilethal	Wild (U.S.)	—
t^{w49}	Semilethal	Wild (Canada)	—
t^{w52}	Lethal	$T/t^{w5} \times T/t^{w5}$	t^9
t^{w71}	Lethal	Wild (Denmark)	t^{w1}
t^{w72}	Lethal	Wild (Denmark)	t^{w1}
t^{w73}	**Lethal**	**Wild (Denmark)**	t^{w73}
t^{w74}	Lethal	Wild (Denmark)	t^{w5}
t^{w75}	Lethal	Wild (D.D.R.)	t^{w5}
t^{w80}	Lethal	Wild (USSR)	t^{w5}

*This was the original t^1 allele (See section 4.4.1).

†This is the present t^1 allele, shown by Silagi (1962) to be identical to t^0.

**Although very precise segregation ratios have been given in previous listings (e.g., Dunn et al., 1962; Bennett et al., 1976), different laboratories have often found different transmission frequencies for some t alleles, probably due in part to the genetic background of the male and the female in the mating. Therefore, we have purposely been less exact in our assignment of transmission frequencies to the alleles. The code that we have used is as follows: $+++$ denotes very high ($\geq$ 90 percent) transmission frequency of the t (or T) alleles; $++$ denotes high (75–90 percent) frequency; $+$ denotes elevated (55–75 percent) frequency; N denotes normal (45–55 percent) frequency; $-$ denotes low (25–45 percent) frequency; $--$ denotes very low ($\leq$ 25 percent) frequency.

The alleles described here form an incomplete list selected because they are discussed in the text. These alleles and others are listed in Dunn et al. (1962), Bennett (1975a), and Bennett et al. (1976). The alleles in boldface type are the prototypes for the six comple-

Segregation Ratio**	Suppression of Recombination	Key References
+++	Yes	Bennett and Dunn (1964)
N	?	Dunn et al. (1962)
–	No	Bennett and Dunn (1964)
N	?	Dunn et al. (1962)
++	Yes	Bennett and Dunn (1964), Hillman and Hillman (1975)
––	No	Bennett and Dunn (1969)
+++	Yes	Bennett and Dunn (1969)
+++	Yes	Bennett and Dunn (1969)
+	Partial?	Bennett et al. (1976)
++	Yes	Dunn and Bennett (1971a), Dunn et al. (1973)
++	Yes	Dunn and Bennett (1971a), Dunn et al. (1973)
++	**Yes**	Dunn and Bennett (1971a), Dunn et al. (1973), Spiegelman et al. (1976)
+++	Yes	Dunn and Bennett (1971a), Dunn et al. (1973)
+++	Yes	Dunn et al. (1973), Bennett (1975a)
+++ ?	?	Dunn et al. (1973), Bennett (1975a)

mentation groups (and the group of T alleles). A question marked entry indicates that there is some uncertainty about the assignment or that the data has yet to be published. The references include those in which the mutant was first described, in which the allele was submitted to genetic analysis, and, where appropriate, in which morphology of homozygous mutant embryos is described.

tives of different lethal classes). To date, six such classes of mutants, or complementation groups, are known. The six representative stages of embryogenesis might be unique in their susceptibility to the lethal effects of *t*-gene lesions; alternatively, further probing might uncover additional classes of *t* mutants. It is clear, however, that certain *t* mutations do arrest embryonic development at particular stages and that the study of these lethal mutants might give further insight into the processes of mammalian development and differentiation.

4.2 Genetics of the T Complex

The multiple phenotypic effects of mutations in the T region of chromosome 17 in the mouse present an interesting and extremely elusive problem to the geneticist. Two facts stand out amid the waves of speculation regarding *t* (or *T*) mutations:

1. At least some of the effects attributed to these mutations are separable by crossing over and thus represent lesions at multiple loci within the region rather than pleiotropic effects of single mutational events.
2. Lethal *t* mutations found in wild populations and laboratory stocks, as well as some derivatives of these, possess a physical alteration in chromosome 17 that probably extends from the centromere well beyond the marker *tf* and is responsible for crossover suppression in this region.

The possible nature of the alteration will be discussed later in this section. First, however, a brief description of the evolution of the problem over the last several decades is warranted.

4.2.1 Isolation and Characterization of *t* Alleles

Background
The first recessive *t* mutations were discovered by Dobrovolskaia-Zavadskaia and Kobozieff (1932a) upon outcrossing heterozygous short-tailed (*T*/+) mice with various laboratory and wild stocks. Three such stocks contained mice heterozygous for a recessive mutation that when combined with *T* caused taillessness. These tailless mice possessed, in addition to the original *T* mutation, a mutation that appeared to be allelic with *T* and, like *T*, was lethal in the homozygous state. The three recessive *t* alleles were maintained in "balanced lethal"

stocks by continuous matings of tailless (T/t) animals inter se
(Dobrovolskaia-Zavadskaia and Kobozieff, 1932a,b; Kobozieff, 1935).
Chesley and Dunn (1936) pursued the study of these balanced lethal
lines. They demonstrated conclusively by exhaustive genetic analysis
that the tailless mice were indeed heterozygous for two lethal mutations, that these mutations were probably alleles because there appeared to be no crossover between them, and that homozygous lethal
embryos of the type T/T or t/t, from matings of T/t inter se, died at
different but characteristic stages of embryogenesis, so that only tailless T/t mice survived. They noted, furthermore, that the recessive t
allele was transmitted by the male parent (either T/t or $+/t$) in somewhat higher than Mendelian ratios of 1:1 (see section 4.3). Transmission through the egg was normal. Although there was some variation in
t transmission frequency among males tested, this variation was independent of whether the t allele had been contributed to the male by his
father or mother. The mating of animals from two of the balanced
lethal lines (e.g., $T/t^x \times T/t^y$) produced some normal-tailed animals
that proved to be compound heterozygotes of the type t^x/t^y, specifically t^0/t^1 (Kobozieff, 1935; Dunn, 1937, 1939; Dunn and Glueksohn-
Schoenheimer, 1943).

When the heterozygous Brachyury $(T/+)$ strain was outcrossed to
representative animals from wild populations, tailless offspring almost
invariably resulted from at least one such mating from each population.
Tests proved these tailless animals to be T/t heterozygotes with characteristics very similar to those obtained from laboratory stocks: the
recessive t mutation appeared to be allelic with T because no evidence
of crossing over between the two loci was observed; the transmission
of the t allele was extremely high when contributed by the male parent;
and, in most cases, the recessive t allele could be maintained in balanced lethal stocks due to the homozygous lethality of both the T/T
and t/t genotypes. Occasionally, the t allele was viable in the homozygous state (later correctly designated semilethal), although it retained
the additional phenotypic effects mentioned above (Dunn and Morgan,
1953; Dunn and Suckling, 1956; Dunn, 1960).

Many different t alleles have arisen from preexisting balanced lethal
stocks as normal-tailed exceptions; that is, among the progeny of T/t^x
$\times T/t^x$ crosses, from which only tailless animals were expected, an
occasional normal-tailed animal was found. Upon testing, these exceptions proved to be heterozygous for two different recessive t alleles

(i.e., t^x/t^y). These rare exceptions, which occurred only in the presence of a t allele, raised two very important points:

1. The frequency of occurrence was approximately 1 of 500 progeny (Dunn et al., 1962). Was this an unusually high frequency of mutation [the normal spontaneous mutation rate in the mouse is 0.8×10^{-5} (Lyon, 1960)], perhaps the result of an inductive phenomenon in the homologous chromosome, or were the rare exceptions a result of an unusual type of recombination?

2. Studies of the phenotypes of rare exceptions most often showed them to differ in many respects from the "parental" strain and suggested a high degree of genetic complexity within the T region.

These observations are discussed in more detail later. The fact that normal-tailed exceptions possess two different t alleles again demonstrates the ability of different t mutations to phenotypically complement each other and to produce viable (although often male sterile) animals.

Terminology

Mutations resulting in short-tailed phenotype in the heterozygous state are designated T and called dominant. Mutations that, when combined with T as a compound heterozygote, cause taillessness (thought to be a result of enhancement of the original T effect) are designated t and are referred to as recessive. Because of the lack of recombination between T and t (except for a very rare instance, which is discussed later) and among the various recessive t's, the term "allele" has been applied. This term is used to identify mutations located within the region of chromosome 17 that characteristically prevent recombination (and thus cannot be tested for true allelism) and that have a number of phenotypic effects, the most important of which, from the standpoint of identification, is interaction with T to cause taillessness.

Recessive t alleles are identified by the following designations: alleles found in wild populations and exceptions derived from them are called t^w and numbered consecutively. Thus, t^{w1} was the first such allele discovered, followed by t^{w2}, etc.; alleles found in laboratory stocks and their derivatives are also numbered in order, but without the w designation, i.e., t^0, t^1, t^2, etc.; finally, exceptions derived from t^6 and characterized by Lyon and co-workers are designated t^h. Representative members of each class are listed in table 4.1.

Dominant T mutations appear to be true alleles and are given a variety of identifying designations (see section 4.4.7).

The last term, which is frequently used in its loosest sense and might be confusing to a student of genetics, is complementation. Classically, complementation tests are used to map genes so closely situated on the chromosome that recombination is extremely rare. If mutant genomes are placed in the same organism and the resulting phenotype is normal, the mutations are said to have occurred in different units of genetic function. Only in rare instances do mutations within a single gene complement each other. In the case of t mutants, "complementation" tests are used to distinguish classes of lethal mutants. A member of one group is said to complement a member of the second group if any progeny of the type t^x/t^y survive. However, complementation is never complete (often viability is low and males are always sterile; see Bennett and Dunn, 1969), so there is at least an overlap of functional loss of genetic information between groups. Lack of complementation within a group of t mutants does not establish genetic identity in the classical sense because variations in phenotype among individual members of a group are possible (see section 4.4). The six known complementation groups are designated according to the first studied representative of the group. These are t^0, t^9, t^{12}, t^{w1}, t^{w5}, and t^{w73}.

Phenotype of t mutants

A complete list of variations in phenotype among t mutants would be confusing and would not add to the content of this section. By way of summary, then, this discussion considers the common characteristics of t alleles classified as lethal, semilethal, or viable.

Lethal t alleles   As mentioned earlier, the majority of t alleles found in wild populations are lethal. Such mutants characteristically exhibit extreme male segregation distortion (see section 4.3) and suppress crossing over in the T region. Complementation studies show that the majority of lethal t alleles in wild populations are members of the t^{w5} group, but included are a few representatives of the t^{w1} group and single exceptions from the t^0 (Dunn and Suckling, 1956) and the newly discovered t^{w73} (Dunn et al., 1973) groups. The possible significance of this distribution of t alleles in feral populations is discussed in section 4.2.4.

Lethal t alleles found in laboratory stocks or derived from preexisting balanced lethal lines (T/t^{wx} or T/t^x) fall into complementation groups t^{12}, t^0, t^{w1}, and t^9 (thus far, t^{w73} is the only representative of its group and t^{w5} appears to be represented only in the wild; see Bennett et al., 1976). Phenotypic variations are relatively great among these t alleles. Members of the t^9 group do not suppress crossing over in the T–tf region and exhibit normal-to-low male transmission frequencies. Transmission frequencies, in general, are lower for alleles found in laboratory stocks and exceptions than for alleles found in wild populations (table 4.1). At least some of the variation in transmission frequencies might be attributable to the background genotype of test animals (J. McGrath and N. Hillman, personal communication).

As mentioned above, lethal t alleles can be classified according to their ability to undergo phenotypic complementation. Another criterion for classification comes from embryological studies on dying t/t homozygous embryos. Both the time of death and physical characteristics of dying embryos appear to be more or less constant among members of a single group. Morphological studies of homozygous lethal embryos are discussed in section 4.4.

Viability and male fertility of mixed or compound heterozygotes containing at least one lethal t allele also depend somewhat on the complementation group of the lethal allele as well as on the phenotype of the second allele in the compound. Viability may also be influenced by non-T-linked genetic factors (Klyde, 1970).

Semilethal t alleles   All but one of the t alleles designated semilethal have been isolated from wild mouse populations. These have characteristics similar to the lethal t alleles in that they suppress crossing over, they have very high male transmission frequencies (transmission is normal in females; see Dunn, 1960), and homozygous semilethals exhibit morphological defects similar to those found in compound heterozygotes containing two different complementary lethal t alleles (e.g., t^0/t^{12} or t^0/t^{w1}; see sections 4.4.8 and 4.4.9). All homozygous semilethals that survive to birth are abnormal (Bennett and Dunn, 1969). Homozygous male semilethals that survive to maturity are sterile as are males heterozygous for two different semilethal alleles (see section 4.3.1).

There seems to be a remarkable similarity between homozygous semilethals and compound heterozygous lethals in that both are male sterile (the former being perhaps the most severely affected because they possess an obvious defect in spermatogenesis; see section 4.3.1) and exhibit varying degrees of lethality. Surviving animals suffer from characteristic abnormalities. Compound semilethal heterozygotes also show some degree of complementation; i.e., compounds sometimes have higher viability than either parent when homozygous. It has been postulated that "semilethal alleles may be members also of the series formed by the lethal alleles . . ." (Bennett and Dunn, 1969).

The one semilethal t allele that arose as an exception in a balanced lethal stock of T/t^{w5}, namely t^{w34}, differed from other semilethals in that it was characterized by a low transmission frequency but maintained a low survival frequency (Bennett and Dunn, 1969). It also did not suppress crossing over in the T–tf interval. Unfortunately, this line was lost before further tests could be done.

Viable t alleles   About the only characteristic viable alleles have in common with lethal and semilethal alleles, and the one that makes their detection and selection possible, is the interaction of the recessive t allele with T to cause taillessness. Viable t alleles make up the majority of the exceptions derived from balanced lethal stocks. In general, they characteristically permit normal recombination in the T region (Dunn and Gluecksohn-Waelsch, 1953b; Dunn et al., 1962), have normal-to-low male transmission frequencies, and are fertile (though to varying degrees) in the homozygous state (Dunn et al., 1962). Although there is no way of determining "identity" with viable alleles as there is with lethals, they are grouped according to transmission ratio characteristics and fertility when combined with a lethal allele in compound heterozygote form.

4.2.2 Recombination in the T Region

Generation of new t alleles: mutation or recombination?
Early studies by Dunn and Caspari (1945) showed that the presence of some t alleles (specifically t^0 and t^1) can inhibit normal crossing over between the markers T and fused (Fu), located approximately 8 map units apart on chromosome 17. Furthermore, exhaustive genetic studies failed to detect recombination between T and any t allele. Lyon and Phillips (1959) introduced the recessive tufted (tf) marker (Lyon, 1956) into the T-bearing chromosome (thus aiding in detection of crossover events by an easily recognized phenotype) and found that the frequency of recombination between T and tf in $T\,tf/t^x$ + heterozygotes (when $t^x = t^6$, a lethal t allele that suppresses crossing over) was

approximately equal to the "mutation" frequency previously observed
in balanced lethal lines that produced new t alleles. Furthermore, recombinational events involving the t allele resulted in a "mutation" of
that allele to a different t. Although it is possible that a t allele somehow induces mutation in the homologous chromosome, the more likely
explanation [proven in later experiments by Lyon and Meredith
(1964a,b,c)] was "that the phenotypic and genetic effects of t-alleles
are due to an abnormal chromosome segment in which rare crossing-
over occurs and, by changing the length or position of the abnormal
region, causes a change in the properties of the t-allele which is observed as a mutation." Grüneberg (1952) had, in fact, suggested this
hypothesis earlier. Subsequently, the tf marker was introduced into
many balanced lethal lines, and it became obvious that all spontaneous
"mutational" events that generated a new t allele were accompanied
by crossing over. This is also true in the case of new t mutants arising
from preexisting t alleles that do not show crossover suppression in the
region $T-tf$ (e.g., t^9, t^4).

Frequency of recombination in the T region

Prior to the work of Lyon and co-workers (Lyon and Phillips, 1959;
Lyon and Meredith, 1964a,b,c), who demonstrated the importance of
using a marker such as tufted to detect recombinational events, the
following procedure was used to study new t alleles arising in T/t-
mutant heterozygotes. Exceptions arising in balanced lethal lines (i.e.,
from the cross of two T/t^x tailless parents) and exhibiting normal-tailed
phenotype were tested by crossing with a $T/+$ heterozygote. If the
exception had arisen from a loss of T (i.e., $+/t^x$), then the cross ($+/t^x$
$\times$ $T/+$) would yield some short-tailed offspring. If, on the other hand,
the exception was a compound heterozygote of the type t^x/t^y, crossing with $T/+$ mates would yield both tailless (T/t^x or T/t^y) and normal-
tailed ($+/t^x$ or $+/t^y$) offspring, but no short-tailed animals. The latter
proved to be the case in each instance tested. This system would fail to
detect other types of possible recombinants (see discussion of t^6 derivatives below), and it is limited by problems of low viability of some,
and male sterility of all, compounds made up of lethal and semilethal t
alleles. These latter exceptions might contribute to frequency calcula-
tions but not to the pool of available t alleles. This factor probably
contributes to the high proportion of viable exceptions compared to

lethals and should be considered when discussing the significance of the various proportions of different t phenotypes arising from preexisting mutants. This is also true when considering the pattern of derivation of t lethals from preexisting lethals. Hence, does t^0 (or members of the t^0 complementation group) fail to give rise to t lethals of other complementation groups or is the probability of obtaining a viable female compound lethal heterozygote much lower than it is in the case of another complementation group (see Bennett et al., 1976)? Some recombinant genotypes are never seen, although they would be detected in this system (e.g., recombination resulting in loss of the T modifying effect but retention of the parental type of lethality and recombination between T and t to yield $T/+$, $+/t$).

Dunn et al. (1962) discussed the probability of detecting rare exceptions in balanced lethal matings. To reiterate their argument, detection depends on the frequency of the event, the probability that the gamete carrying the new allele will combine with the appropriate partner [due to high male transmission frequencies characteristic of most lethal t mutations, this value depends on whether the new allele arises in the female (roughly 0.9) or the male (0.5)], and the probability that the new compound heterozygote will survive (as low as 15 percent in the case of some mixed lethal heterozygotes or as high as 100 percent in the case of lethal-viable combinations).

Exceptions arising from t^6

The most enlightening information concerning the genetic structure of the T region has been derived from the recombination experiments of Lyon and Meredith (1964a,b,c). They obtained a series of t alleles from a balanced lethal stock containing t^6 and T (tf) on homologous chromosomes (T tf/t^6 $+$) during studies on effects of x rays on mutation in the T region. They found exceptions of the type t^6 $+/t^{hn}$ tf and T tf/t^{hn} tf (i.e., normal-tailed nontufted and tailless tufted from the original tailless nontufted stock) as well as one exception (t^{h7}, described below) of the type T tf/t^{h7} $+$ which could have been an x-ray-induced mutation because there was no evidence of recombination. These exceptions (excluding t^{h7}) were similar to the majority of exceptions obtained from balanced lethal stocks (table 4.1); that is, all were viable, all failed to suppress recombination in the T–tf interval (except for t^{h5}, which showed a partial suppression of recombination), all had normal or low transmission frequency, and homozygous males were fertile. The retention of the ability to cause taillessness in combination with T (hereafter referred to as "T modification") with loss of lethality and sterility can logically be explained in either of two ways:

1. Lethality and sterility are specified by genetic elements (or a single pleiotropic element) located distal to T (because loss of lethality and sterility was accompanied by recombination with tf) with an intervening abnormal stretch of chromosome. Normally, the intervening region would link the T complex into a single functional unit. The rare recombinational events under discussion would then disrupt this unit into individual elements.

2. Alternatively, sterility and lethality might be the result of a given length of abnormal chromosome; if the recombinant segment does not contain a minimum length of the abnormal region, then sterility and lethality will be lost (see Lyon and Meredith, 1964a).

Additional derivatives of t^6 (t^{h18}) and t^{h7} (t^{h13}, t^{h14}, t^{h15}, and t^{h16}) provided still more information about the nature of this chromosomal region. The t^{h18} allele arose from the cross $+ tf/t^6 + \times T tf/+ tf$ as a single short-tailed nontufted exception $T tf/+(t^{h18}) +$ (designated "$+(t^{h18})$" because it fails to cause T modification). Not only does t^{h18} permit recombination between T and tf but it also recombines with T in the cross $T tf/+(t^{h18}) + \times + tf/+ tf$ yielding the genotype $T(t^{h18}) +/+ tf$ at a frequency of 6.2 percent. Furthermore, t^{h18} retains the lethality of t^6. This represents the opposite type of recombination to that previously observed arising from lethal t alleles in that now the T modification and recombination suppression effects are lost, but lethality is retained. This type of recombinant was never detected in *balanced lethal* crosses because the resulting genotype $[T(t^y)/T$ or $T(t^y)/t^x]$ would undoubtedly be lethal.

The model proposing that lethality and T modification are located at somewhat opposite ends of an abnormal chromosome segment is supported by t^{h18}. Additional characteristics of this unusual mutant allele include absence of male segregation distortion and the ability to recombine not only with T, but also with other t^h alleles that retain the T-modification effect but not the t^6 lethality. Recombinants of this type $[T tf/t^{hn}(t^{h18}) +$, where $n = 2$, 6, or 11] were identical to heterozygotes of the type $T(t^{h18}) +/t^{hn} tf$ in that they were tailless and found to contain t^6 lethality upon further testing. In addition, the reconstituted chromosome was still not identical to the parental t^6 chromosome because it did not suppress recombination in the t region, consistent with the idea that a "midpiece" is responsible for the latter effect.

The t^{h7} allele, a *normal-tailed*, nontufted exception from the balanced lethal T tf/t^6 + stock, had the genotype T tf/t^{h7} +. This *suppressive* effect on the T Brachyury allele (it does not cause taillessness in combination with T but actually causes a lengthening of the tail) was not a result of mutation elsewhere in the genome because it was linked to, and did not undergo normal recombination with, tf. The t^{h7} allele carries the t^6 lethal factor and recombination suppression but has normal male segregation. Among the exceptional recombinants derived from t^{h7}, one (t^{h16}) was identical to t^6, but another (t^{h14}) retained only the normal T-modification effect. These two recombinants, as well as the phenotype of t^{h7}, could be explained by proposing that the T region of t^{h7} is identical to that of t^6 except for a duplication in the T-modification region (see Lyon and Meredith, 1964b). This duplication more than compensates for the T-gene deficiency (resulting in a normal-tailed phenotype), whereas the single mutated region present in t^6 might be at least partially T-deficient, a trait that would enhance the Brachyury effect (i.e., taillessness). Recombination at the duplication would give rise to the t^{h14} genotype (carrying a single dose of the T-modification region, thus interacting with T to give taillessness, with all else wild type) and the t^{h16} (or t^6) genotype. The only additional piece of evidence that this model does not incorporate is the loss of segregation distortion due to the duplication (a phenotypic characteristic of t^{h7} and t^{h14} but not t^{h16}).

Two additional alleles derived from t^{h7} resembled the two types of complementary recombinants obtained from t^6, as though recombination had occurred between the duplication and the lethal end of the region. Thus t^{h15} (like t^{h7}) suppresses the T phenotype, but it is viable and permits crossing over. On the other hand, t^{h13} has t^6 lethality and permits crossing over, but it has no effect on T.

It appears from these studies that, indeed, at least some of the phenotypic effects of mutations in the T region are due to alterations in discrete genetic loci, principally lethality and the interaction with the T allele to cause taillessness. The property of crossover suppression, on the other hand, appears to extend over a larger portion of the chromosome segment, as it would if it were the result of chromosomal abber-

ration rather than a discrete mutational event. Only the property of
segregation distortion remains to be explained on the basis of this (or
any!) model.

4.2.3 Structural Considerations

The multiple phenotypic effects coupled with crossover suppression
due to recessive lethal mutations in the T region of chromosome 17 and
the additional phenomenon of the conversion of one *t* phenotype to
another as a result of a rare recombinational event have engendered
wide acceptance for the theory that at least the original recombinational or mutational event that gave rise to such a mutant was accompanied by a gross alteration in the physical structure of the chromosome in that region. The alteration has been preserved, along with the
other phenotypic effects of the mutation, throughout at least recent
evolution of the mouse because of its ability to suppress recombination
and because of its accompanying [and possibly primary (see Lyon,
1964b)] effect on male segregation ratios.

In genetic studies, Hammerberg and Klein (1975a) mapped four
genes on chromosome 17 with respect to the centromere:

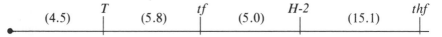

They demonstrated that not only do lethal *t* mutations suppress crossing over in the *T–Fu* (Dunn and Caspari, 1945) or *T–tf* (Lyon and
Phillips, 1959) interval, but also that suppression often extends beyond
the *tf* locus to the H-2 region. They concluded that "almost the entire
chromosomal arm from the centromere to the H-2 complex is for all
practical purposes excluded from the normal process of recombination." There is even some evidence that recombination suppression
extends beyond H-2 to the *thf* locus (15 map units from *H-2*), the t^{w5}
allele having the greatest effect (Hammerberg and Klein, 1975a).
Although this evidence suggests that a structural change might have
occurred over this extensive chromosomal region, recombination, when
it does occur outside the *T–tf* interval in the presence of a *t* allele, is
not accompanied by an alteration in that *t* allele. Furthermore, all
phenotypic effects attributable to *t* mutations are located in the *T–tf*
region. It seems likely, then, that the aberrant chromosome structure

is limited to the T region (between T and tf), but nevertheless influences recombination for some further distance.

What structural changes in the chromosome might result in the observations mentioned above? Dunn et al. (1962) discused two hypotheses that would explain not only crossover suppression but also the variations in phenotype of t mutations that always accompany a rare recombinational event and the interesting facts that there are only six known lethal complementation groups and that members of different groups of lethal t alleles at least partially complement each other when combined in a compound heterozygote. The first hypothesis incorporates the fact that chromosomal inversions alter recombination frequencies and suggests that "a series of spatially separated inversions each associated with a different t lethal" might explain these observations. The second hypothesis predicts "a series of deficient or deleted sections which suppress recombination in regions adjacent to them."

Several attempts have been made to identify the chromosomal region in question and to determine cytologically whether or not a gross structural change has taken place. Unfortunately, Mus musculus chromosomes are all acrocentric, present a continuous size gradation, and cannot be easily distinguished with routine cytological techniques. Bennett (1965) attempted to identify chromosomes carrying t mutations, but found no abnormalities that could be associated with them. Womack and Roderick (1974) concluded from their cytological studies that if an inversion is indeed responsible for the t phenotype, it is not a sizable portion of the chromosome. Similar conclusions had been reached by Jaffe (1952).

By placing linkage group IX (chromosome 17) of Mus musculus on the metacentric chromosome of Mus poschiavinus, Klein (1971) determined that the map distance between the centromere and the H-2 complex is close to the value obtained for the T to H-2 interval. He suggested that "the physical proximity of the repetitive DNA sequence in the centromeric heterochromatin could be responsible for at least some of the strange phenomena associated with the t-alleles."

In conclusion, although the genetical evidence strongly suggests that an extensive region of chromosome 17 is physically altered in some way, there is no cytological evidence that this is the case. Perhaps the problem should be reassessed with the more sophisticated chromosome banding techniques now available.

4.2.4 Evolution and Population Genetics of the T Region

How are recessive lethal and semilethal t mutations maintained as poly-morphisms in wild populations? It is interesting that the vast majority of t alleles found in wild populations in this country and in others (table 4.1) are members of the t^{w5} complementation group, although these populations include a small sampling of the t^{w1} group and single representatives of the t^0 and t^{w73} groups (Dunn and Suckling, 1965; Dunn et al., 1973). Furthermore, Bennett et al. (1976) pointed out that although t^{w5} can give rise to representatives of all other complementation groups as a result of rare, unequal recombinational events, a member of this group has never been generated from a preexisting allele. Hammerberg and Klein (1975a) found that t^{w5} suppresses recombination over the longest chromosome distance (beyond H-2) of any t allele tested. Thus, based on the assumption that the initial event that gave rise to a t mutant (an extremely rare mutational event because a wild-type-to-t transition has never been documented) was the result of a gross structural change in the chromosome (Lyon and Meredith, 1964b), it is logical to expect that the t^{w5} allele, having all the inherent properties required to ensure its survival and propagation in nature, came about first.

The high male segregation distortion of t^w alleles has been shown to be sufficient to ensure the persistence of the large proportion of t-carrying mice (up to 50 percent) in feral populations in spite of selective pressure against t alleles due to lethality (see Dunn, 1964). Furthermore, the initial t-generating event presumably would have resulted also in extreme recombination suppression, which ensures inheritance of the entire block of phenotypic effects, including lethality and T modification (and perhaps even H-2 haplotypes; see Hammerberg and Klein 1975a), which appear to be at opposite ends of the chromosome region in question (Lyon, 1964b). Thus, once the initial mutational event occurred (resulting in the t^{w5} genotype), its survival was ensured, not due to any possible selective advantage of T modification, lethality, or even linkage to H-2, but fortuitously as a result of the initial structural change in the chromosome.

The limited number of t^w phenotypes can be easily explained again on the basis of the structural change hypothesis. A recombinational event giving rise to a viable t allele is generally accompanied by loss of recombination suppression and segregation distortion. The new t allele usually reduces fertility in homozygous males or compound lethal-viable males. Semilethal recombinants, although found in wild populations are also at a slight disadvantage because homozygous males are sterile as are compound lethal-semilethal males. It has been shown by Lewontin (1962) that male sterility is more disadvantageous than lethality in terms of gene pool reduction and might even cause extinction of the whole population in the case of small groups. Recombinants that retain extreme crossover suppression and segregation distortion might then be the only exceptions likely to survive the process of natural selection; t^{w1} and t^{w73} are possible examples.

In summary, it is possible that the existence of t allele polymorphism in nature is a result of the combined selective advantage of high transmission frequency and crossover suppression. These are, in turn, due to an alteration in a chromosome segment that fortuitously includes genetic loci responsible for enhancement of the T-mutation effect on tail length and lethality (Lyon and Meredith, 1964b).

4.3 The Effect of t Mutations on Male Reproduction

The mechanism of segregation distortion of t mutations remains almost as much a mystery today as it was when Chesley and Dunn first reported its occurrence in 1936. Because female heterozygotes ($+/t$, T/t, or t^x/t^y) are fertile and transmit the t allele in the normal Mendelian ratio, this discussion will consider only possible aberrations in the male reproductive system of mutant animals that might account for both the observed deviation from Mendelian inheritance and the paradoxical effect of some t mutations that cause sterility when combined with another t mutation in a compound heterozygote. For example, t^x and t^y might represent two recessive lethal mutations. In heterozygous ($+/t$) males, each might be transmitted to greater than 90 percent of the offspring. Yet, if two heterozygous animals are mated ($+/t^x \times +/t^y$), progeny that are of the compound heterozygous type t^x/t^y might be viable, but all surviving males will be sterile. Therefore, it appears as though some process confers a selective advantage in the transmission of either one of the t mutations alone, whereas that property, or some other one, results in complete inhibition of transmission of both mutations when combined in the same animal.

4.3.1 Sterility in t-Mutant Males

Sterility, or reduced fertility, associated with t mutations appears, at least superficially, to be of two types. The type caused by homozygous effects of many semilethal t mutations (e.g., t^{w2}) appears to result from aberrant spermatogenesis. Sperm number, morphology, and motility are all affected in homozygous t semilethal males (Bennett and Dunn, 1971; Dooher and Bennett, 1974). Compound heterozygotes possessing two different semilethal mutations are sterile, but they have not been studied further.

The second type of sterility does not appear to be accompanied by an obvious defect in spermatogenesis, and early experiments (Glucksohn-Schoenheimer et al., 1950; Dunn, 1952) ruled out the possibility that apparent sterility was due to prenatal death of embryos. Males homozygous for viable t mutations often have reduced fertility and might even be completely sterile (Braden and Gluecksohn-Waelsch, 1958). Males heterozygous for a viable allele and a lethal or semilethal t mutation are, with rare exceptions, relatively infertile, or "quasi-sterile" (Dunn and Bennett, 1969), and lethal-lethal or lethal-semilethal compounds are always sterile (Bennett and Dunn, 1967). The number of sperm found in ejaculates of such compound heterozygous mutant males appears to be normal, and although the mean proportion of sperm with morphological abnormalities appears to be higher compared to those from males of genotype $+/+$, T/t, or $+/t$ (Bryson, 1944; Braden and Gluecksohn-Waelsch, 1958; Olds, 1971), there is no strict correlation with the degree of infertility in an individual animal, nor is it thought to be sufficiently high to account for the sterility observed in many cases. Indeed, Hillman and co-workers in extensive EM studies on sperm and progenitor cells at various stages of differentiation have been unable to detect *any* specific effect of T or t mutations on sperm morphology (N. Hillman, personal communication). However, Bryson (1944) did report that t^0 and t^1 (lethal t mutations that at that time were members of different complementation groups) when in combination had a definite affect on spermatogenesis. In general, such compound heterozygous males had reduced testis size and weight and a high proportion of sperm with abnormal heads. In addition, by studying the rate of dispersion of sperm into nutrient medium in vitro, he concluded that sperm from sterile t^0/t^1 males exhibited reduced motility.

The question of whether sperm motility is affected in the case of quasi-sterile or sterile t mutants (other than the semilethals mentioned above) has not been answered satisfactorily. Braden and Gluecksohn-Waelsch (1958) found that sperm recovered from the uterus several hours after mating with a sterile male (carrying two different t alleles, e.g., t^1/t^3) had an average "motility rating" lower than that obtained with males carrying only one or no t gene. However, in vitro tests did not show any differences in "the degree or duration of motility" of sperm from these same males. They also found greatly reduced numbers of sperm in the fallopian tubes of females mated to sterile males and suggested that sterility was a result of the inability of sperm to traverse the uterotubal junction. Olds (1970) also found in similar matings that sperm from sterile males were present in normal numbers in the uterus, but were absent or present in low numbers in the oviducts. On the other hand, Bennett and Dunn (1967) reported that sperm from sterile males were found in the oviducts (often in normal numbers), but had reduced motility as did sperm recovered from the uterus. They concluded that motility was not necessary in order for sperm to enter the fallopian tubes, but it might be required for them to *remain* at the site of fertiliztion because at a later time they did observe a difference in the number of sperm remaining in the oviduct of females mated with sterile males compared to normal matings.

There has been much speculation as to the cause of sterility and reduced fertility in the cases of lethal-lethal, lethal-semilethal, lethal-viable, and viable-viable compound heterozygous and t-homozygous viable males. Olds (1970) suggested that something present in semen of sterile males causes a closure of the uterotubal junction or that a reaction takes place in the uterus that causes coagulation of the semen. Both conditions would prevent sperm from reaching the site of fertilization. However, the experiment of Braden and Gluecksohn-Waelsch (1958), in which sperm from sterile and fertile males were mixed and used to artificially inseminate females did not lead to a decrease in the ability of + sperm to fertilize, as would be expected if Olds's theory were correct. In a later experiment, Olds (1971) injected sperm from sterile mutant (t^{w18}/t^{w32}), fertile mutant (T/t^{w18} or T/t^{w32}), and wild-type males into the ovarian bursae of females just after ovulation. She concluded that sterility was not due solely to the inability of sperm to reach the site of fertilization, because essentially no eggs were fertilized by sperm from the sterile male (a very small number, 2–5 percent, being questionable) and in no case did she observe these sperm in the eggs or perivitelline space, although a minimum of 10,000 sperm were injected per bursa. Furthermore, it is known that the interaction of sperm and egg during the fertilization process is both highly specific and complex; close proximity of two gametes is not sufficient to ensure penetration. Perhaps even a slight reduction in motility or sperm viability results in inability to penetrate. In addition, aberrations in sperm physiology could result in insufficient capacitation, but they would not be obvious in the preceding experiments.

What does appear obvious from a survey of the literature is that there is a correlation between the "strength" of the t mutation and fertility. Thus, homozygous semilethals, compound semilethals, compound lethal-semilethals, and compound lethals are always sterile; lethal-viable compounds are usually sterile; and viable-viable compounds and homozygotes are sometimes sterile (e.g., Bennett and Dunn, 1971). Spermatogenesis could very well be affected in all such animals although it is obvious only in the case of the most severely affected mutants. Furthermore, with combinations of alleles resulting in quasi-sterility, individual males can vary in fertility. For example, males of genotype t^0/t^{38} can be 100 percent fertile or show degrees of reduced fertility as low of 11 percent (Dunn and Bennett, 1969). It appears, therefore, as though factors besides genotype within the T complex affect male fertility. Dunn and Bennett (1969) also pointed out that in the case of mixed compound heterozygotes, there appeared to be no correlation between the transmission frequency of the individual t alleles involved and male fertility.

4.3.2 Segregation Distortion in t-Mutant Males

The possibility that sterility, especially in the case of the lethal-lethal compounds, is caused by abnormalities arising during spermatogenesis has received little recognition, perhaps because of the paradoxical effect of these same mutations, which cause *increased* transmission frequencies of the t allele when it is present in a single dose. Setting aside, for the moment, the problem of sterility, let us discuss the possible mechanisms involved in segregation distortion. Recessive t alleles that are transmitted in a ratio higher than normal (50 percent) include most of the known lethals, except t^4, t^9, and t^{w18}, and all of the semilethal t mutations discovered in wild mouse populations (see table 4.1).

The dominant T allele does not show segregation distortion (Kobozieff, 1935).

Two types of mechanisms have been proposed to account for segregation distortion in the t mutants. The first of these, referred to generally as *meiotic drive*, includes any process during spermatogenesis that results in an unequal distribution of a particular chromosome at the spermatid stage. This process might involve selective destruction of non-t-bearing cells or enhancement of differentiation of t-bearing spermatocytes. The second mechanistic consideration involves *postmeiotic gene activity*, i.e., any process that either alters the original 50:50 ratio of t-bearing and non-t-bearing sperm or confers on t-bearing sperm a physiological superiority and thus increases its effectiveness at the time of fertilization.

As early as 1944, Bryson reported that he found no evidence of abnormal spermatogenesis or meiotic disturbances in heterozygous t^x/t^y or $+/t^x$ males. The meiotic drive phenomenon could be achieved by extra division of spermatids receiving the t allele, but this possibility was ruled out by determining the size distribution of spermatid nuclei: if division took place with no further growth, then a class of spermatids with small nuclei should be present. Division accompanied by growth should result in an increase in the diameter of the seminiferous tubules and testis weight. Neither of these alternatives was observed. Hammerberg and Klein (1975b) have eliminated other possible mechanisms of unequal chromosome distribution, such as early meiotic pairing and somatic cell segregation in the developing testis. Using mice carrying a translocation in chromosome 17, they were able to follow the fate of t-bearing and non-t-bearing chromosomes during meiosis [i.e., males heterozygous for translocation 7 and a t allele ($T\ 7/t^x$)]. The results of these studies indicated that there is little or no deviation from the normal distribution of chromosomes during meiosis, and it could be concluded that segregation distortion of t alleles is not a result of meiotic drive. (The small deviation from normal distribution that was detected in some cases was attributed to the presence of the translocation rather than to the t mutation.)

Attempts to determine directly whether or not postmeiotic gene function occurs in sperm generally have yielded negative conclusions. On the other hand, several lines of indirect evidence suggest that the effect responsible for altered transmission frequency in t-mutant mice is postmeiotic. For example, Braden (1958) reported that when the time of mating is delayed to coincide more closely with the time of ovulation, the transmission of t alleles from a $+/t$ or T/t male can be reduced and, in some cases, approaches Mendelian ratios. He concluded that the mechanism of segregation distortion has a physiological basis that is manifest during the period between ejaculation and fertilization. Yanagisawa et al. (1961) reported a late mating effect for t^1, but not for two t alleles with extremely high transmission ratios (t^{w6}, t^{w10}). Later, Dunn and Bennett (1971b) indicated that even the effect on t^1 transmission is not readily reproducible. The most recent results have swung back in favor of late mating. Braden (1972) has reproduced and extended his original observations, whereas Erickson (1973) and J. McGrath and N. Hillman (personal communication) have observed a reproducible delayed mating effect on the transmission frequency of the t^6 allele. Both Braden and McGrath and Hillman have additionally found that the genetic background of the female in some way influences the extent of the late mating effect.

In a different approach to the problem of increased transmission frequency of t alleles, Ginsberg and Hillman (1974) reported that the presence of a high-ratio t allele in a pooled sperm sample could be correlated with increased oxygen uptake and a reduction in the NADH:NAD ratio, both of which reflect an increase in overall metabolic rate. These observations are consistent with the idea of postmeiotic gene function. Sperm surface antigen studies by Bennett and co-workers also present a case in favor of postmeiotic expression of genes in the T complex. These experiments are discussed in section 4.5.1.

Because the evidence presented here suggests that the non-Mendelian inheritance of certain t alleles has a physiological basis and is not due to meiotic drive, the term "segregation distortion" perhaps is inappropriate. Thus, in future discussions we shall refer to this phenomenon more specifically as "transmission ratio distortion."

4.3.3 Overview

The data described above lead to several questions concerning the effects of t mutations on sperm function, none of which has been satisfactorily resolved:

　　1. Is transmission ratio distortion of the t mutation due to a positive

t-specified effector that confers a particular advantage on t-bearing sperm (e.g., Ginsberg and Hillman, 1974), or is it a negative t-specified factor that is disseminated by t-bearing sperm and inactivates the remainder of the sperm population (e.g., Braden and Gluecksohn-Waelsch, 1958)? The aforementioned experiments by Ginsberg and Hillman (1974) seem to favor the former mechanism, though this model does not readily explain the sterility of mixed t compound heterozygotes. Gluecksohn-Waelsch (1972) has favored the latter hypothesis; according to her recent theory, t-mutant sperm express on their surfaces a gene product that in some way inactivates non-t-bearing sperm, presumably by affecting their motility or ability to capacitate. In a t^x/t^y compound heterozygote, the t^x and t^y sperm inactivate each other, resulting in a sterile male. Aside from the problem that Gluecksohn-Waelsch pointed out, namely, preventing self-destruction (i.e., t sperm in a $+/t$ heterozygote must only kill $+$ sperm), her hypothesis is inconsistent with her earlier finding (Braden and Gluecksohn-Waelsch, 1958) that the presence of sperm from several types of sterile males did not interfere with the fertilizing capacity of $+/+$ sperm after artificial insemination. Furthermore, Bennett and Dunn (1971) discovered that in compound heterozygotes carrying both a high- and a low-transmission-ratio t allele, the high-ratio allele does not necessarily fertilize with its usual high frequency. In fact, in a study with t^{w2}/t^{38} males, almost 80 percent of the offspring (resulting from crosses with $T/+$ females) carried t^{38}, despite the fact that the normal transmission frequencies for t^{w2} and t^{38} are 0.85 and 0.13, respectively! Bennett and Dunn concluded that the most likely explanation of their results was some interaction between the two t-mutant alleles during spermatogenesis, and this hypothesis raises the following question, which is also unanswered at this time.

2. Is transmission ratio distortion a result of premeiotic or postmeiotic gene expression or a combination of both? Because there has been so little evidence of postmeiotic gene expression in the past, the experiments described above, although they support the idea of a postmeiotic influence on transmission of t alleles, often receive guarded reactions. On the other hand, it is difficult to conceive of ways in which premeiotic gene expression could explain the observed results.

3. Are transmission ratio distortion and sterility occurring by the same mechanism? Although it would undoubtedly simplify matters if

this were the case, hypotheses attempting to explain one of these phenomena inevitably do not apply well to the other. Indeed, much of the available evidence supports the involvement of different mechanisms. For example, although sterility can be correlated with lethality and severity of the t mutations involved, there is little correlation with transmission frequency. Homozygous viable (t^3, t^{17}, etc.) mutants and compounds show, in some instances, extreme reduction in fertility (Braden and Gluecksohn-Waelsch, 1958) although their transmission frequencies are low or normal. The lethal alleles in the t^9 complementation group likewise result in sterility when combined with other alleles (see Bennett and Dunn, 1971), but again they do not have an effect on transmission frequency (table 4.1). This, incidentally, argues further against Gluecksohn-Waelsch's hypothesis concerning adverse effects of t^x sperm on wild-type or t^y sperm, because one would then expect the same alleles that cause transmission ratio distortion in simple heterozygotes to be sterile as compounds, and vice versa.

4. Are wild-type T-complex gene products necessary for fertilization? Because t-mutant sperm are so effective at fertilization, it would be logical to conclude that the wild-type product(s) of the T complex is (are) not essential for sperm function. On the other hand, it is difficult to imagine ways in which homozygosity for many t alleles results in death to embryos but actually confers an advantage to sperm bearing the mutation over wild-type sperm. Two sets of observations are pertinent here. First, Lyon and co-workers (1972), using a translocation, showed that sperm could be made completely deficient not only for the T region of chromosome 17, but also for *low* and *H*-2, and still retain normal function. This would seem to confirm that gene activity essential for somatic cell survival is unnecessary in haploid sperm cells. However, it would be incorrect to propose that expression of wild-type T-complex genes was also unnecessary during spermatogenesis. The observation that homozygous t semilethal males show abnormal sperm production (Bennett and Dunn, 1971; Dooher and Bennett, 1974) speaks against such an idea.

Consequently, we are left with the rather complex picture that normal T-complex gene products are essential, at least in some cases and possibly in all instances, during spermatogenesis in order to ensure normal sperm function. In postmeiotic stages, the T region is not needed by fertilizing sperm possibly because T-gene products formed prior to meiosis are still available. On the other hand, it appears that expres-

sion of the premeiotic T-complex genes is irrelevant in the explanation
of transmission ratio distortion, although postmeiotic gene function is
probably involved. Of course, the best way to assess the validity of
these speculations is to characterize and analyze the products of genes
in the T region (see section 4.5).

4.4 Morphological Analyses of T-Complex Mutant Embryos

Detailed morphological studies, usually in the form of analyses of sectioned and stained embryos, have been carried out on at least one
mutant type from each of the six complementation groups made up of
recessive lethal t alleles. These studies are reviewed here, not necessarily in historical order, but rather on the basis of the time of lethality
of the various mutants. The abnormalities that are observable in compound heterozygotes and semilethals also are considered. Because
these descriptions summarize the work of many laboratories, different
conventions were used to stage the embryos and to determine embryonic age. In an effort to present these data in a uniform manner, every
attempt has been made to stage and age early embryos according to the
description of Snell and Stevens (1966) and embryos beyond the ninth
day according to Theiler (1972), except that their "n days" is expressed as "the $(n + 1)$th day," in accordance with the convention
used throughout this book (see the glossary). Table 4.2 lists the embryonic stages and gestation ages that are considered in this section.

4.4.1 Complementation Group t^{12}

t^{12}/t^{12} embryos
Embryos homozygous for the t^{12} mutation or for other mutations in
this complementation group die prior to implantation. Consequently,
these mutations are among the earliest-acting lethals in mammals.
Smith (1956), Mintz (1964a,b), and Calarco and Brown (1968) have
studied t^{12}/t^{12} embryos at the light microscope level. Ultrastructural
analyses have also been carried out by Calarco and Brown (1968) and
by Hillman et al. (1970).

  In her histological analyses, Smith (1956) noted that development of
t^{12}/t^{12} embryos to the morula stage was normal; mutant morulae could
not be distinguished from normals by overall morphology, by histology,
or on the basis of cell number. However, t^{12}/t^{12} embryos failed to form

Table 4.2. Developmental Stages during Which t and T Mutations Are Expressed

Stage of Development	Age [a]	t^{w32}	t^{12}	t^{0}	t^{6}	t^{w73}	t^{w5}	$t^{9};$ t^{4}	t^{w18}	t^{w1}	T	T^{c}	T^{OR}	T^{h}
1-Cell	1													
2-Cell	2	○	○											
		○	○											
4-Cell	E-M3	○	○											
		○	○											
8-Cell	M-L3	◐	◐											
		●	●											
Morula	L3-E4	●	●											
			●											
Blastocyst	4													
				○										
Implanting blastocyst	M5			○										
				○										
Short egg cylinder; proamniotic cavity	E-M6			●	●									
				●	●	●					●	●		
Elongated egg cylinder	L6-E7				●	●		○			●	●		
							●	○						
Mesoderm; primitive streak	M-L7						●	○						
							●	○						
Head process; notochord; neural groove; amnion	E-M8							●						
								●	●					

Table 4.2 (continued). Developmental Stages during Which t and T Mutations Are Expressed

Stage of Development	Age[a]	t^{w32}	t^{12}	t^0	t^6	t^{w73}	t^{w5}	t^9; t^4	t^{w18}	t^{w1}	T	T^c	T^{OR}	T^h
Early somite; head fold	L8							●	●					
									●		●	●		
7–11 Somites	E-M9							●	●		●	●		
								●	●		●	●		
Turning of embryo; neural tube closure; 12–20 somites	L9-E10								●		●	●		
									●		●	●		
Posterior neuropore; forelimb bud; 21–29 somites	M10-E11								●		●			
									●					
Head limb and tail bud, 30–34 somites	E-M11								●					

[a]Gestastion age is given as the day of pregnancy; E = early, M = mid, L = late.

The filled circles represent the stages at which abnormalities or obvious retardation of growth of the various t and T mutations are apparent. The open circles represent the earliest stages at which mutant embryos can be distinguished from normals, although the criteria used do not qualify as overt abnormalities or severe retardation of development. The affected embryos do not advance beyond the stage indicated by the last closed circle except in the case of t^{w1}, some of which develop to substantially later stages (see text). Also, the table does not indicate the time of death of the embryos, only the latest developmental stage reached. In some cases, where indicated, the mutant embryos advance beyond a particular stage, but do not fully attain the next one. Staging and aging are according to Snell and Stevens (1966) and Theiler (1972). The t^{w5} assignment applies to the other four members of that complementation group analyzed at the same time. The periods of developmental arrest assigned to T^c and T^h are tentative. The available information in the literature is inadequate for assignments to T^{hp} and T^{Orl}.

blastocysts. Prior to death, the embryos appeared to persist for several hours in a state of developmental arrest. Thus, whereas all embryos from the control matings had reached the blastocyst stage by the afternoon of the fourth day of gestation, only 60 percent of the experimental embryos had done so at this time. Most of the remaining embryos, presumably t^{12}/t^{12} mutants, were scored as normal morulae. Their nucleoli remained rounded rather than transforming to an elongated state, and there was no intensification of staining with Azure B, events that Smith reported to occur normally in the morula-to-blastocyst transition. However, by the end of the fourth day of gestation, the only evidence of mutant embryos was a small number of abnormal morulae; the majority had either degenerated or disappeared. More recently, Mintz (1971) has provided evidence that the rapid disappearance of the abnormal t^{12}/t^{12} embryos at this time is due to the removal of their zona pellucidas by a uterine zonalysin.

Mintz's studies (1964a,b) were in general agreement with those of Smith. Aside from nucleolar morphology and Azure B staining, she showed that t^{12}/t^{12} morulae failed to undergo properly two other morphological changes that are preludes to blastocyst formation: the conversion from a granular to a translucent cytoplasm and the appearance of discrete cytoplasmic vacuolations, which presumably contribute subsequently to the formation of the blastocoelic cavity. Mintz (1964b) further demonstrated that in chimeras between wild-type and t^{12}/t^{12} embryos, although blastocysts could result, the mutant cells could still be distinguished from normals by their rounded nuclei and less intense staining with Azure B. The t^{12}/t^{12} cells were also larger, suggesting a decline in mitotic rate. Mintz concluded that association of mutant cells with normal cells did not improve the developmental capacity of the former.

Calarco and Brown (1968) reported that although normal and mutant morulae showed no *overall* difference in Azure B staining, *individual* blastomeres in normal, but not mutant, morulae did vary in staining intensity. Furthermore, these investigators pointed out that cell-cell associations appeared to be more tenuous in t^{12}/t^{12} than in wild-type embryos, and that mutant embryo cells appeared to be more rounded. These observations were consistent with earlier reports that the outside cells of mutant embryos, although they initially flattened during the characteristic smoothing out of the morula surface, tended to round up again at later stages (Smith, 1956) and that bonds between adjacent mutant cells in the trophectoderm layer of chimeric blastocysts occasionally broke, leading to herniation of cells of the inner cell mass (Mintz, 1964b). The weakening of cell-cell association is presumably not due to the failure of the outer cells of mutant embryos to form the focal tight junctions that appear normally at the morula stage (Calarco and Brown, 1968).

Calarco and Brown reported that t^{12}/t^{12} and wild-type morulae were very similar ultrastructurally. They did note, however, that only mutant morulae contained dense cytoplasmic inclusions, resembling crystalloids found in normal and mutant morula cells, but lacking periodicity. They suggested that these might be crystalloid precursors visible only in t^{12}/t^{12} cells because they were being processed more slowly than is usual. They noted also that the nuclei of t^{12}/t^{12} cells contained several small (0.1–0.7 μn) agranular inclusions with the same density as that of the much larger nucleoli. Calarco and Brown drew attention to the fact that similar structures exist in anucleolate amphibian and plant cells.

Calarco and Brown concluded that all the observed differences between wild-type and t^{12}/t^{12} morulae were not necessarily a direct result of altered expression of genes in the T complex, but could have been due to "degenerative changes of metabolically declining cells." Hillman et al. (1970), who strongly embraced this argument, observed the same abnormalities in mutant embryos. They also found that binucleate cells were common in mutant morulae. Hillman et al. reported that death of cells in mutant embryos was asynchronous and, unlike earlier workers, claimed that small numbers of mutant embryos could develop to the early blastocyst stage. On the other hand, they also reported that developmental arrest of some t^{12}/t^{12} embryos took place as early as the 8-cell stage and was preceded by the appearance at even earlier stages of the nuclear fibrillo-granular bodies first reported by Calarco and Brown and of nuclear lipid droplets and excessive cytoplasmic lipid accumulations. These authors ascribed most of the abnormalities observed in t^{12}/t^{12} embryos at the late morula stage to secondary degenerative changes and suggested that production of altered T-complex gene products begins during early cleavage.

t^{w32}/t^{w32} embryos

Hillman and Hillman (1975) recently studied t^{w32}/t^{32} embryos. They reported that the phenolethal period of homozygous mutant embryos is somewhat earlier than that of t^{12}/t^{12} embryos; most embryos were reported to be developmentally arrested between the 8-cell and early morula stage. Furthermore, ultrastructural analyses indicated that although t^{w32}/t^{w32} embryos possessed most of the same abnormalities as t^{12}/t^{12} embryos (excess lipid, abnormal nucleoli, binucleate cells, early death of individual blastomeres), there are two clear differences. Whereas t^{w32}/t^{w32} mutant embryos did not possess the nuclear fibrillo-granular bodies found in t^{12}/t^{12} blastomeres, the mitochondria in the former (but not in the latter) were distinctly abnormal: as early as the 8-cell stage, t^{w32}/t^{w32} embryos contained round, electron-dense mitochondria with abnormal cristae and often with crystalline inclusions. On the basis of differences in ultrastructure and in the period of developmental arrest, Hillman and Hillman proposed that t^{w32} and t^{12} are separate alleles. However, these workers rightly pointed out that this conclusion was somewhat equivocal until both mutations could be studied on an isogenic background.

t^1/t^1 embryos

In the early literature, Gluecksohn-Schoenheimer (1938b) inspected uteri on the eighth day of pregnancy from heterozygous $+/t^1$ mice mated inter se and found no evidence of resorption sites, although litter size was already reduced. She concluded that t^1/t^1 embryos died prior to implantation. Presumably, then, the t^1 mutation belonged in the t^{12} complementation group. However, a morphological analysis of these embryos was never carried out, and by 1962 it was discovered that the original t^1 line had been lost. The present t^1 line is apparently identical to t^0 (see Silagi, 1962).

4.4.2 Complementation Group t^0

t^0/t^0 embryos

In their initial studies, Chesley and Dunn (1936) concluded that resorption of t^0/t^0 embryos began as early as the seventh day of pregnancy and was complete by the eighth day. The first detailed morphological studies on t^0/t^0 embryos were carried out by Gluecksohn-Schoenheimer in 1940. She was able to find histological evidence of abnormal development early on the sixth day, at the short egg-cylinder stage, but not earlier. However, because the number of embryos scored as abnormal was some 25 percent less than that found late on the seventh day, it seems that the time of expression of mutant properties by the homozygotes is somewhat variable. Characteristic of those embryos arrested at earlier stages was the absence of discrete separation of the ectoderm layer into embryonic and extraembryonic portions; instead, a single clump of ectoderm cells persisted. The visceral endoderm layer also appeared to be abnormal in many of the earlier stage arrests in that the cells were larger and more irregularly aligned about the ectoderm.

By the beginning of the seventh day of gestation, the arrested embryos were clearly much shorter than the normal egg cylinders, which had elongated at this age. The arrested t^0/t^0 embryos had apparently failed to form a proamniotic cavity. The visceral endoderm cells became more irregular and multilayering about the ectoderm occurred. In some cases, the interaction between visceral endoderm and ectoderm seemed to be disrupted and the two layers were detached. Although these seventh-day t^0/t^0 embryos were developmentally arrested, some cells were still viable because mitotic figures could be seen. By the end of the seventh day, however, degeneration was usually complete.

t^1/t^1 embryos
As mentioned in section 4.4.1, the t^1 allele is presently assigned to the t^0 complementation group. Silagi (1962) compared t^0/t^0 and t^1/t^1 embryos and concluded that they were identical.

t^6/t^6 embryos
The t^6 allele was reported by Dunn and Glueksohn-Schoenheimer (1950) to be a recurrence of t^0. Carter and Phillips (1950) and Lyon and Meredith (1954b) confirmed that t^6/t^6 embryos died shortly after implantation. Nadijcka and Hillman (1975a) recently analyzed t^6/t^6 embryos at both light and electron microscope levels. In general, a comparison of their studies with those of Glueksohn-Schoenheimer indicates that embryos homozygous for t^0 and for t^6 behave in a relatively similar manner, although some differences may exist. For example, Glueksohn-Schoenheimer reported that t^0/t^0 embryos arrested at the short egg-cylinder stage and failed to form discrete embryonic and extraembryonic ectoderm moieties or a proamniotic cavity, but Nadijcka and Hillman reported that 35 percent of t^6/t^6 embryos did reach the elongated egg-cylinder stage by the seventh gestation day. Most of these had formed proamniotic cavities, although in only about

half the cases did the cavity extend into both the embryonic and extra-embryonic regions. In this respect, the apparently longer survival time of t^6/t^6 compared to t^0/t^0 embryos might be analogous to similar findings when t^{w32}/t^{w32} and t^{12}/t^{12} embryos were compared (see section 4.4.1).

Nadijcka and Hillman noted abnormalities in the ectoderm and visceral endoderm layers of t^6/t^6 embryos, as had Gluecksohn-Schoenheimer for t^0/t^0 embryos. However, the former investigators also observed that the parietal endoderm layer of t^6/t^6 mutant embryos was also often irregular, overgrowing into folds that in some cases almost occluded the yolk-sac cavity. This phenomenon was not reported by Gluecksohn-Schoenheimer for homozygous t^0 embryos, although she did not discuss the morphology of the parietal endoderm layer.

At the ultrastructural level, Nadijcka and Hillman (1975) noted that cells in t^6/t^6 embryos contained excessive numbers of cytoplasmic lipid droplets. These droplets were observed prior to arrest of the embryos. In fact, excessive lipid accumulation was observed in more than half of the mutant embryos before the end of the fifth day of gestation (late blastocyst stage). The investigators pointed out that these inclusions were also characteristic of earlier t-mutant embroys (t^{w32}/t^{w32} and t^{12}/t^{12}; see section 4.4.1). The lipid droplets were present in all of the embryonic and extraembryonic cell types (although they are normally seen only in trophoblast) and increased in numbers as degeneration of the embryos proceeded. Parietal endoderm cells in mutant embryos were also unusual in that they possessed large vacuoles filled with cellular debris. Mitochondrial crystalloid inclusions, similar to those described in t^{w32}/t^{w32} cells, were found in all cell types in t^6/t^6 egg cylinders, but were not present at earlier stages. The nuclei of t^6/t^6 embryos did not have the granulofibrillar inclusions found in t^{12} homozygotes (M. Nadijcka and N. Hillman, personal communication).

There is a resemblance between t^6/t^6 embryos and t^{12}/t^{12} and t^{w32}/t^{32} embryos in that premature death of individual cells occurs prior to gross arrest. The earliest cases appeared in late blastocysts. Cell death occurred in all parts of the blastocyst, but was most common in trophectoderm cells at the embryonic pole. At later stages, death of t^6/t^6 cells was asynchronous, as observed by Gluecksohn-Schoenheimer with t^0/t^0 embryos (1940), and was apparently random, i.e., in general, no single cell type appeared to be dying at an earlier stage

than any others. Nadijcka and Hillman did note, however, that trophoblast cells in t^6/t^6 conceptuses that survived to later stages did appear
normal throughout the period of study.

4.4.3 Complementation Group t^{w73}

To date, there is but a single member in this mutant "group" (Bennett,
1975a). It was detected in a wild mouse population (Dunn and Bennett,
1971). Mutant embryos have been analyzed by Spiegelman et al. (1976).
By the beginning of the seventh day of pregnancy, about half the
t^{w73}/t^{w73} mutants could be identified as abnormal. These were arrested
at the elongated egg-cylinder stage and were much shorter than
normal littermates. The cells in the mutant were crowded and occasionally contained dense granules. Visceral endoderm cells overlying the
embryonic ectoderm remained cuboidal instead of becoming squamous.
The mutant embryos could remain arrested in this state for several
days before they degenerated. In these respects, the mutants resembled
t^6/t^6 embryos (see section 4.4.2). However, there was a remarkable
difference: in t^{w73}/t^{w73} embryos, the *extraembryonic cell types*, particularly trophoblast cells, were most abnormal. Late on the sixth day,
trophoblast cells did not show good contact with uterine tissues, and
the ectoplacental cone (EPC) was absent or abnormally small. It is not
clear whether implantation had begun normally and had reversed, because earlier stages were not analyzed. However, it was noted that
uterine epithelial cells in the area surrounding the conceptus had
eroded (M. Spiegelman, personal communication), as occurs normally
during implantation. With increasing time, trophoblast and EPC cells
still failed to penetrate the uterus. The trophoblast cells did not become
giant, but scattered about as clusters of small cells. Even by the ninth
day of pregnancy, only occasional trophoblast giant or multinucleate
cells were observed, and attachment of the conceptus to the uterus was
tenuous. It is particularly notable at this time that the trophoblast layer
surrounding normal embryos was replete with sinusoids containing maternal erythrocytes and occasional lymphocytes. In contrast, the trophoblast region of t^{w73}/t^{w73} conceptuses was infiltrated by lymphocytes
and polymorphonuclear leukocytes.

Less dramatic abnormalities were also evident in the parietal endoderm layer. These cells produced only scanty amounts of Reichert's
membrane. This malfunction, along with the lack of a substantial troph-

oblast layer, caused the parietal endoderm cells in many instances to break contact with the placenta and to obscure part of the yolk-sac cavity.

4.4.4 Complementation Group t^{w5}

At least 19 members of the t^{w5} complementation group are known to exist (Bennett, 1975a). All were derived from wild populations and are, therefore, on various genetic backgrounds. Embryos homozygous for five members of this group—t^{w5}, t^{w6}, t^{w11}, t^{w13}, and t^{w17}—were analyzed histologically in a single study (Bennett and Dunn, 1958). This appears to be the only published morphological analysis to date of mutant embryos from the t^{w5} complementation group. Ultrastructural studies have yet to be carried out.

Bennett and Dunn were unable to find any histological evidence of abnormality of t^{w5}/t^{w5} (or related) mutant embryos up to the time of organization of embryonic and extraembryonic ectoderm in the egg cylinder (sixth day of gestation). The proamniotic cavity was formed normally. Although some pycnotic cells were found in the embryonic ectoderm layer, these were also present, presumably in equal numbers, in normal littermates. At about the time of mesoderm formation in normal embryos (mid- to late seventh day), however, abnormalities were often seen in the mutants, simultaneously in the embryonic ectoderm and visceral endoderm regions. The embryonic ectoderm began to undergo substantial degradation, but the cells of the visceral endoderm layer failed to differentiate in the extraembryonic region from cuboidal cells to the characteristic high columnar secretary epithelium. Pycnosis of ectodermal cells seemed to be restricted to the embryonic, as opposed to the extraembryonic, area.

At the time when normal embryos were forming the notochord and head process (early to mid-eighth day), degeneration of ectoderm in mutant embryos continued. By this time, it was also apparent that all endodermal derivatives were abnormal: the entire visceral endoderm was developmentally arrested as cuboidal cells and did not differentiate into the high columnar and squamous cells in the extraembryonic and embryonic regions, respectively. Even the parietal endoderm layer was aberrant; it formed closely applied, rounded cells instead of widely spaced, squamous ones. Also, the formation and proliferation of mesoderm were retarded. By the end of the eighth day of gestation, homo-

zygous mutant embryos were found in various stages of developmental
arrest, ranging from abnormal elongated egg cylinders, almost completely devoid of embryonic ectoderm, to further developed structures
containing mesoderm and abnormally infolded embryonic and extraembryonic ectoderm. By the ninth day, some of the conceptuses were
obviously totally degenerate. It was striking, however, that others
showed almost completely normal development of chorion, yolk sac,
and allantois, as well as tubes of extraembryonic ectoderm, although
most or all cells of the embryo proper were pycnotic or had disappeared. Detailed analyses of later stages were not available, but it was
indicated by Bennett and Dunn that the lethal period extended to the
eleventh day of gestation.

Although the above description applies to all of the five mutant embryo types that Bennett and Dunn studied, these investigators reported
that there were variations among the five. For example, t^{w6}/t^{w6} embryos appeared to be the most severely affected and were generally
arrested at earlier stages than the other embryos: loss of embryonic
ectoderm was always complete, and normal formation of the extraembryonic membranes was rare. On the other hand, although some
abnormalities were also present in t^{w11}/t^{w11} or t^{w13}/t^{w13} embryos as
early as those detected in t^{w6}/t^{w6} embryos, about 25 percent of the
former still contained distinct differentiated embryonic ectoderm late
on the eighth day of gestation. Intermediate effects appeared in t^{w5}/t^{w5}
and t^{w17}/t^{w17} embryos. It was stressed that this variation could have
been due to the different genetic backgrounds of the embryos under
analysis.

4.4.5 Complementation Group t^9

Bennett (1975a) has listed five members of the complementation group
t^9: t^4, t^9, t^{w18}, t^{w30}, and t^{w52}. Of these, t^4 and t^9 homozygous mutants
were characterized morphologically by Moser and Glueksohn-Waelsch
(1967), and Bennett and Dunn (1960) described defects in t^{w18}/t^{w18} embryos. More recently, Spiegelman and Bennett (1974) and Spiegelman
(1976) have analyzed t^{w18}/t^{w18} embryos by light and electron microscopy. Moser and Glueksohn-Waelsch reported that t^4 and t^9 mutants
were isolated independently, but the two alleles were identical genetically and embryologically. Consequently, they pooled the data obtained
with t^4/t^4 and t^9/t^9 embryos, and they are considered together in this

presentation. Although it has also been claimed that t^9 and t^{w18} alleles are identical (Van Valen, 1964; cited by Spiegelman and Bennett, 1974), some differences appear to exist between embryos homozygous for one allele or the other. To our knowledge, these differences have not been resolved in the literature, and we shall therefore describe t^4/t^4 and t^9/t^9 embryos separately from t^{w18}/t^{w18} embryos.

t^4/t^4 and t^9/t^9 embryos

According to Moser and Gluecksohn-Waelsch (1967), abnormalities in t^4/t^4 or t^9/t^9 homozygous embryos are detectable as early as the elongated egg-cylinder stage, late on the sixth day of gestation; death and resorption occur on the ninth or tenth day. The early signs of abnormality, apparently seen in about 70 percent of mutant embryos, were (1) a degree of disorganization of the germ layers, (2) the presence of dead cells in the proamniotic cavity, (3) the failure of visceral endoderm cells overlying the embryonic ectoderm to become transformed from cuboidal to squamous, and, rather uniquely, (4) the fusion of these visceral endoderm cells with the antimesometrial parietal endoderm or trophoblast (incorrectly referred to as decidua by the authors). This bipolar attachment of the egg cylinder has not been described in other t-mutant phenotypes and appears not to have any obvious relationship to the subsequent abnormalities observed in these embryos.

Late-seventh-day mutant embryos failed to form an organized mesoderm layer. On the other hand, the proamniotic cavity appeared to be partially occluded by an outgrowth of undifferentiated tissue from the primitive streak area. After another day of pregnancy, a number of gross abnormalities could be seen: the embryo was abnormally short, the head folds were overdeveloped and there was a lack of somites and a general failure of the mesoderm to differentiate normally. Although many homozygous mutant embryos died early on the ninth day of gestation, the survivors showed irregular head folds and duplication of neural folds. Neural groove duplication also appeared in some cases; in these embryos, the underlying mesoderm was necrotic. Somites were still absent. The allantois was often unattached and occasionally duplicated. Finally, misplaced tissue that histologically resembled primitive streak had grown between the yolk sac and the amnion, or through it and into the amniotic cavity.

t^{w18}/t^{w18} embryos

The study by Bennett and Dunn (1960) on t^{w18}/t^{w18} embryos was somewhat clearer and more detailed than that of Moser and Gluecksohn-Waelsch on t^4 and t^9 homozygotes. This study and the more recent analyses by Spiegelman and Bennett (1974) and Spiegelman (1976), provide a rather complete picture of the aberrations caused by this mutation. Bennett and Dunn did not observe in t^{w18}/t^{w18} mutants the abnormalities in elongated egg cylinders recorded by Moser and Gluecksohn-Waelsch. By the eighth day of gestation, however, mutant embryos had visibly slowed in their development, having failed to produce several axial structures: head process, notochord, archenteron, and neural groove. There was also a prominent thickening in the primitive streak area. The mutant embryos continued to lag behind normals without overt signs of degeneration through the eighth day of pregnancy and into the ninth. By this time, axial structures had begun to appear, though only the most advanced t^{w18}/t^{w18} embryos had reached the somite stage. The protuberance in the primitive streak area was striking and in most cases bulged far into the proamniotic cavity. The presumptive neural ectoderm cells bounding the outgrowth were healthy and present in normal amounts. The internal cells accounting for the bulk of the population were generally somewhat pycnotic, but on the whole they did not differ dramatically from normal primitive streak cells. Consequently, due to the unusual topology in the primitive streak area, two neural grooves formed instead of one when the mutant embryos reached the appropriate stage.

By the beginning of the tenth day of gestation, at which time normal embryos had 12–20 somites and were turning and forming neural tubes, the t^{w18}/t^{w18} embryos were relatively more retarded. They contained no more than 10–12 somites, and the neural grooves had not yet closed. The first signs of overt degeneration and pycnosis were visible, particularly in the primitive streak area. In many embryos, a lump of disorganized tissue was seen adherent to, and partially outside, the amnion. Bennett and Dunn proposed that this growth, subsequently noted also by Moser and Gluecksohn-Waelsch (1967) in t^4/t^4 and t^9/t^9 embryos, might have been derived, then broken off, from the primitive streak and become trapped during closure of the amnion.

In the most advanced t^{w18}/t^{w18} embryos to develop, duplicated closed neural tubes were observed; in some cases the two were joined, in others they remained separate. Blood vessels, when present, were greatly dilated and contained relatively few blood cells. It should be noted that even the best-developed t^4/t^4 or t^9/t^9 embryos described by Moser and Gluecksohn-Waelsch failed to produce somites or closed neural tubes.

In the studies described above, it was assumed that the major abnormalities of the mutant homozygotes were caused by overgrowth of primitive streak cells and/or their derivatives. This was somewhat disturbing, because all other t-mutant embryos suffered from underdevelopment, rather than overproduction, of various cell types. The careful studies by Spiegelman and Bennett (1974) and Spiegelman (1976) of t^{w18}/t^{w18} embryos (denoted as t^9/t^9 by the authors, who assumed the two alleles to be identical) seems to have clarified the situation. These investigators observed that the protuberance in the primitive streak area was not due to overgrowth of cells, but rather to failure of primitive streak cells to differentiate to normal mesoderm cells, which migrate to other areas of the embryo. The cells, therefore might have been arrested as primitive streak cells or at some intermediate stage. The mutant primitive streak cells appeared by light microscopy to be smaller than normal and to have irregular nuclei, greater cytoplasmic density, and a greater degree of cytoplasmic vacuolation, compared to normal counterparts. Whereas mesoderm cells in late-eighth-day normal embryos were stellate and loosely packed, "mesoderm" cells in mutant homozygotes were more rounded and tightly packed. Furthermore, the mutant cells did not acquire filopodia (thin intercellular cytoplasmic bridges), but instead formed lobopodia (wide rounded projections). At this time, histologic abnormality appeared to be restricted to cells in the mesoderm area; cells in the presumptive neural ectoderm layer, for example, were identical in both normal and mutant embryos.

The electron-miscroscopic analyses of Spiegelman and Bennett confirmed and extended their light-microscopic observations. They showed that normal embryonic mesoderm cells had several characteristic ultrastructural features: the filopodia connecting adjacent cells had dense contact zones, with small focal junctions, longer (up to 1 μm) junctions, and occasional desmosomes all in evidence. The apposing membranes were usually 5–20 nm apart and were often separated by an electrondense secretion. Characteristically, the filopodia were rich in subsurface microfilament lattice arrays that presumably participated in some aspect of cellular migration. The "mesoderm" cells in the homozygous mutant embryo were not so tightly apposed and showed only occasional signs of amorphous deposits between the cell membranes with few or no subsurface microfilaments in evidence. Instead, ribosomes appeared in subsurface cytoplasmic spaces; they were not seen in this location in normal cells.

Other cytoplasmic organelles were normal in mutant "mesoderm" cells, although they were clustered in a perinuclear position in contrast to their random location within wild-type cells. Curiously, mutant cells appeared to contain more ribosomes and polysomes overall than their counterparts in normal embryos.

4.4.6 Complementation Group t^{w1}

In 1959, Bennett et al. described the morphology of embryos homozygous for the t^{w1} allele and for three other alleles in the same complementation group (t^{w3}, t^{w12}, t^{w20}). Embryos homozygous for these alleles showed an extremely variable lethal period, ranging from the ninth to the twentieth days of gestation. There was also a very notable difference among the alleles in the proportion of early- and late-lethal embryos: for example, virtually all t^{w20}/t^{w20}, but only 65 percent of t^{w1}/t^{w1}, embryos had died by the seventeenth day of gestation. The t^{w3}/t^{w3} and t^{w12}/t^{w12} mutants were intermediate in severity (Bennett et al., 1959a). Surprisingly, although no less than 65 percent of the homozygous mutant embryos (and 95 percent of t^{w20}/t^{w20} embryos) in this first study were determined to have died by the tenth day of gestation on the basis of the size of the placental remnant ("mole"), the early morphology of these mutant embryos has yet to be described. Instead, Bennett and co-workers (1959a,b) chose to concentrate upon the abnormalities of the relatively few mutant embryos that survived the thirteenth day of gestation. The defects routinely observed in these late-lethal homozygotes were similar for the four alleles and were described as follows (Bennett et al., 1959a,b).

1. *Reduced size*. By the beginning of the ninth day, about 70 percent of the mutant embryos could be scored as abnormal; of these, 15 percent were already dead. The surviving abnormals were generally only one-half to three-quarters normal size. Even the longest surviving embryos (sixteenth to twentieth days) were smaller by one-third to one-quarter compared to normal littermates.

2. *Abnormalities in nervous system structure.* In the earlier stages, homozygous mutants showed pycnosis of neural cells, primarily in the hindbrain region and tending to extend into the spinal cord. Damage in this region was most severe on the eleventh and twelfth days of gestation. Characteristically, the ventral portion was much more pycnotic and necrotic than the dorsal area. By the thirteenth day of gestation, only half of the mutant embryos had survived. Interestingly, in those still alive, pycnosis of the brain was much less intense than previously, although the hindbrain was deficient in numbers of neural cells. By the fifteenth day, 30 percent had survived, and these showed pycnosis most prevalently in the forebrain and midbrain regions. The spinal cord was reduced in size and the lumen was occluded. It appears, then, that the mutant embryos could almost be split into two groups: the first half of the embryos showed severe hindbrain necrosis and died before the fourteenth day; the remainder survived this critical stage, but subsequently suffered defects in other portions of the nervous system. The degenerative process actually appears to move in a bidirectional wave radiating from the hindbrain toward the forebrain and the spinal cord; the older the embryo under study, the closer to the extremities of the nervous system were the degenerative processes taking place. Thus, in the oldest surviving mutant embryos, pycnosis was confined to the extreme posterior section of the spinal cord and to the more anterior regions of the brain. Once the wave had passed a particular area, pycnosis diminished there; however, total recovery did not occur as evidenced by a lowered cell density in that area.

3. *Skeletal abnormalities.* Overall, the skeletal structure was not overtly abnormal, but was retarded from the earliest stages with respect to chondrification and subsequently to ossification. The area in which developmental retardation and size reduction were most noticeable was the chondrocranium, the cartilaginous floor of the skull that is replaced later in development by the basal skull bones. From the early stages of chondrocranium formation (fourteenth day), at which time the three component elements—nasal septum, hypophyseal cartilage, and parachordal cartilage—are still separate, the mutant embryos showed retardation or even absence of chondrification. As these structures interconnected and developed to form a continuous plate in the normal embryo, the mutant's chondrocranium was always either immature or arrested in development. Characteristically, the area of the nasal sep-

tum was most reduced in size in homozygous mutant embryos when compared with normals.

4. *Edema*. Edema was only occasionally noted in mutant embryos on the thirteenth day, but within the next day of development all mutant embryos were overtly edematous. Probably related to this defect was pericardial enlargement.

The relationship between the various defects is not completely clear. Bennett et al. (1959b) pointed out that axial cartilage formation is induced by spinal cord tissues. However, their preliminary in vitro experiments suggested that t^{w1}/t^{w1} spinal cord effectively induced cartilage formation. They noted as well that a variety of agents that induce edema in mammalian embryos also produce brain abnormalities, such as the microcephaly and hydrocephaly seen in t^{w1}/t^{w1} embryos. On the other hand, Bennett et al. mentioned that at least one mouse mutant, ragged, caused edema but did not show nervous tissue abnormality.

4.4.7 "Complementation Group" *T*

Dobrovolskaia-Zavadskaia (1927) isolated the original *T* (Brachyury) mutant mouse from breeding stock. Since that time, descendants of this mouse have played a pivotal role in the isolation and testing of recessive *t* mutations. For many years, this line was the only source of the dominant *T* mutation, but since 1944, a variety of *T*-mutant mice have been discovered or induced (see table 4.1; Bennett et al., 1975). It is apparent from embryological and genetic studies that a number of these mutations are allelic with, but not identical to, *T* Brachyury; in the sense that $T^x T^y$ is lethal (see below) and $T^x t^y$ mice are viable, we shall refer to T as a complementation group. To date, the embryological defects caused by the newer *T* mutations have not been analyzed nearly so well as those of the original *T* mutation. However, after a consideration of the effects of the original *T* on embryogenesis, a comparison with other *T* mutations can be made on the basis of available information.

T/T embryos

In 1935, Chesley provided a thorough description of *T/T* embryos. According to him, and subsequently to Spiegelman (1976), normals, *T/t* heterozygotes, and *T/T* homozygotes were indistinguishable prior to the early somite stage (ninth day of gestation). At that time, however, distinctions were possible even by external observation. Characteristically, T/T embryos developed paired or unpaired vesicles or blisters on the sides of the midline. Grüneberg (1958) and Spiegelman (1976) also documented this phenomenon. Less conspicuous at this time were distortion of somites and waviness in the neural tube.

By the tenth day (10- to 12-somite stage for normals), the blisters had disappeared, but examination of the intact embryos revealed further irregularities in the neural tube and somites. One day later, normal embryos had developed both anterior and posterior limb buds, but only the former were apparent in T/T embryos. In fact, the entire posterior region of the embryo appeared to be greatly reduced in size and structure. Enlargement of the pericardial cavity was also characteristic of T/T embryos at this stage.

Chesley (1935) carried out histological analyses of T/T over this period of development as well. He reported that prior to the ninth day of gestation, the primitive streak area appeared to be morphologically normal in T/T embryos. At that time, he noted that notochord production from primitive streak cells had failed to take place, although it did occur at this stage in normal embryos. In fact, Spiegelman (1976) pointed out that although the mid-ninth-day primitive streak is normally a relatively small body of cells restricted to the posterior extremity of the trunk, the primitive streak in T/T embryos appeared to persist as a much larger structure, perhaps, as in the case of t^{w18}/t^{w18} embryos (see section 4.4.5), because primitive streak cells failed to develop and to move out of that area. Chesley maintained that one of the structures that did not form normally from the primitive streak was notochord. Gluecksohn-Schoenheimer (1938a) subsequently argued that notochord cells were never present in T/T homozygotes, but Grüneberg (1958) and later Spiegelman (1976) pointed out that this is incorrect. In fact, notochord cells are produced, but they do not form a characteristic definitive structure. Instead, the cells either remained fused to the dorsal roof of the gut or formed atypical associations with the neural tube and other neighboring tissues.

Chesley also reported that the neural tube showed abnormalities as early as the middle of the ninth day. This structure was asymmetric and branching and, according to Spiegelman, slow in closing. By the tenth day, the branches had become larger than the primary neural tube in some regions, and neural tissue in the extreme posterior part of the embryos was a totally disorganized mass. Chesley also characterized the blisters that appeared on the sides of the neural tube and disappeared one day later as being bounded in most cases by an ectodermal layer, with an occasional contribution of mesoderm. Although somite formation began normally, these structures became increasingly abnormal; somite tissue was no longer recognizable by the eleventh day. Spiegelman showed that posteriorly, the somite and notochord region were often occupied by tubular epithelial structures, which presumably had been formed from cells derived from the primitive streak area. These structures appeared not to be connected to the less aberrant axial structures in the anterior regions of the embryo. Finally, Chesley noted a variety of less striking abnormalities, such as the enlarged, blood-filled pericardial cavity and retarded heart development, but concluded that these defects were secondary to the aberrant notochord and neural tube structures. Gluecksohn-Schoenheimer (1944) pointed out that the allantois in T/T embryos failed to develop properly and did not make contact with the EPC. Consequently, umbilical vessels did not develop and communication with the maternal circulation was poor. Gluecksohn-Schoenheimer proposed that it was for this reason that T/T embryos died abruptly on the eleventh day of gestation.

Spiegelman (1976) recently carried out an ultrastructural analysis of cells in the notochord and neural tube regions of T/T embryos on the ninth and tenth days of development. These studies confirmed and extended light-microscopic observations. Spiegelman found that in normal embryos on the ninth and tenth days of gestation, neural tube and notochord cells were separated by both the basal lamina of the former, as well as by a small space. In T/T mutants, however, the basal lamina of the neural tube cells were discontinuous in some areas; rounded projections emanated from these regions, and often the membranes bounding these protuberances came into close contact with notochord cells. In fact, spacing was much closer than normal, and cellular junctions such as those usually occurring within a single tissue were observed between neural tube and notochord cells. Spiegelman has proposed that these junctions developed because both types of cells had failed to differentiate to the degree that the cells would no longer recognize each other as "self."

$T/+$ embryos

Unlike the recessive t alleles, heterozygotes carrying the T allele are recognizable in utero by alterations in the posterior extremity that subsequently lead to the shortened-tail or *Brachyury* phenotype. $T/+$ embryos have been described by Chesley (1935) and in an excellent and thorough study by Grüneberg (1958). Chesley distinguished $T/+$ embryos from wild type by external characteristics as early as the twelfth day of gestation, following a period in which tail growth was rapid. The heterozygote was obvious because of a sharp constriction in the tail at this time. Tail growth ceased posterior to the constriction and in fact that area atrophied and often fell off before birth. Histologically, Chesley identified some $T/+$ embryos on the basis of notochord abnormalities at the 10-somite stage, but only in the most extreme cases; a clear diagnosis was usually possible on the tenth day, by the 22-somite stage. The irregularities were observed in the posterior and ventral regions of the notochord. The formation of abnormal diverticula was most common. The neural tube in the posterior extremities was sometimes misshapen. However, because notochord was *always* abnormal, whereas neural tube structure was not so dependably aberrant, Chesley felt that the primary defect was in the notochord and that the effect on the neural tube was secondary. Grüneberg (1958) concurred with this view. On the other hand, Chesley also considered the possibility that the T mutation exerted its initial effects at the level of the primitive streak prior to formation of definitive notochord and neural tube cells. This view has been espoused by Spiegelman (1976).

Grüneberg (1958), like Chesley, was unable to discriminate between $+/+$ and $T/+$ embryos on the basis of external characteristics prior to the twelfth day of gestation. He carried out a detailed study of $T/+$ embryos from the tenth to the fifteenth day. He found that initially the notochord at the growing point was wider than usual and often dangled as an appendage of the neural tube. Anterior to this spot, the notochord had not broken its association with the dorsal roof of the hindgut and cloaca. In some regions, the notochord had fused into the base of the neural tube and the roof of the gut, rather than appearing as a discrete and independent rod-shaped structure throughout its length, as it is in $+/+$ embryos. Although the notochord did eventually break these associations in $T/+$ animals, fusion often recurred later in gestation. This phenomenon led Grüneberg to conclude that notochord cells in $T/+$ and T/T embryos were "stickier" than normal. This concept is consistent with the aforementioned EM analysis by Spiegelman.

Grüneberg also noted that the notochord in the tail region of $T/+$ animals tended to possess a central lumen. Often these lumina were short and discontinuous. The lumina were bounded by a simple layer of notochord and, as such, give the impression of a tubule surrounded by epithelial cells. Perhaps the large tubules in the notochord and somite area observed by Spiegelman in T/T embryos were more extreme examples of this phenomenon.

Finally, Grüneberg confirmed Chesley's observation that the notochord was often branched in $T/+$ embryos. Branching occurred mainly in the region of the cloaca and proximal parts of the tail. The branches could be quite short or relatively long (250 μm). The diverticula usually ended blindly, but occasionally coalesced with the primary notochord.

Although Chesley noted some abnormalities in neural tube structure in $T/+$ heterozygotes, Grüneberg did not comment on the neural tube other than to note that it did not appear to be overly distorted at sites where it had incorporated notochord. The significance of the recent report by Yanagisawa and Kitamura (1975) that the mitotic index in the middle and ventral neural tube of tenth day $T/+$ embryos is significantly greater than that of T/T or $+/+$ embryos is obscure at this time.

Other T-mutant embryos

In 1959, Lyon isolated a short-tailed mouse from a control population in a mutagenesis experiment. The mouse was designated $T^h/+$. Embryological analyses suggested that T^h/T^h embryos arrested much earlier in development than T/T: on the ninth day, conceptuses were less than half normal size and appeared to be grossly disorganized. Bennett (1975a) indicated that the embryos died by the eighth day; characteristically, embryonic ectoderm ceased proliferation and degenerated. On the other hand, T/T^h embryos were phenotypically indistinguishable from T/T embryos.

In 1966, Searle isolated a radiation induced mutation, curtailed. In the heterozygote ($T^c/+$), the mutant appeared to be more severe than T, causing several skeletal abnormalities not observed in $T/+$ mice. In fact, in many instances, the phenotype of $T^c/+$ mice was difficult to distinguish from tailless and crossing with viable t-mutant mice gave offspring which usually died within two weeks of birth.

By the eleventh day of gestation, T^c/T^c embryos showed a more dramatic curtailment of the posterior part of the body than T/T embryos. In fact, even the forelimbs failed to develop normally. Furthermore, the neural folds in the trunk region did not close, and there were kinks in the spinal chord. Finally, distention of the pericardial cavity was much more pronounced than in T/T embryos. Interestingly, T^c/T^h embryos resembled T^c homozygotes, whereas T^cT embryos showed adverse effects intermediate between T/T and T^c/T^c.

Johnson (1974) has described yet another allele, T^{hp}. The $T^{hp}/+$ mice have short tails that either are blunt or end in an acute bend, which gives it a hairpinlike appearance. No T^{hp}/T^{hp} embryos have been characterized. Johnson has raised the possibility that these embryos die prior to implantation. There was a curious maternal effect with this allele such that even $T^{hp}/+$ embryos died, but only if the mutant gene was maternal. Thus, in $T^{hp}/+$ embryos in which the mutant gene was sperm-derived, the familiar pattern of abnormal development in the posterior region occurred between the twelfth and fifteenth gestation day and gave rise to the short-tailed phenotype. On the other hand, $T^{hp}/+$ from the reciprocal cross had much more severe defects from the eleventh day, including a condition described as duplication of the neural tube, which in reality might have been the abnormal notochord tissue-derived tubules like those described by Spiegelman in T/T mutants, extreme edema, inflation of the pericardial cavity, and polydactyly. The actual time of death was not noted. Johnson suspected that T^{hp}/T heterozygotes were viable and short-tailed, but Bennett and Spiegelman (cited in Bennett et al., 1975) found that T^{hp}/T was lethal and, like most other T^x/T compound heterozygotes, behaved like T/T.

Bennett et al. (1975) studied seven irradiation-derived short-tailed mutant mice produced at Oak Ridge National Laboratory. Six of these were found to have the genotype $T^{OR}/+$ (we shall use the superscript designation OR instead of Or, which Bennett et al. used, to avoid confusion with the T^{Orl} mutant mice that were produced in Orleans, France) and two that were studied embryologically appeared to be affected identically. The T^{OR}/T^{OR} embryos appear to resemble those of Lyon carrying T^h/T^h (described above); that is, they died much earlier than T/T. Analysis indicated that T^{OR}/T^{OR} showed developmental arrest at the elongated egg-cylinder stage. On the seventh day of gestation, the embryonic ectoderm was pycnotic and degenerate, but the endoderm cells overlying the ectoderm had failed to undergo the transition from cuboidal to squamous. Overall, T^{OR}/T^{OR} were only about one-quarter the size of normal littermates. These embryos never progressed to the stage of mesoderm and primitive streak formation, but degenerated almost completely by the end of the eighth day of gestation, leaving behind much less affected extraembryonic structures, including EPC, trophoblast giant cells, and parietal endoderm. In all these reports, T^{OR}/T^{OR} embryos were somewhat similar to those homozygous for t^{w5}. However, the time of death of T^{OR}/T^{OR} embryos was less variable than that of t^{w5}/t^{w5} embryos. Also, T^{OR}/t^{w5} compound heterozygotes were viable. On the other hand, T^{OR}/T embryos were lethal and behaved like T/T rather than T^{OR}/T^{OR} embryos.

Bennett et al. (1975) have reviewed previous reports concerning various T mutations, a number of which appeared to be identical with the original T Brachyury. One other mutation, T^{Orl}, was produced by irradiation by Moutier in Orleans in 1973 [cited by Bennett (1975a) and Bennett et al. (1975)]. It appears to be similar to T^{hp} (Bennett, 1975a), except for some genetic differences (see Bennett et al., 1975). The embryological effects of T^{Orl} have not been published, although Bennett et al. have noted that T^{Orl}/T embryos appear to be indistinguishable from T/T homozygotes.

4.4.8 Compound Heterozygotes

T/t^x embryos

With very rare exceptions (see section 4.2.2), T/t^x animals are born tailless or with a vestige composed mainly of skin. The initial studies of T/t^x were carried out by Glueksohn-Schoenheimer (1938a). Grüneberg (1958) studied them in more detail and compared them with $T/+$ embryos (see section 4.4.7). Grüneberg concluded that the differences between the two were ones of degree rather than kind. Thus, whereas $T/+$ embryos showed a thickening in the notochord only at the growing point, T/t^6 embryos had thickened notochords right through the tail area and into the cloacal and lumbar regions. Grüneberg speculated that this extensive thickness might have resulted from incorporation of "extraneous" cells into the notochord. The notochord in these regions also often possessed large central lumina. Finally, by the eleventh day, the notochord had fused completely into the neural tube in the tail area; Gluecksohn-Schoenheimer (1938a) also noted connections between the notochord and the hindgut. Pycnosis in the tail area was evident in T/t embryos before it began to occur in $T/+$ embryos.

t^x/t^y embryos

As mentioned earlier (section 4.2.1), new t alleles are assigned to complementation groups on the basis of their ability or failure to produce live young in a heterozygous state with other characterized t alleles. Thus, an *intracomplementation* group cross produces no live t^x/t^y young, whereas an *intercomplementation* group cross produces t^x/t^y viables with normal tails. However, in most cases, this procedure has been used strictly as a genetic tool, and analysis of the viable compound heterozygotes is rare. Furthermore, even in intercomplementation group crosses, it is common that only a fraction (as low as 1 percent) of t^x/t^y offspring are actually born. Klyde (1970) showed by selection for high or low viability of t^0/t^{w12} heterozygotes that other factors are involved in survival of intercomplementation group embryos. Very few papers have appeared for the sole purpose of characterizing morphological defects in t^x/t^y heterozygotes that fail to survive. The small amount of information that is available is reviewed here.

Intracomplementation group heterozygotes   As mentioned above, the purpose of crosses leading to t^x/t^y heterozygote formation was to prove complementarity. In a number of such studies (e.g., Bennett and Dunn, 1958, 1964; Bennett et al., 1959b), it was merely mentioned that the compound heterozygote died at the same stage or showed the same type of abnormality as the homozygote t^x/t^x already known to belong to a particular complementation group. In only two instances were careful morphological analyses of intracomplementation group compound heterozygotes carried out: t^0/t^1 by Silagi (1962) and t^{12}/t^{w32} by Hillman (1975). It was, in fact, in Silagi's investigations of the offspring of $T/t^0 \times T/t^1$ mice that it was revealed that the original t^1 mutation (presumably in the t^{12} complementation group; Gluecksohn-Schoenheimer, 1938b) had been lost. The "new" t^1 mutation belonged to the t^0 complementation group because, as Silagi reported, t^0/t^1 embryos behaved identically to t^1/t^1 embryos (see section 4.4.2).

Hillman (1975) investigated t^{w32}/t^{12} compound heterozygotes because her previous ultrastructural analyses indicated that t^{w32}/t^{w32} embryos were somewhat different from t^{12}/t^{12} homozygotes (see section 4.4.1).

The t^{w32}/t^{12} heterozygotes showed the earlier lethal period of t^{w32}/t^{w32} embryos, mid, as opposed to late, morula, but contained the fibrillo-granular bodies characteristic of t^{12}/t^{12} cells. The cells of the heterozygote contained the excessive lipid accumulations and double nuclei seen in t^{12} and t^{w32} homozygous embryos. On the other hand, t^{w32}/t^{12} heterozygotes were unique in that they possessed abnormal mitochondria from the 2-cell stage. Instead of the dense ovoid bodies seen in normal embryos at this stage, the compound heterozygote had mitochondria that were rounded, swollen, and lacking the characteristic dense matrix. This abnormality appears to be different from that observed in mitochondria at t^{w32}/t^{w32} embryos later in development (see section 4.4.1). It remains to be seen whether t^0 and t^9 intracomplementation group compound heterozygotes will also show differences from the homozygotes, because there might also be variation in embryonic development among members of these groups (see sections 4.4.2 and 4.4.4).

Consideration was given to T^x/T^y compound heterozygotes in section 4.4.7.

Intercomplementation group heterozygotes   In 1937, Dunn indicated that t^0/t^1 (the latter being the original t^1 allele) compound heterozygotes were all viable and that females were normal, but males were sterile. However, the realization that t^0 and t^1 showed transmission ratio distortion made it apparent that many t^0/t^1 heterozygotes from $T/t^0 \times T/t^1$ crosses were failing to survive. Dunn and Gluecksohn-Schoenheimer (1943) later confirmed this fact by showing that the t^0/t^1 embryos scored as abnormal on the eleventh to sixteenth days of pregnancy were often reduced in size, microcephalic, and/or microphthalamic.

In 1956, Smith indicated that t^1/t^{12} heterozygotes died mainly between the thirteenth and twentieth days of gestation (this study postdated the loss of the original t^1, so the t^1 mutant used here was from the t^0 complementation group). This statement was challenged by Silagi (1962), who has published the only comprehensive study of compound heterozygotes to date. She found that t^0/t^{12} (or t^1/t^{12}) heterozygotes first showed abnormalities on the eighth day of gestation. Characteristically, the embryos resembled those of t^{w18} in that they showed growth retardation except for an apparent overgrowth in the primitive streak area. The adversely affected compound heterozygotes differed from t^{w18}/t^{w18} embryos in that they did not subsequently form duplicate neural grooves (see section 4.4.5). Instead, by the ninth day, the most severe mutants had produced the kind of "extraembryonic organism" seen with t^{w5}/t^{w5} homozygotes prior to resorption (see section 4.4.4). However, Silagi indicated that t^0/t^{12} heterozygotes can be categorized in four classes, of which the above-described mutants were the least normal. The next most seriously defective class had failed to develop from the previous day. These embryos still had large primitive streak areas. Subsequently these embryos degenerated and died without forming somites. Death of this class of embryos and the more abnormal one previously described occurred abruptly on the twelfth and thirteenth days of gestation. From one-third to one-half of the t^0/t^{12} embryos died at this time. The period of death, although close to that of T/T homozygotes, was not due to the same cause (supposedly failure of connection of the allantois with the EPC; see section 4.4.7) because the extraembryonic membranes were less defective than the embryo proper.

The less dramatically affected classes of t^0/t^{12} embryos included a group that was only slightly retarded developmentally on the ninth day of gestation, but was reduced in size. Eventually this group developed forebrain abnormalities and often died on the twelfth or thirteenth day of pregnancy. The remaining class of heterozygous embryos was initially almost normal in size, but slightly retarded in development. These embryos apparently fared better than the other heterozygotes. Those that survived the "lethal crisis" on the twelfth and thirteenth days generally survived at least until birth. By that time, however, there was again a large proportion of embryos with abnormalities of varying severity, the defects ranging from unilateral anophthalmia to microcephaly to headless monsters. Only half of the remaining heterozygotes were born live. Some of the newborns carried the less severe deformities and all were reduced in size to 50–90 percent compared to littermates. The final lethal crisis occurred in the neonatal period during nursing, presumably due to unsuccessful competition with healthier littermates. Finally, the few mice (4 percent of the total t^0/t^{12} group) that did survive weanings had short lifespans. As had been reported earlier (Dunn, 1937), the male compound heterozygotes were sterile, although at least half the females bore litters. The offspring were not described at any length.

In a brief review of the subject of intercomplementation group compound heterozygotes, Bennett (1964) indicated that offspring from other crosses suffered the same kinds of abnormalities as those described by Silagi, namely, microcephaly, microphthalmia, and micrognathia.

4.4.9 t Semilethals

In 1969, Bennett and Dunn reported on four t alleles—t^{w2}, t^{w8}, t^{w36}, and t^{w49}—derived from wild mice that they categorized as semilethals. Of the four, t^{w2} was the least severe mutation; it caused embryonic death of half of the homozygotes. By comparison, more than 98 percent of t^{w49}/t^{w49} embryos failed to survive. The t^{w8} and t^{w36} homozygotes were intermediate, resulting in 12 percent and 20 percent liveborn, respectively.

Although comprehensive analyses of the embryological developmental failures of these mutants were not carried out, the information provided for t^{w36}/t^{w36} and t^{w49}/t^{w49} homozygotes, as well as t^{w36}/t^{w49} compound heterozygotes, gives the distinct impression that the semilethals showed the same patterns of deformations and abnormalities described by Silagi (1962) for intercomplementation group compound heterozygotes. In other words, the embryos that were scored as t^{w36}/t^{w36} or t^{w49}/t^{w49} on the fifteenth to seventeenth days were characteristically retarded and showed in many cases forebrain malformations and jawlessness. Two differences were apparent, however. First, some of the embryos also showed herniation of internal organs through the abdominal wall, not reported by Silagi for t^0/t^{12} embryos. More importantly, by the fifteenth day, the only remaining evidence of 49 percent of the t^{w36}/t^{w36} embryos and 60 percent of the t^{w49}/t^{w49} embryos was the presence of moles. Although Bennett and Dunn argued that many of the embryos died on the thirteenth day, as Silagi reported for t^0/t^{12} embryos, it is quite likely that many embryos died at earlier stages. In fact, in her recent review, Bennett (1975a) indicated that some semilethals die shortly after implantation, but others show abnormalities resembling those of t^{w5}/t^{w5} and t^9/t^9 embryos.

4.4.10 Overview of Morphological Analyses

The morphological studies described above have obviously been instrumental in revealing the phenocritical period of t-mutant embryos. On the other hand, there are some limitations to the amount of useful information that can be obtained from analyses such as these. For example, it is very difficult to fix a time of death for the affected embryos. Different laboratories have used different criteria: the time of extreme pycnosis, the time at which the entire embryo has been resorbed, the time at which mitotic figures are no longer evident, etc. But often only some parts of the embryo show necrosis and other sections are relatively unaffected, as the "extraembryonic organisms" discussed previously exemplify most strikingly. Problems result from the use of mitotic figures as criteria of viability, because these have been seen in tissues containing extensive amounts of pycnosis. However, this aspect of the argument is somewhat academic in view of the fact that t-mutant embryos usually spend a good deal of time in developmental arrest prior to death (see below). Determination of the actual time of death assumes importance only when one is interested in the direct cause, e.g., Gluecksohn-Schoenheimer's proposal that T/T embryos died due to failure of the allantois.

A much more critical issue in this respect concerns the *earliest* time at which adverse effects of T-complex mutations can be detected. If it should happen that expression of T-complex genes occurs substantially earlier in development than the first detectable morphological defect, as is often the case, for example, with temperature-sensitive mutants in Drosophila (see Suzuki, 1970), then efforts to characterize the primary lesion caused by t mutations could be unnecessarily complicated by the presence of many secondary aberrations. In fact, the problem of cause and effect is always a thorny one, and in at least one case, the apparent confusion between the two has led to a dead-end pathway of research (studies on ribosomal RNA production by t^{12} embryos; see section 4.6).

One way to protect against being misled about the time of action of mutant t genes from morphological studies would be to analyze embryos carefully at stages *prior* to those in which overt defects are observed. Such analyses should be done at the ultrastructural level. Table 4.2 shows rather clearly that in a number of cases careful analysis of earlier stages had led to the discovery of abnormalities, at least in some cells, long before the time of obvious developmental retardation or arrest. In other instances, however, only a simple statement exists to the effect that studies of earlier stages did not produce any evidence of abnormality. There is no way to know just how carefully and conscientiously these studies were carried out. On the other hand, the kind of thorough analyses that might be necessary to detect subtle early defects, particularly in later acting t mutations when the embryos are quite large and many sections would have to be studied, could require a tremendous amount of work. Furthermore, complications might arise, including confusion with unrelated abnormalities that undoubtedly occur in a small number of non-t-mutant littermates and the fact that even in normal embryos, isolated cell death is usual and characteristic of a number of periods in development, beginning as early as the blastocyst stage (see, e.g., El-Shershaby and Hinchliffe, 1974). Nevertheless, observations such as Spiegelman's report [see the discussion following Spiegelman (1976)] that mesoderm cells in T/T embryos possess excessive lipid accumulations similar to those observed by Hillman and co-workers in t^{12}, t^{w32}, and t^6 homozygous embryos (see sections 4.4.1 and 4.4.2) may be taken as a suggestion that later-acting t mutations possess the same "early warning signals" as the earlier ones.

One of the consistent characteristics of t-mutant embryos is that of developmental arrest. In most cases, t/t embryos become progressively retarded, not only in size, but also in the degree of development, when compared to normally developing littermates. In some instances (e.g., t^6, t^{w73}, and less strikingly, t^{12}), mutant embryos appear to spend extended periods of time at a final developmental stage before degeneration becomes apparent. From morphological studies it is not possible to decide whether arrested embryos failed to proceed in development because they were missing information necessary to undergo differentiation to the cell types characteristic of the succeeding stages (developmental failure hypothesis) or whether the stage of arrest reflected a "plateau" level that they were unable to exceed because energy or metabolic requirements could not be satisfied (metabolic barrier hypothesis). In other words, the question raised is whether the T complex specifies information directly related to differentiative events, an idea put forward by Dunn, Bennett, Gluecksohn-Waelsch, and others, or whether this region of the chromosome encodes products essential for normal cellular metabolism, disturbances that will result in the arrest of embryonic development at a particular stage, depending upon the severity of the mutation. This idea has been considered previously by Silagi (1962) and by Hillman (1975). Although an intensive consideration of these hypotheses is reserved until other pertinent experiments have been described (see section 4.8), it is appropriate to deal briefly here with two corollaries of the developmental failure hypothesis.

The first corollary states that developmental arrest takes place at
specific and sharply demarcated stages in development during which
important organizational events occur. The data in table 4.2 describe
the stages at which the various homozygous t and T embryos show
developmental arrest. Taken together, these data are really quite sug-
gestive of a continuum in the occurrence of abnormalities from the
earliest to the latest acting t mutations, particularly when the early,
more subtle effects (open circles in the table) are taken into account.
As mentioned above and as others have noted previously (see especially Klyde, 1970), differences in the genetic background on which the
various t mutations occur may be very influential in determining the
stages at which the embryos are first affected and finally arrested.
Indeed, genetic variation unrelated to the T complex may account in
large part for the overlaps in the t-mutant effects. Once again, this
problem can be satisfactorily investigated only by comparing the various t and T alleles on isogenic backgrounds. After this comparison has
been done, if the merging of the effects of the various mutations is still
evident, the idea that t mutants "strike" only at discrete periods in
development will have to be reassessed. Although a continuum of t-
mutant effects would not necessarily invalidate the developmental failure hypothesis, it would certainly not be difficult to explain the effect
by the metabolic barrier hypothesis (differential abilities to bypass or
temporarily delay the metabolic block depending upon other enzyme
levels would allow development to proceed to varying degrees for each
t allele).

The second corollary, that only one or a few cell types in the embryo
shows a primary effect of the t or T mutation, perhaps deserves more
careful scrutiny than it has been given. In some cases, such as that of
notochord cells in T/T, this corollary clearly holds. On the other hand,
in mutants such as t^6/t^6 and t^{w5}/t^{w5}, both ectoderm and endoderm cells
show obvious defects and at about the same stage; in fact, in t^6/t^6
embryos only trophoblast cells seem initially to escape the t^6/t^6 effect
in utero, and this is not even the case in vitro (see section 4.7). That
ectoderm cells degenerate and die prior to endoderm cells in t^6 and t^{w5}
homozygotes does not necessarily indicate that it is this cell that is
primarily affected; it might merely be more severely affected by the t
lesion at that stage. In any event the conclusion that only certain cells
in the embryo are subject to the primary effects of t or T mutations
does not necessarily hold for all the complementation groups. At the

least, morphological analyses of t^{12}/t^{12}, t^6/t^6, and t^{w73}/t^{w73} embryos
have clearly invalidated the original proposal of Glueksohn-
Schoenheimer (1940), reiterated as recently as 1970, that "mutations at
the T locus find a common denominator in the abnormalities of the
notochord-mesoderm material" (Glueksohn-Waelsch and Erickson,
1970). A further consideration of whether t mutations are generalized
cell lethals is presented in section 4.8.

4.5 Surface Antigens and the T Complex

Glueksohn-Waelsch and Erickson (1970) and Bennett et al. (1972a)
proposed that T-complex-gene products are components of the cell surface. From this line of reasoning, one could ultimately explain why
t-mutant sperm possess a selective advantage at fertilization. At the
same time, altered cell surface components could be predicted to interfere with normal cell-cell interaction, resulting in organizational failure
and death of t-mutant embryos. Bennett and her colleagues thus endeavored to search for T-complex-specified surface antigens and reported their first success in 1972. Since that time, surface antigens have
occupied a central role in the conceptualization by many investigators
of the function of the T region.

4.5.1 T-Complex Antigens on Sperm

In 1972, Bennett, Boyse, and Old reported that sperm from animals
bearing the T mutation carried a surface antigen not present on sperm
from wild-type animals. This observation was made possible by the
adaptation of an immunological cytotoxicity assay for use with sperm
(Goldberg et al., 1970, 1973). The strategy in these experiments was to
multiply inoculate mice with sperm preparations from T/t^{w2} donors.
After an appropriate series of injections, Goldberg et al. collected serum and absorbed it with sperm from wild-type ($+/+$) animals in order
to remove non-T-complex-related sperm autoantibodies. The resultant
antiserum was cytotoxic against sperm from $T/+$ animals but not
against wild-type ($+/+$) sperm. Furthermore, the antiserum killed about
half the sperm from $T/+$ animals, the fraction that might have been
expected if there were equal numbers of $+$ and T sperm. It soon
became apparent, however, that this latter quantitative relationship was
not generally observed.

In a more detailed series of experiments, Bennett et al. (1972b) studied T specific antisera from T/t^{w2} sperm, as described above, and also from $T^J/+$ sperm (T^J is an independently observed T mutant originating in K. Hummel's inbred BALB stock at Jackson Laboratories; it is apparently identical to T Brachyury). The latter antisera were obtained both from BALB and BALB/T^J mice. The anti-T/t^{w2} antiserum was cytotoxic for sperm from $T/+$ and $T^J/+$ (BALB) mice, but when values for nonspecific cytotoxicity were subtracted, much less than half the sperm were affected. On the other hand, when sperm from T/t^{w2} mice were subjected to assay, about twice as many sperm were specifically killed by the antiserum. This effect presented the likelihood that *both* T and t^{w2} were specifying surface antigens. The anti-$T^J/+$ (BALB) antisera raised in BALB, and even BALB/T^J, mice also effectively killed T-bearing sperm, establishing the T antigens as a specific category of sperm autoantigens. Furthermore, the fact that T antigen was not expressed on lymphoid or epidermal cells of T-mutant animals, supported the view that the antigen is restricted to sperm [although it was predicted that the appropriate embryonic cell types (i.e., notochord and probably neural tube) would be found to express T antigen; see section 4.5.3]. These results, then, provided qualitative evidence that sperm produced specific T antigen; absolute quantitation was not possible because of the relative weakness and insensitivity of the assay. Thus, although half or fewer $T/+$ sperm showed reactivity with various anti-T antisera by the cytotoxicity assay, the temptation to conclude that expression of T antigen on sperm was a postmeiotic event (see section 4.3) was resisted. In fact, Bennett et al. noted that because there was apparently little synthetic activity in spermatogenic cells, it was likely that there was at least some premeiotic input into sperm surface antigens, as appears to be the case with the male-specific H-Y antigen (Goldberg et al., 1973). Consequently it was suggested that the distribution of T and $+^T$ antigens on the sperm surface was relative rather than absolute, i.e., both T and $+$ sperm would contain both types of antigen, but there would be more T than $+^T$ antigen on T sperm and, conversely, more $+^T$ than T antigen on $+$ sperm.

Yanagisawa et al. (1974a) extended this kind of approach to the recessive t mutants and confirmed suspicions that the sperm surface could also express t-mutant surface antigens. This paper described the production of a series of antisperm antisera: anti-T/t^1, anti-T/t^0, anti-T/t^{w1}, anti-T/t^{w5}, and anti-T/t^{w32}, as well as anti-$+/+$, the latter being produced by injecting wild-type sperm into wild-type mice. Overall, the data were quite consistent: antisera of the type T/t^x were cytotoxic to sperm from $T/+$, $+/t^x$, and T/t^x mice, but not to those from $+/t^y$ or $+/+$ mice. Furthermore, about twice as many T/t^x sperm were killed as $T/+$ or $+/t^x$, as would be expected if T and t^x antigens were being recognized. Again, presumably due to technical problems (which apparently are still unresolved), the cytotoxic index (fraction of sperm specifically killed by the antisera) never exceeded 0.6 and was usually between 0.2 and 0.4.

Two other significant pieces of information were reported. First, t^0- and t^1-containing sperm were both killed by anti-T/t^0 or anti-T/t^1 antiserum, demonstrating that these two alleles, which are apparently identical (see table 4.1) although they have been maintained in separate balanced lethal stocks for twenty years, showed the same surface antigen specificity. The second point of interest was that $+/+$ sperm were capable of producing an antiserum in $+/+$ mice. After absorption with T/t^0, T/t^{w1}, and T/t^{w32} sperm, it was still cytotoxic to $+/+$ and, to a lesser extent, to $T/+$ or $+/t^x$ sperm, but it had no effect upon T/t^x sperm. Although the latter observation is satisfying at first glance, it actually presents a paradox, which is discussed below.

In a subsequent communication, Yanagisawa et al. (1974b) attempted to use the system that they had developed to test the inference from "late-mating" experiments (Braden, 1958; see section 4.3.2) that sperm carrying t alleles that show transmission frequency distortion live longer than their wild-type counterparts. Consequently, they tested sperm from $+/t^{w1}$ mice incubated in vitro for increasing periods of time with anti-T/t^{w1} antiserum. They reported that for up to four hours incubation time, an increasing fraction of the surviving sperm showed cytotoxicity when treated with the antiserum (see, e.g., table 4.3, Experiment 1A). In the fifth hour there was a dramatic and unexplained drop in the cytotoxic index, back to the one-hour value. As a control, $T/+$ sperm were tested in the same way (Experiment 1B), and there was no apparent increase in the fraction of viable T sperm according to the cytotoxicity assay (as would be expected because T sperm do not show transmission frequency distortion). The converse experiment, involving $+/+$ antiserum gave consistent results: the fraction of viable $+$ sperm in the $+/t^{w1}$ population fell with time in culture (table 4.3, Experiment 2A), but there was no change in the cytotoxic index of the $T/+$ sample (Experiment 2B). The data thus appear to be consistent with the hy-

Table 4.3. Effect of Antisera on t^{w1} and + Sperm with Increasing Time in Culture

			Cytotoxic Index	
Experiment	Antiserum	Sperm Donor	1-hr Culture	4-hr Culture
1A	T/t^{w1}	$+/t^{w1}$	0.15	0.40
B		$T/+$	0.30	0.25
2A	$+/+$	$+/t^{w1}$	0.40	0.15
B		$T/+$	0.10	0.15

These data are taken from Yanagisawa et al. (1974b). The numbers are approximate because they were taken from graphical presentations. The cytotoxic indices represent the fraction of sperm that were alive at the time of assay but were subsequently killed by exposure to antiserum.

pothesis that t^{w1} sperm survive longer than + sperm. These experiments would have been more convincing, however, if the starting (1 hr) values for A and B in both Experiments 1 and 2 had been similar. Thus, the initial value in Experiment 1B is 0.3, close to the maximal attainable value (0.4) with the T/t^{w1} antiserum used (the latter value has been determined from other data in the paper). Therefore, technical limitations presumably would have prevented the detection of a sharp increase in the percentage of viable T sperm with increasing culture time. Similarly, in Experiment 2B, the initial cytotoxic index, 0.1, was at or close to a control (no antiserum or no complement) value level; again, a detectable drop would not have been measured. Thus, although the experimental data look attractive, the control data are questionable.

More recent sperm antigen assays have revealed that the entire system is a good deal more complicated than originally supposed (Bennett, 1975a,b; Artzt and Bennett, 1977). Sperm absorption studies have shown that each t allele (or *haplotype* as it is now called by Bennett and Artzt) is represented on the sperm surface by a set of specificities, rather than by a single species. Thus, although t^{w18}, t^0, and t^{w2} sperm have one or more unique specificities apiece, there are some other shared specificites between t^0 and t^{w18} sperm, or t^0 and t^{w2} sperm, and possibly one common to all three alleles. On the basis of these data, Artzt and Bennett (1977) have drawn an analogy with the "private" and "public" specificities of the H-2 complex (see Klein, 1975).

Assays of sperm from two viable recombinants derived from t^0 and one from t^{w2} indicated that they contained at least some of the shared specificities, but each lacked the unique specificity of the sperm from the parental line. Artzt and Bennett have argued, perhaps somewhat prematurely, that the unique specificities "have a functional role responsible for lethality or semilethality," whereas common specificities might be related to factors causing taillessness in combination with T. Having obtained these data on unique and shared specificities, these investigators have repeated their earlier studies (Yanagisawa et al., 1974a), and presumably due to the more sensitive assay procedures now being used, they have been able to find shared specificities among the alleles previously reported to express unique antigens. Furthermore, Bennett (1975a) has reported that even members of the same complementation group (e.g., t^{w1} and t^{w12}) can have a different array of specificities.

Finally, in analysis of one of the T^{OR} mutants (see section 4.4.7), Bennett et al. (1975) reported that T^{OR} sperm did not contain antigens cross-reacting with their T antisera, nor were they able to generate a T^{OR}-specific antiserum. They concluded that T^{OR} was a deletion. They did not, however, eliminate the possibility that T^{OR} expresses $+^T$ antigens and perhaps possesses instead altered antigens similar to those on t^0 or t^{w5} sperm (T^{OR}/T^{OR} embryos show some similarities to t^0/t^0 and t^{w5}/t^{w5} embryos with respect to time of development arrest and morphological abnormalities; see section 4.4.7).

In summary, it now appears as though at least some part of the T region is involved in the specification of sperm surface antigens. Antigens appear to be produced or modified by T, t, $+^T$, and $+^t$ alleles. Both unique and common specificities are probably expressed by each class of alleles. The available evidence favors the idea that the deposition of these antigens on the sperm surface is largely a postmeiotic event (in fact, solid evidence in favor of any premeiotic contribution has yet to be presented), unlike the expression of H-Y (Goldberg et al., 1973) and probably H-2 antigens (Goldberg et al., 1971). It has been proposed that these T- and t-specified antigens are involved in t-mutant embryo lethality and taillessness. One may also speculate that they play a role in transmission ratio distortion (Gluecksohn-Waelsch and Erickson, 1970).

There are still aspects of the sperm antigen story that are unclear. Some of these, such as the question of whether the array of T (or t) antigens on the sperm surface takes place prior to or following meiosis will probably be definitively answered only when *all* sperm carrying mutant antigens can be detected by cytotoxicity or some other immunoassay procedure. It now appears likely that T (or $+^T$) and t^x (or $+^{t^x}$) "alleles" occupy different regions of the chromosome (see section 4.2.2). If so, the results obtained with the $+/+$ antisera (Yanagisawa et al., 1974a) are particularly perplexing because t^x-mutant sperm would presumably have the genotype $+^T t^x$ and T-mutant sperm would carry $T +^{tx}$. Whether expression of the T-complex antigens were premeiotic or postmeiotic, the total sperm population ($+^T t^x$ and $T +^{tx}$) should express equal amounts of mutant and wild-type antigens. Consequently, when Yanagisawa et al. absorbed $+/+$ antiserum with T/t^x sperm, *all* antibodies to T-complex antigen should have been removed, but there was still a reaction with $+/+$ sperm.

Artzt et al. (1974) attempted to explain this paradox in the following way. They proposed that different t "alleles" were at different sites in the T region. However, a t mutation would not only cause an alteration at its own site but also would result in the conversion of $+^T$ to t^T. Thus, the genotype of a T/t^x mutant might be:

$$\frac{T +^{t^x} +^{t^y} +^{t^z}}{t^T t^x +^{t^y} +^{t^z}}$$

There would then be no $+^T$ allele in the mice used as a source of the absorbing sperm by Yanagisawa et al., and the absorbed anti-$+/+$ antiserum would, in reality, contain only anti-$+^T$ antibodies. Although this scheme readily explains the observed immunological observations, it is rather difficult to understand on genetic grounds how a single mutation might affect both a t site and the T site. This would be especially hard to conceptualize in the case of a t^z mutation, according to the scheme written above: presumably, the $+^{t^z}$ and the $+^T$ sites would have to be altered without any change to the intervening $+^{t^x}$ and $+^{t^y}$ sites, because the latter would be expressing wild-type surface antigens.

An alternate explanation of the data of Yanagisawa et al. is that t and T surface antigens might effectively mask wild-type T-complex antigens, either physically or else by being produced in greater numbers. It will be interesting to see how this problem is resolved.

4.5.2 F9 Antigen and the T Complex

As indicated above, the study of T-complex antigens on sperm has
progressed through three levels of sophistication: (1) the demonstration
that the T region plays a role in the specification of some antigens on
the sperm surface; (2) the revelation that sperm carrying different *t*
alleles possess a different array of *t*-specified surface antigens; and
(3) the discovery of unique and shared *t* specificities. As far as parallel
studies on embryos are concerned, the work has advanced so slowly
that the investigators involved are still struggling at the first level.
There are many reasons for this lack of progress. A number of them
are technical and include difficulties in collecting enough material to
generate antibodies and problems involved in detecting specific antigens in sectioned material. At present, most of the information available on the expression of T-complex antigens on embryonic cells is
somewhat indirect. It involves the F9 surface antigen of teratocarcinoma cells and non-T-complex antigens present or absent during early
stages of embryogenesis. Because various aspects of these subjects are
treated in detail in chapters 5 and 7 of this monograph, the discussion
here includes only data pertinent to the T complex.

In 1971, Edidin et al. reported that when injected into rabbits, mouse
embryoid bodies, the ascitic form of teratomas, generated antisera that
cross-reacted with early mouse embryos. Following this report, Artzt
et al. (1973) generated antisera by injecting irradiated cells from a cultured teratocarcinoma stem cell line, F9 (see chapter 7) into syngeneic
strain 129 mice. These antisera, which were cytotoxic to F9 cells but
not to a large variety of normal adult or tumor cells, were considered
to be detecting autoantigens on the F9 cell surface.

The anti-F9 antisera gave a positive reaction when tested against
embryos. The exact schedule of F9 antigen appearance on, and disappearance from, the embryo surface is still unclear, and the different
results apparently depend upon the kind of immunoassay used. Initially, unfertilized and fertilized 1-cell eggs were found to be unreactive
and 4- and 8-cell embryos were positive, although the reaction at the
later stages was not homogeneous for all the cells in the embryo (Artzt
et al., 1973). Subsequently, Babinet et al. (1975) reported that although
unfertilized and fertilized eggs were still negative, F9 antigen could be
detected from the 2-cell embryo right through to the blastocyst stage,

at which time both trophectoderm and ICM cells were positive. On the other hand, Bennett indicated that her laboratory had failed to detect F9 antigen on intact blastocysts [J. Caldwell (unpublished results), cited by Bennett (1975a)]. In the most recently published account, the Jacob laboratory claims that F9 antigen is discernible on the surface of the egg following fertilization. Antigen is expressed not only on the surface of the blastocyst, but also on embryonic ectoderm cells right through the ninth day of gestation [M. H. Buc-Caron (unpublished results), cited by Nicolas et al. (1976)].

In a parallel series of experiments, F9 antigen has also been detected on the sperm surface (Artzt et al., 1974; Babinet et al., 1975; Fellous et al., 1974; Vitetta et al., 1975). It is from these studies that relationships between F9 antigen and T-complex-gene products have been drawn. Artzt et al. (1974) found that anti-F9 antiserum showed a decreased potency against F9 cells following absorption with sperm. Importantly, sperm from mice carrying the t^{w32} allele (i.e., T/t^{w32} or $+/t^{w32}$) were only half as effective as wild-type sperm in absorbing F9 antibodies. Sperm from other t-mutant mice (T/t^{w1}, $T/+$) absorbed the F9 antibody as effectively as wild-type sperm. From these observations, Artzt et al. concluded that F9 antigen was the product of the $+^{t^{w32}}$ (or $+^{t^{12}}$) allele, with the single reservation (Bennett, 1975a) that the observed results could have been due to "rearrangements of antigenic sites" (masking of wild-type antigens?) on the sperm surface by the presence of t^{w32}.

By indirect immunofluorescence and cytotoxicity analyses, Buc-Caron et al. (1974) and Fellous et al. (1975) provided direct confirmation that F9 antigen was present on mouse sperm cells. Furthermore, a cross-reacting specificity was detected on human sperm. Fellous et al. (1975) proposed that antigen was localized primarily on the post-acrosomal cap of the sperm. Babinet and co-workers (1975) carried out a study of F9 antigen expression on the male germ line and found that they could detect the antigen on cells at all stages of spermatogenesis: gonocytes, spermatocytes, and spermatids were all positive, but the structural cells of the testis did not appear to possess F9 antigen. The latter point was demonstrated conclusively with Sl^d/Sl^d mice, which lack primordial germ cells and their derivatives.

Artzt et al. (1974) have also tested F9 cells for their expression of T-complex surface antigens by the use of sperm antiserum. They found that anti-$+^T/+^t$ antiserum was cytotoxic to F9 cells; however, after absorption with T/t^{w1}, T/t^{w5} and T/t^{w32} testicular cells, the antiserum was noncytotoxic against F9 cells [although, as Yanigasawa et al. (1974a) had reported earlier, this absorbed antiserum still reacted with +/+ sperm]. Artzt et al. offered two possible explanations for this observation. The first, that F9 cells expressed non-T-related auto-antigens, was subsequently proposed to be unlikely (Fellous et al., 1975). The second explanation recalls the argument put forward at the end of the last section, namely, that the $+^T$ and $+^t$ "alleles" actually occupy different sites in the T region, and therefore mice that are genotypically T/t^x also produce $+^{t^x}$ gene products. In this way, anti-F9 antibody, presumed to be identical to anti- $+^{t^{w32}}$, would be absorbed out by sperm from T/t^{w32} (or any T/t^x) mice.

4.5.3 T-Complex Antigens on Embryos

At the time of this writing, there has been no published report to indicate that any of the T-complex-specific antisera derived from sperm inoculations cross-reacts with any embryonic cell types. Nor have any of the antisera produced from inoculations of embryonic or extra-embryonic tissues (see chapter 5) been implicated in carrying specific-ities against T-complex gene products. We have learned, however, that it has been demonstrated with some success that t^{w32}-mutant embryos express t^{w32} surface antigens (F. Jacob, personal communication).[2]

4.5.4 Relationships between T and H-2

Erickson and Gluecksohn-Waelsch (1970) and Artzt and Bennett (1975) have compared the T and the H-2 regions; the latter have decided that the similarities warrant the conclusion that H-2 might have been an evolutionary precursor of T or vice versa, or that both might have been

---

[2]Recently, Kemler et al. (1976) reported that cleavage-stage embryos bearing t mutations reacted with antisera prepared against sperm from animals of the homologous t genotype. Their results demonstrate the presence on cleavage-stage embryos heterozygous for the alleles t^{w32}, t^{w5}, t^0, and t^{w18} the same altered antigenic specifities found on t-mutant sperm. Therefore, the time of expression of the mutant antigens is not necessarily coinci-dent with the lethal period of the mutant phenotype. These investigators also demon-strated that F9 antigen is absent from embryos homozygous for either t^{w32} or t^{w5} but not from T/T or t^{w18}/t^{w18} embryos. The relationship between F9 and t antigens now appears to be more complex than was first proposed (Artzt et al., 1974), and although these antigens may be genetically related, their coexistence or mutual exclusion on the surface of cleavage-stage embryos appears to be determined by some factor(s) other than simple genetic allelism.

products of a common, more primitive gene. It was noted in favor of this argument that the two regions are linked on chromosome 17 and that each occupies a large portion of the chromosome (although the size of the T region has not been measured, it is presumed to be large). It should be indicated, however, that there are genes between T and H-2 that seem not to be obviously related functionally to T or H-2, for example, tufted.

The most important shared property of T and H-2 is the nature of the gene products. The H-2 complex specifies histocompatibility surface antigens; at least part of the T complex is also involved in surface antigen expression. In both cases, the antigens are complex in that there are unique as well as shared specificities. Recently, Vitetta et al. (1975) isolated and compared H-2 and F9 (putative $+^{t^{12}}$) surface antigens, and they found a number of common features: both appear to be dimers of 90,000 molecular weight with subunits interconnected by disulfide bridges. Both are also noncovalently associated with one or more proteins in the molecular weight range of β2-microglobulins (c. 12,000 daltons).

Artzt and Bennett considered three other correlations of T and H-2 genes. Just as the importance of the H-2 system is underscored by analogous genes in other species (such as the HL-A region in humans), the cross-reactivity of anti-F9 antiserum with human and bovine sperm (Buc-Caron et al., 1974) possibly indicates that gene complexes akin to T are also present in other mammals. Furthermore, Artzt and Bennett have paralleled the fact that the H-2 region contains two serologically detectable genes (H-2^D, H-2^K) with their present hypothesis, as yet unproven, that the T region includes a set of genes controlling lethality and a second region that influences tail size. Finally, a number of groups have been investigating the possible reciprocal relationship between the the expression of H-2 and F9. Briefly, undifferentiated teratocarcinoma cells produce F9 antigen until they differentiate, at which time there appears to be a transition to H-2 expression (see chapter 7). According to Artzt and Bennett, the same kind of transition occurs in the case of the embryo. However, we feel that it is premature to draw such a conclusion, because the timing of expression of both H-2 and F9 antigens on embryonic cells is uncertain. The recent observations that blastocysts can transiently express H-2 antigens in ovariectomized hosts (Håkansson et al., 1975) should provide an ideal test for the proposed reciprocal relationship of H-2 and F9 antigens, because the latter should be diluted out during this period. Clearly, any consideration of the relationship between expression of H-2 and T-complex-specified surface antigens should be reserved until the latter have been identified definitively on the surface of embryonic cells. Finally, Artzt and Bennett pointed out that all adult cells other than sperm express H-2, but sperm express only F9. The latter part of the statement, which stems from the work of Vitetta et al. (1975), is directly contradictory to the earlier reports of histocompatibility antigen expression on sperm in mice (H-2: Vojtiskova et al., 1969; Goldberg et al., 1970) and in humans (HL-A: Fellous and Dausset, 1970). No explanation has been given to account for this discrepancy.

Overall, we are inclined to agree that the T complex is in some way related to surface antigen expression on sperm and probably on embryonic cells as well. However, there are inconsistencies yet to be resolved and critical experiments still to be done in order to assess the role of surface antigens in T-gene function(s) with any assurance. Nevertheless, we shall add our own speculations concerning the T-complex antigens to those already described at the end of this chapter (see section 4.8.1).

4.6 Biochemical Studies of T-Complex Mutant Embryos

As has often been the case in other areas of developmental biology, biochemical or molecular biological analyses have lagged far behind morphological, histological, and genetic studies on the T complex. In this instance, the system is particularly refractory to such experimentation because:

1. Embryos are difficult to obtain in large enough numbers for routine biochemical analyses at the time of the early acting *t* lethals.
2. At the critical times of the later lethals, the embryos are relatively small but complex, and it is difficult to isolate the particular tissues that should be studied (e.g., notochord or neural tube).
3. Because the mutants can be obtained only in heterozygote crosses, even when taking transmission ratio distortion into account, at least half the embryos will be phenotypically normal, so any biochemical variation of *t* mutants in pooled samples will tend to dilute out.
4. If mutants are selected at late enough stages in development such that morphological abnormalities can be detected (so as to avoid pool-

ing with normal embryos), possibly misleading secondary biochemical effects almost certainly will have taken place. In most cases the early, more subtle morphological aberrations characteristic of certain t mutations are not visible in intact embryos, but only in sectioned material.

Despite these difficulties, some biochemical studies have been carried out, though these have been restricted to early-acting t lethals.

There is a single biochemical aspect concerning T-complex mutants that has received considerable attention by investigators in the field. It is the synthesis of RNA, particularly rRNA, by homozygous t^{12} embryos. Initially, Smith (1956) and Mintz (1964 a,b) reported, on the basis of histochemical and autoradiographic studies, that t^{12} mutant morulae showed signs of underproduction of cytoplasmic RNA. Taken together with the abnormalities in development of the nucleoli in these embryos (see section 4.4.1), inadequate rRNA synthesis was an obvious candidate for the arrest of t^{12}/t^{12} embryos. Further support was gained from a preliminary note (Klein and Raska, 1968) claiming that rRNA genes were underrepresented in DNA from $+/t^{12}$ mice and from parallels drawn between structural aberrations in t^{12}/t^{12} embryos and anucleolate Xenopus embryos (Calarco and Brown, 1968; Gluecksohn-Waelsch and Erickson, 1970). However, more careful analyses of t^{12}-mutant embryos demonstrated clearly, as Calarco and Brown (1968) first suspected, that abnormal RNA metabolism was a secondary, degenerative effect rather than the primary lesion. Hillman et al. (1970) supported this interpretation with their ultrastructural analyses on t^{12}/t^{12} mutants (see section 4.4.1), and Hillman (1972) and Hillman and Tasca (1973) confirmed it in subsequent experiments. With the use of high-resolution autoradiography following incubation of embryos with ^{3}H-uridine, Hillman demonstrated that t^{12}/t^{12} morulae continued to produce normal amounts of RNA well after ultrastructural abnormalities (condensation of the pars fibrosae and subsequent rounding of the nucleoli) appeared. No differences were evident between normal and mutant embryos in pulse or pulse-chase experiments in the number or distribution (nucleolus, nucleoplasm, cytoplasm) of silver grains. Even in more degenerate cells showing breakdown of polysomes and rough endoplasmic reticulum, RNA synthesis was taking place in the nucleoli, although the RNA synthesized appeared to be retained there. Grain density was too heavy to allow quantitative comparisons with normal cells.

Hillman and Tasca carried out biochemical studies of ^{3}H-uridine up-take and incorporation into RNA by late morulae from $(+/t^{12} \times +/t^{12})$ and wild-type crosses. No differences between experimental and control populations were observed after short (3 hr) or long (45 hr) incubations with the labeled nucleoside. Furthermore, the electrophoretic profiles of the RNA synthesized were virtually superimposable. Although the data of Klein and Raskas (1968) have never been directly refuted or confirmed, it seems justified to conclude from the studies by the Hillman laboratory that faulty RNA synthesis is not the primary lesion in t^{12}/t^{12} embryos and that the T complex is presumably not involved with rRNA synthesis.

In an effort to further characterize early biochemical differences between t^{12}/t^{12} and normal embryos, Erickson et al. (1974) and Hillman (1975) studied protein synthesis at preimplantation stages. As was the case with RNA synthesis, no large differences were obvious between wild-type or heterozygous embryos and t^{12}-mutant embryos with respect to uptake or incorporation into protein of ^{3}H-leucine. Furthermore, Erickson et al. found that the electrophoretic profiles of the labeled proteins were virtually identical.

These data led to the idea that t^{12}/t^{12} embryos show normal development to the late morula stage, at which time they are abruptly arrested. This hypothesis is not completely consistent with the observations by Hillman et al. (1970) of ultrastructural differences between normal and mutant embryos long before the late morula stage (see section 4.4.1). Our laboratory has begun to study this question more closely by measuring enzyme activity in single embryos (Wudl and Sherman, 1976). In this way, the problem of obscuring small differences between normal and mutant embryos that result from sample pooling is obviated. Data from a comparative study of β-glucuronidase activities in individual normal and mutant embryos are shown in figures 4.1 and 4.2. Statistical analyses revealed that there was no difference between the distribution of enzyme activities (figures 4.1c and 4.2c) in 4-cell embryos from $(+/+ \times +/t^{12})$ and $(+/t^{12} \times +/t^{12})$ crosses, respectively (the latter cross should contain 46 percent t^{12}/t^{12} embryos, whereas the former contains no t^{12}/t^{12} embryos, but 92 percent $+/t^{12}$). On the other hand, by the midmorula stage (figures 4.1b and 4.2b), there was already a significant difference (at the 0.05 level) between the variance in enzyme activities in the two populations. Twenty-four hours later, control blastocysts had shown about a tenfold increase in β-glucuronidase activity (figure

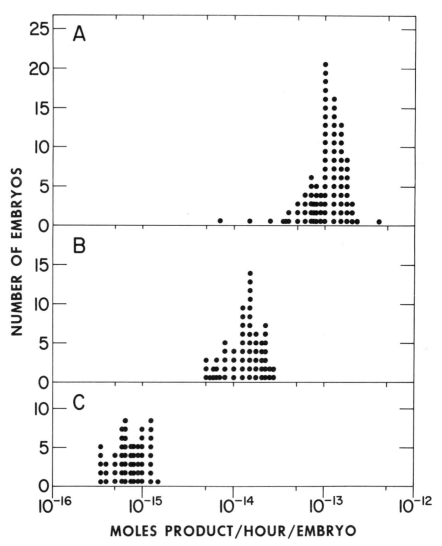

4.1 β-Glucuronidase activities of normal preimplantation embryos. Two-cell embryos were removed from +/+ mothers mated with +/t^{12} males, placed in culture, and individually assayed by a microfluorometric technique at the expanded blastocyst stage (a), morula stage (b), and 4-cell stage (c) as described by Wudl and Sherman (1976). Note that the activity rises approximately tenfold at each of these stages. Each circle denotes the enzyme activity for a single embryo. From Wudl and Sherman (1976).

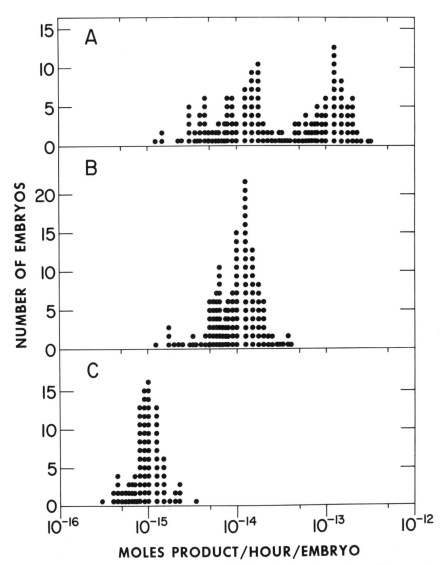

4.2 β-Glucuronidase activities of normal and t^{12}/t^{12} preimplanation embryos. The experiment was carried out as in figure 4.1, except that embryos were derived from a $(+/t^{12} \times +/t^{12})$ cross. The expected proportion of t^{12}/t^{12} embryos from this cross is 45.9 percent. The stages are as in figure 4.1, except that the embryos in (a) are a mixture of morulae and expanded blastocysts. From Wudl and Sherman (1976).

4.1a), and only wild-type and heterozygous embryos from the experimental cross had progressed in a normal manner, both morphologically and biochemically. In fact, the presumptive mutant embryos showed no increase at all in enzyme activity in the 24-hour period (figure 4.2a). These data are consistent with the assertion by Hillman that at least some embryos begin to die by the midmorula stage (as reflected especially by the 15 percent of experimental embryos with activities lower than any of the controls at the midmorula stage; cf. figures 4.1b and 4.2b). They also attest to the dramatic and abrupt cessation of metabolism in all t^{12}/t^2 embryos at the late morula stage. It should be noted, however, that experiments such as these serve only as indicators of cellular metabolic levels; there are as yet no indications of any specific enzymes that might have a primary involvement in T-complex effects. Furthermore, these assays are useful primarily for enzymes like β-glucuronidase, which can be detected in single embryos and which rise (or fall) dramatically during the stage of development at which the particular t mutation is suspected of being expressed.

Some recent biochemical studies by Ginsberg and Hillman (1975) and Nadijcka and Hillman (1975b) dovetail nicely with the ultrastructural analyses carried out by that laboratory. Ginsberg and Hillman compared ATP metabolism in embryos derived from $(+/t^{12} \times +/t^{12})$, $(+/t^{w32} \times +/t^{w32})$, and $(+/+ \times +/+)$ crosses. They found dramatic differences, particularly at the 2-cell stage. Pooled samples containing mutant embryos had higher levels of ATP and much higher ADP contents than the control wild-type population. Furthermore, the rates of ATP synthesis and turnover were higher in populations containing t^{12}/t^{12} and t^{w32}/t^{w32} embryos than in the wild-type sample after the first cleavage. These differences between mutant and wild-type embryos were observed until the 8-cell stage for embryos containing t^{w32} homozygotes and until the early morula stage for the t^{12}-containing population. At the morula stage, ATP and ADP levels in mutant embryos were similar to those of controls, as Erickson et al. (1974) had found earlier. However, Ginsberg and Hillman demonstrated that the synthetic and turnover rates of ATP in mutant embryos had in fact dropped below control levels. At the late blastocyst stage, by which time homozygous mutant embryos had died and were removed from the pools, ATP metabolism was, as would be expected, comparable in all cases. Although these experiments would have been even more impressive had control and mutant embryos been carried on similar genetic backgrounds or if controls contained $+/t$ as well as $+/+$ embryos, the data nevertheless support the views that normal and t^{12}- or t^{w32}-mutant embryos differ as early as the 2-cell stage and that t^{12} and t^{w32} homozygotes are not the same in every respect. In fact, the earlier drop in ATP metabolism by t^{w32} embryos, compared to those carrying the t^{12} mutation, is consistent with the earlier time of death of the former (see section 4.4.1).

In another study, Nadijcka and Hillman (1975b) carried out high-resolution autoradiographic analyses of the incorporation of precursors seen in the abnormal lipid droplets in t^{12}/t^{12} and t^{w32}/t^{w32} embryos. They first showed that pyruvate was incorporated into the lipid accumulations and thereby confirmed that the embryos were making the material rather than absorbing it from the external milieu. Subsequently, they showed that although phospholipid precursors (ethanolamine and choline chloride) were not selectively associated with the lipid droplets, palmitic acid clearly was. They could not eliminate the possibility of cholesterol accumulation in these cells (because they were unable either to take up labeled mevalonic acid or to convert it to cholesterol), but it was clear that the excessive numbers of lipid droplets contained neutral lipid, at least in part. Ginsberg and Hillman (1975) noted that this situation very easily could have resulted from the diversion of the excessive levels of ATP in the mutant embyros to acetyl-CoA, a substrate for neutral lipids. In fact, a further correlation in this respect has been drawn with t^6 embryos [Ginsberg and Hillman (unpublished), cited in Ginsberg and Hillman (1975)], in which there appear to be abnormally high levels of ATP metabolism, but only at the late blastocyst stage, coincident with the time that lipid accumulations are first seen in these mutants (see section 4.4.2). Finally, Ginsberg and Hillman have cited evidence that accumulation of crystals in mitochondria, due to the excessive uptake of inorganic ions by those organelles, occurs only in the presence of elevated levels of ATP and ADP. This phenomenon, of course, might explain the appearance of crystalline inclusions in t^{w32} and t^6 mitochondria (see sections 4.4.1 and 4.4.2). However, it is not clear why t^{12} homozygotes, which have generally higher ATP and ADP levels than t^{w32}/t^{w32} embryos at equivalent stages (Ginsberg and Hillman, 1975), do not also show evidence of such inclusions. Nevertheless, we feel that the observations of Hillman and co-workers have introduced a new and important dimension to studies on early-acting t mutations, and we shall consider more fully the possible significance of their results later (see section 4.8).

4.7 In Vitro Studies of T-Complex Mutant Embryos

Some forty years ago, Ephrussi (1935) reasoned that "the application of tissue culture to the study of lethal embryos . . . is based on the assumption that this technic should permit the differentiation of the potentialities of the cells of an embryo which is non-viable in vivo." Despite the remarkable beginning that Ephrussi achieved with his studies on T/T embryos, there has been little inclination until very recently to study t-mutant embryos in vitro. In the past, there have been two major barriers to such experimentation. First, it was necessary either to select the mutant embryos in utero and place them in culture prior to their death or alternatively to culture all embryos from mutant mice individually and select those that could subsequently be shown to be t-mutant. Second, conditions had to be appropriate for culture of embryos at the various stages. Recent developments have potentially removed both of these barriers, and the time is ripe for incisive in vitro studies of t-mutant embryos.

Although relatively few culture (or ectopic site) studies on t-mutant embryos have been carried out, at least some experiments have been performed on members of four of the six complementation groups, as well as T Brachyury. These will be reviewed in order of time of death of the mutant embryos in utero (see section 4.4).

4.7.1 t^{12} Embryos

The greatest number of in vitro experiments on t-mutant embryos have been carried out with t^{12} embryo populations. In 1963, both Mintz and Silagi reported that presumptive t^{12}/t^{12} morulae died in vitro at the morula stage, just as they do in utero. This observation reduced the possibility that mutant embryos died as a result of abnormal fetal-maternal interaction. Mintz (1964a) analyzed arrested t^{12}/t^{12} morulae histologically and, in a comparison with Smith's (1956) studies on the same mutants taken directly from the genital tract, concluded that the abnormalities in vitro were the same as those in utero. Hillman et al. (1970) and Hillman and Hillman (1975) subsequently confirmed this finding by demonstrating that t^{12}/t^{12} and t^{w32}/t^{w32} mutant embryos showed the same ultrastructural defects whether removed from the genital tract and processed immediately for electron microscopy or previously cultured from the 2-cell stage (see section 4.4.1). In a definitive series of experiments, Mintz (1964b) constructed $t^{12}/t^{12}\leftrightarrow$ normal chimeras from midcleavage embryos. She found that although several of the chimeras reached the blastocyst stage, the t^{12}/t^{12} cells therein did not develop beyond the stage at which they normally arrest (see section 4.4.1). The conclusion from these experiments is clearly that t^{12} is a cell lethal, i.e., all blastomeres in the morula succumb to its adverse effects.

4.7.2 t^6 Embryos

Erickson and Pedersen (1975) compared the behavior in vitro of blastocysts from experimental ($+/t^6 \times +/t^6$) and control ($+/t^6 \times T/+$) crosses. They observed that 25 percent more outgrowths in the experimental cross contained only trophoblast cells, i.e., all inner cell mass (ICM) derivatives were absent compared to the control cross. Of those embryos containing ICMs, no statistically significant differences were present between controls and experimentals with respect to the degree of organization of the ICM (i.e., two-layered, two-layered but disorganized, totally disorganized, etc.). These authors concluded that the excess 25 percent of the embryos containing only trophoblast cells were t^6/t^6, and from this they proposed that trophoblast cells were relatively unaffected by the t^6 mutation. One could, in turn, conclude from this that the t^6 mutation is not a generalized cell lethal, but selectively kills ICM derivatives.

For reasons that will become apparent in section 4.7.4, we felt it unlikely that trophoblast cells are excluded from the effects of t^6. Consequently, we carried out experiments similar to those of Erickson and Pedersen, but we came to a very different conclusion. We agree that presumptive t^6/t^6 blastocysts are more likely to give rise to outgrowths of trophoblast lacking ICM derivatives; however, a careful inspection suggested that the trophoblast cells from presumptive t^6/t^6 embryos were arrested in their development before the eighth equivalent gestation day (EGD) (L. R. Wudl and M. I. Sherman, unpublished observations). Our criterion of trophoblast development was the diameter of the nuclei. We have shown previously that mouse trophoblast cells are polyploid and that the extent of polyploidization increases as the cells develop. Trophoblast cells undergo polyploidization in vitro (Barlow and Sherman, 1972), and the resulting DNA contents are reflected by proportionate increases in nuclear diameter of the cultured cells (Sherman et al., 1977).

Table 4.4 and figure 4.3 show the nuclear diameters measured in experimental (from $+/t^6 \times +/t^6$ crosses) and control (from $+/+ \times +/t^6$ crosses) outgrowths that contained only trophoblast cells on the eighth EGD. It is apparent that even by the eighth EGD, the average nuclear diameter in the presumptive t^6/t^6 trophoblast population was less than that of control trophoblast population. In fact, only 37.5 percent of the experimental nuclei had diameters exceeding that found for normal diploid nuclei in the G2 phase of mitosis (see Sherman, 1975); the comparative figure for the control trophoblast population is 67 percent (figure 4.3). Furthermore, although the nuclear diameters of control trophoblast cells increased during the next four days in culture, there was no concomitant increase in the presumptive t^6/t^6 trophoblast nuclear diameters (table 4.4 and figures 4.3 and 4.4). By the eleventh EGD, trophoblast cells in the control outgrowths still looked healthy (e.g., figure 4.4c), whereas many t^6/t^6 trophoblast cells had died and the remainder were unhealthy in appearance; of those surviving, more than half still had nuclei with diameters in the G2 diploid range (figures 4.3c and 4.4f). In an extension of these studies, we made trophectodermal vesicles (see Sherman, 1975a) from t^6/t^6 and normal embryos. These structures, made by culturing isolated midcleavage blastomeres, never had ICMs, but they gave rise to outgrowths of giant trophoblast cells apparently identical to those formed from normal blastocyst outgrowths. We observed the arrest of trophoblast cells in outgrowths from t^6/t^6, but not $+/t^6$ trophectodermal vesicles (L. R. Wudl and M. I. Sherman, unpublished results). Although we are attempting to confirm these observations by studying other parameters of trophoblast cell development, we feel that the data already recorded strongly suggest that both ICM and trophoblast cells are affected by t^6 and that the mutation is therefore a generalized cell lethal.

4.7.3 t^{w73} Embryos

Embryos bearing this recently discovered allele have not yet been studied in vitro. It would be extremely interesting, however, to determine how t^{w73}/t^{w73} embryos behave in culture. Phenomena analogous

Table 4.4. Average Nuclear Diameters of Cultured Trophoblast Cells from Normal ($+/t^6$ or $+/+$) and Presumed t^6/t^6 Blastocysts

Presumed Genotype	Average Nuclear Diameter (μm) on EGD[a]			
	8	9	10	11
$+/t^6$ (or $+/+$)	31 (26)	—	39 (22)	42 (17)
t^6/t^6	24 (40)	24 (39)	—	24 (30)

[a] Equivalent gestation day, i.e., the age of the embryo had it been left in utero.
Blastocyst outgrowths from ($+/+ \times +/t^6$) and ($+/t^6 \times +/t^6$) crosses were selected for those containing only trophoblast cells (no ICM). Outgrowths were photographed with phase contrast optics. Nuclear diameters (averages of the largest and smallest diameter of each nucleus) were determined from the fourth to the eighth days of culture (eighth to eleventh EGD). The number of nuclei measured in each case are shown in parentheses. See also figures 4.3 and 4.4.

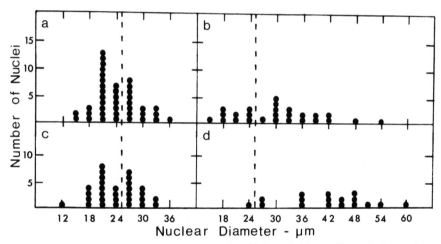

4.3 Nuclear diameters of normal and presumptive t^6/t^6 trophoblast cells on the 8th and 11th EGD. Nuclear diameters [(smallest + largest diameter)/2] were measured in figures 4.4a, c, d, and f. Diameters to the left of the vertical dotted lines fall within the range found for diploid embryonic cells measured in the same way (see Sherman, 1975).
(a) Presumptive t^6/t^6 trophoblast outgrowth, 8th EGD; (b) normal trophoblast outgrowth, 11th EGD; (c) presumptive t^6/t^6 trophoblast outgrowth, 8th EGD; (d) normal trophoblast outgrowth, 11th EGD.

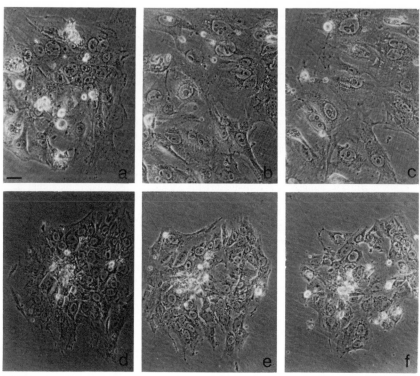

4.4 Behavior of normal and presumptive t^6/t^6 trophoblast outgrowths in culture. Fourth-day blastocysts from a $(+/+ \times +/t^6)$ cross were cultured individually. [see Sherman (1976) for details of culture conditions]. By the 4th day of culture (8th EGD), the outgrowth in (a) contained only trophoblast cells, i.e., had lost all ICM derivatives. The development of this outgrowth is shown in the 10th (b) and 11th (c) EGD. The development of a similar outgrowth from a $(+/t^6 \times +/t^6)$ cross blastocyst is shown on the 8th (d), 9th (e), and 11th (f) EGD. Scale marker in (a) = $50\mu m$.

to implantation of the blastocyst take place under normal culture conditions (see Sherman and Wudl, 1976). Consequently, it could be learned whether trophoblast cells are capable of implanting but then detach or whether they fail to attach altogether, a point not clear from the studies of Spiegelman et al. (1976) and difficult to determine in utero. Normally, trophoblast cells are much more resilient than ICM derivatives in vitro (see discussion section in Erickson and Pedersen, 1975); if trophoblast cells are selectively affected by the t^{w73} mutation (the eventual demise of t^{w73}/t^{w73} embryos in utero being due to placental failure), then it might be possible to generate long-term cultures of ICM derivatives that are genotypically t^{w73}/t^{w73}.

4.7.4 t^{w5} Embryos

A procedure has been developed (Sherman, 1976) for the culture of single blastocysts with a high survival rate after two weeks in vitro (usually 60 percent or more for wild-type blastocysts), and we have assessed the survival of t^{w5}-bearing blastocysts after several days in vitro (Wudl and Sherman, 1976). The results are clearly consistent with the idea that t^{w5} is a generalized cell lethal: blastocysts from the experimental cross ($+/t^{w5} \times +/t^{w5}$) showed a higher mortality rate than control ($+/+ \times +/t^{w5}$) blastocysts, the increment equal to the expected number of t^{w5}/t^{w5} homozygotes in the former. Furthermore, the simplest interpretation of data from studies on survival of t^{w5}-embryo-containing chimeras was that t^{w5}/t^{w5} cells are not rescued by normal neighbors, just as Mintz (1964b) found for t^{12} cells. Finally, β-glucuronidase analyses similar to those carried out on t^{12}-containing embryos (section 4.6) have established that the time of arrest of t^{w5}/t^{w5} embryonic cells is very similar in utero and in vitro (Wudl and Sherman, 1976).

4.7.5 t^{w18} Embryos

Although t^{w18}/t^{w18} embryos have not yet been studied in culture, Artzt and Bennett (1972) have investigated the behavior of presumptive t^{w18} homozygotes under the testis capsule. These authors obtained embryos on the ninth day of pregnancy and, after scoring them as t^{w18}/t^{w18} (on the basis of reduced size and absence of head folds) or normal, transplanted them to related males. Virtually all of the controls formed organized, well-developed embryomas, containing representative tissue types of the three germ layers, but only 60 percent of the putative t^{w18}/t^{w18} embryos gave rise to growths upon inspection between 28 to 40 days posttransplantation. Of those persisting, about 15 percent were scored as misclassed, i.e., having derived from retarded or abnormal non-t^{w18}/t^{w18} embryos, because they resembled tumors formed by normal embryos. The remaining tumors were quite different: all contained ectodermal derivatives, but 40 percent lacked mesodermal cell types and an equal number lacked endoderm. In fact, 2 of the 26 embryos formed tumors containing only ectodermal cell types. Furthermore, about one-third of the tumors were classified as malignant by the histological criteria of invasiveness and the presence of "primitive, undifferentiated cells." These tumors resembled neuroblastomas or medulloblastomas. Similar tumors were also obtained when eighth-day embryos were transplanted under the testis capsule. None of the malignant tumors grew as secondary subcutaneous transplants, even in antithymocyte-serum–treated hosts. Artzt and Bennett suspected that this was due to immunological considerations. Because t^{w18} has since been backcrossed onto genetically defined lines (K. Artzt, personal communication), it will be interesting to see whether transplantable tumors can be obtained.

4.7.6 T Embryos

Ephrussi (1935) dissected T/T embryos from uteri on the tenth day of gestation. By this time, mutant embryos are recognizable by the absence of somites and, indeed, of the posterior end of the body (section 4.4.7). Ephrussi showed that heart organ cultures showing contractions and slow growth could be maintained for up to two months, as long as was possible with hearts from normal embryos. Whole embryos were also maintained in culture and showed signs of growth for two weeks; the cells by that time had thus survived beyond the gestation period. Finally, embryos were fragmented into either small or large pieces. From the former were generated long-term (1 to 2 months) cultures of epithelioid and fibroblastic cells. The latter gave rise to complex organized structures, including cartilage. Ephrussi emphasized the significance of this observation, because T/T embryos die in utero prior to cartilage formation. It should be pointed out, however, that Bennett (1958) was unable to induce cartilage formation by T/T somites in coculture with tenth-day T/T neural tubes. This defect was attributed to T/T somites because T/T neural tubes could induce cartilage in somites from normal embryos or in T/T limb buds. Gluecksohn-Schoenheimer (1944) transplanted a population of eighth-day embryos from $T/+$ inter se crosses into the extraembryonic coelom of the chick. After 18–47 hr, she sectioned the embryos. An appropriate fraction (25 percent) of these embryos showed abnormalities characteristic of T/T embryos in utero: absence of posterior body structures and typically abnormal allantois.

On the basis of these experiments, it can be concluded that the abnormal developmental pattern of T/T embryos is "determined" several hours prior to the time that it takes place. On the other hand, the T mutation appears to be quite selective in the cell types that it affects, unlike t^{w5} (section 4.7.4). Some T/T cells not only can survive substantially longer in vitro than they do in utero, but also can differentiate to a greater extent. These studies provide compelling evidence of the demands that the uterus imposes upon the embryo: abnormal development of only a fraction of the cells can lead to prompt rejection and degeneration of the entire embryo in vivo.

4.8 Conclusions

A model that properly describes the complex T region and explains satisfactorily the properties of its mutations has yet to be put forward, some fifty years after work on this region of the chromosome began. This is partly because certain critical information about the T region is unavailable, but it also stems from the problem of having to reconcile some of the apparently contradictory facts. Thus, a model describing the nature and number of genes present in the T region should explain (1) suppression of recombination in most, but not all, t lethals and semilethals; (2) separation of tail modification effects from lethality and sterility; and (3) failure of recombination to give rise to a wild-type genotype. Explaining cohesively the effects of mutation in the T region requires consideration not only of perturbations of developing embryos (lethality in homozygotes during various stages in development, semilethality, and complementation in compound heterozygotes), but also of alterations of sperm properties and function (high transmission frequency of some t alleles, but normal or even low transmission frequency of others; sterility in compound heterozygous and homozygous semilethal males; and changes in the sperm surface detected by surface antigen studies). Finally, once all these properties have been satisfactorily accounted for, the next step will be to understand the interplay of genetic background on function of T-complex genes, including some effects that are apparently maternal (Gluecksohn-Waelsch and Erickson, 1970; Johnson, 1974; Mickova and Iványi, 1975). Indeed, the more one learns about the T region, the more complicated the story becomes!

4.8.1 What Are the Products of the T Complex?

The most popular theory to date is that at least some genes in the T region specify surface antigens present on embryos and on sperm (but presumably on no other adult cell types). This hypothesis draws support from the immunological studies described in section 4.5, although better characterization of altered antigens on the surface of t/t embryos is to be desired. It would not be difficult to imagine how such alterations in the cell surface could dramatically disrupt embryogenesis in t/t homozygotes. Bear in mind, however, that $+/t$ heterozygotes appear to develop in a completely normal manner, even though fully half of their T-complex-specified surface antigens are presumably aberrant. Once again, although altered sperm surface antigens could be responsible for transmission ratio distortion in $+/t$ males *or* sterility in t^x/t^y males, it is a different matter to explain how the same sperm alterations could cause both effects. Finally, the map of the T region constructed by Artzt et al. (1974; see section 4.5.1) on the basis of surface antigen studies can be reconciled in most respects with that proposed by Lyon and Meredith (1964b; see section 4.2.2) by means of genetic considerations: (1) by inserting a length of chromosome involving recombination suppression between the T and t^x t^y, t^z, etc., loci in the map of Artzt et al., and (2) by splitting the "lethality" segment of the T region in Lyon and Meredith's model into six "sites," one for each of the complementation groups. However, even this combined model would still leave the problem of explaining how the t^{w5} mutation, for example, could be characterized by physical alterations of the chromosome at (1) the $+^{t^{w5}}$ site (giving rise to lethality in homozygotes), (2) the $+^T$ site (causing modification of tail size with T), and (3) some site between the T and the t loci in order to suppress recombination, all without altering the other intervening sites, such as $+^{t^{12}}$, $+^{t^9}$, etc. (because these alleles presumably would still be expressing their wild-type surface antigenic products).

Largely on the basis of studies by Hillman and her co-workers, a second hypothesis has been put forward to explain the effects of T-region mutations, namely a disturbance of some essential cellular metabolic function (see Hillman, 1975). Such a lesion could disrupt normal intermediary metabolism and affect both embryogenesis (e.g., Ginsberg and Hillman, 1975) and sperm function (e.g., Ginsberg and Hillman, 1974). Thus, embryogenesis would be disrupted at various stages in development that require particularly high levels of energy expenditure and/or cellular growth. The severity of alteration of the T-complex-gene product(s) would determine the stage that an embryo homozygous for a particular *t* allele could reach before succumbing to the lesion. Sperm function could be acted upon in a way that would explain transmission ratio distortion, e.g., by superior aerobic metabolism in *t* sperm.

The two hypotheses are not as incompatible as they might seem to be, because altered surface antigens can easily be responsible for perturbing cellular metabolism, or vice versa. Thus, the primary role of the *t*-specified surface antigens might not be to facilitate cell-cell recognition and interaction, as most often suggested (e.g., Artzt and Bennett, 1975), but to perform some other function, e.g., in metabolite uptake or hormone binding. A precedent for such a possibility might be drawn from the apparent relationship between H-2 genotype and cyclic AMP levels (Meruelo and Edidin, 1975). Furthermore, as Meruelo and Edidin (1975) have pointed out, heterozygotes at the H-2 complex can have cyclic AMP levels characteristic of the parent with the higher cyclic AMP values, rather than showing intermediate levels. By analogy, heterozygous *t*-mutant embryos might also show such an effect in resembling the wild-type parent. Bennett (1975b) has, in fact, raised the possibility that membrane elements, presumably including those specified by the T complex, "may serve as receptor sites for humoral, cellular or 'microenvironmental' factors."

One should also consider the converse relationship between a metabolic aberration and altered surface antigens, namely, that a primary effect disrupting normal metabolic processes might lead secondarily to interference with the production of characteristic surface proteins. For example, if the lesion were at the level of some enzyme involved in intermediary metabolism, the carbohydrate moieties of surface glycoproteins might be altered, and this change could then lead to altered antigenicity of these proteins (Boyd, 1966). In this case, the production of wild-type enzyme from the nonmutant gene would explain normalcy of heterozygous embryos. Hillman and co-workers have reported aberrations in some t-mutant embryos long before signs of overt degeneration become apparent (Hillman et al., 1970; Hillman and Hillman, 1975; Hillman and Nadijcka, 1975; Ginsberg and Hillman, 1975). But, because it is not yet known just when various T-complex-related surface antigens might appear on embryonic cells, we cannot say whether metabolic abnormalities in t-mutant embryos precede or follow surface alterations.

Finally, it has been mentioned that the T region appears to occupy a large segment of chromosome 17. There is no reason to suppose that genes within this region could not specify *both* surface antigens and particular enzymes required for normal cell metabolism.

4.8.2 t Mutants: Organizational or Cellular Lethals?

Section 4.4.10 discussed the developmental failure and metabolic barrier hypotheses. To reiterate briefly, the former suggests that t/t (or T/T) embryos die because one or a few cell types lack information necessary to complete a critical differentiative step. The result in the demanding uterine milieu would be failure and subsequent degeneration of the entire embryo, a condition designated "organizational lethality." In contrast, the latter hypothesis stipulates that homozygous t embryos might fail to reach a particular developmental stage due to some metabolic block in the production of required energy or metabolites. It is feasible that *all* cells in the embryo would suffer from this lesion and the result would be embryonic death due to generalized cell lethality.

In support of the organizational lethality proposal are the observations of disturbed cell-cell interactions in t^{w18}/t^{w18} and T/T embryos (see sections 4.4.5 and 4.4.7), as well as the studies suggesting that some cell types in t^{w18}/t^{w18} and T/T embryos die in ectopic sites or in vitro, whereas others survive for extended periods of time (section 4.7). On the other hand, in vitro studies indicate that t^6 and t^{w5} are generalized cell lethal mutations (section 4.7). (One cannot draw any conclusions from t^{12}/t^{12} embryos in this case because there is only a single cell type at the time of death.) Furthermore, although most recent reviews indicate that aberrations are initially restricted to the ectoderm layer in t^6 and t^{w5} homozygous embryos, the original analyses clearly indicated that endoderm cells failed to develop into the characteristic phenotypes (secretary columnar cells in the extraembryonic region and squamous cells in the embryonic region; see sections 4.4.2 and 4.4.4). In fact, Bennett and Dunn (1958) initially suggested that failure of embryonic ectoderm cells in t^{w5}/t^{w5} embryos was an effect secondary to the inability of endoderm cells to properly assume their nutritive roles.

We shall assume, therefore, that t^6 and t^{w5} homozygous embryos are generalized cell lethals although this is not the case for t^{w18} and T homozygotes. It must be, then, that some cells in the latter types escape the adverse consequences of a t (or T) mutation (1) because t^{w18} and T are not so severe in their effects as are t^6 and t^{w5} or (2) because t (or T) mutations affect only relatively undifferentiated cells and some cells reach a degree of differentiation advanced enough to allow them to escape the adverse effects of altered t^{w18}- or T-gene products by the time in development at which those genes are required. Note that neither alternative stipulates a particular time of *activation* of the T and t genes. If alternative (1) is correct, then all t (and T) genes could be, but need not be, expressed at the same early time in development so that embryos carrying the more severe mutations capitulate earlier than those bearing milder versions. According to possibility (2), either the t (and T) alleles could be expressed shortly before overt signs of abnormality are observed or all t genes might be active throughout embryogenesis but their gene products would not be required until developmental arrest takes place.

Whether alternative (1) or (2) is correct, if it turns out that the expression of the various t- (and T-) gene products occurs at progressively later times in development (a particular time of expression characteristic of each of the six classes of t alleles), then the differential susceptibility of specific cell types could be attributed to their relative differentiative states. Thus, in t^{w5}, embryonic ectoderm might be more dramatically affected (pycnosis) than visceral endoderm (developmental arrest) because the latter has progressed farther along in its differentiative pathway. Similarly, neural derivatives might be more "mature" than mesodermal ones at the time of expression of the t^{w18} gene, and the result would be escape from the effect by neural derivatives but not by mesodermal ones, as found in ectopic growths by Artzt and Bennett (1972). This pattern appears not to apply, however, to the t^{w73} mutation, wherein trophoblast cells, which seem always to be least susceptible to perturbation, at least in culture (see discussion in Pedersen and Erickson, 1975), is primarily affected (section 4.4.3). However, this mutation is clearly a rare one. Only a single example of this kind has been found, and its properties must be assessed further.

Two other points should be made about alternatives (1) and (2). The first is that both alternatives would be equally acceptable if the primary t effect were on surface antigens or on some enzyme involved in general cell metabolism. Second, whichever of the alternatives is correct, the observations described suggest that the early-to-mid-eighth-day embryo is the first to contain cells, principally neural ectoderm, which for one reason or another are immune to the effects of t mutations. Perhaps embryos at this stage should be studied more carefully with this point in mind.

4.8.3 Are t Mutations Truly "Developmental" Mutations?

We began this review by expressing some uncertainty as to whether t mutations were indeed developmental in nature. If the primary gene products of the T complex were surface antigens and if these antigens became expressed and/or functional at various times in development, then the T-complex genes would clearly be developmental in nature. If, on the other hand, the T complex specified only one or more enzymes that, if altered, resulted in a block of embryogenesis as a consequence of a running-down in general cell metabolism, then as far as we are concerned, the T-complex genes would not warrant the status of developmental genes. Of course, this argument is largely semantic. The critical fact is that a complete unraveling of the mysteries of the complex T region will undoubtedly provide an improved understanding of control mechanisms in mammalian embryogenesis.

Acknowledgments This manuscript has benefited substantially from discussions, suggestions, and exchanges of ideas with Nina Hillman. We want to thank her as well for allowing us to describe the results of unpublished experiments underway in her laboratory. We are also grateful to Karen Artzt, Dorothea Bennett, Hedwig Jakob, and Martha Spiegelman for providing preprints of their work. Finally, we thank Pat Perkowski for typing this manuscript.

References

ARTZT, K, and BENNETT, D.
(1972).
A genetically caused embryonal ectodermal tumor in the mouse. J. Nat.
Cancer Inst. *48*, 141–158.

ARTZT, K., and BENNETT, D.
(1975).
Analogies between embryonic (T/t)
antigens and adult major histocompatibility (H-2) antigens. Nature *226*,
545–547.

ARTZT, K., and BENNETT, D.
(1977).
Serological analysis of sperm of anti-
genically cross-reacting *T/t*-haplo-
types and their interactions.
Immunogenetics, in press.

ARTZT, K., BENNETT, D., and
JACOB, F. (1974).
Primitive teratocarcinoma cells express a differentiation antigen specified by a gene at the T-locus in the
mouse. Proc. Nat. Acad. Sci. USA
71, 811–814.

ARTZT, K., DUBOIS, P.,
BENNETT, D., CONDAMINE, H.,
BABINET, C., and JACOB, F.
(1973).
Surface antigens common to mouse
cleavage embryos and primitive ter-
atocarcinoma cells in culture. Proc.
Nat. Acad. Sci. USA *70*, 2988–2992.

BABINET, C., CONDAMINE H.,
FELLOUS, M., GACHELIN, G.,
KEMLER, R., and JACOB, F.
(1975).
Expression of a cell surface antigen
common to primitive mouse teratocarcinoma cells and cleavage embryos
during embryogenesis and spermatogenesis. *In* Teratomas and Differentiation, M. I. Sherman and D. Solter,
eds. (New York: Academic Press),
pp. 101–107.

BARLOW, P. W., and SHERMAN,
M. I. (1972).
The biochemistry of differentiation of
mouse trophoblast: studies on polyploidy. J. Embryol. Exp. Morph. *27*,
447–465.

BENNETT, D. (1958).
In vitro study of cartilage inducation
in T/T mice. Nature *181*, 1286.

BENNETT, D. (1964).
Embryological effects of lethal alleles
in the *t*-region. Science *144*, 263–267.

BENNETT, D. (1965).
The karyotype of the mouse, with
identification of a translocation. Proc.
Nat. Acad. Sci. USA *53*, 730–737.

BENNETT, D. (1975a).
The T-locus of the mouse. Cell *6*,
441–454.

BENNETT, D. (1975b).
T-locus mutants: suggestions for the
control of early embryonic organiza-
tion through cell surface components.
In The Early Development of Mam-
mals, M. Balls and A. E. Wild, eds.
(London: Cambridge University
Press), pp. 207–218.

BENNETT, D., BADENHAUSEN,
S., and DUNN, L. C. (1959b)
The embryological effects of four late-
lethal *t* alleles in the mouse, which
affect the neural tube and skeleton. J.
Morph. *105*, 105–144.

BENNETT, D., BOYSE, E. A., and OLD, L. J. (1972a).
Cell surface immunogenetics in the study of morphogenesis. *In* Proc. Third Lepetit Colloquium, L. G. Silvestri, ed.(Amsterdam: North-Holland), pp. 247–262.

BENNETT, D., and DUNN, L. C. (1958).
Effects on embryonic development of a group of genetically similar lethal alleles derived from different populations of wild house mice. J. Morph. *103*, 135–158.

BENNETT, D., and DUNN, L. C. (1960).
A lethal mutant (t^{w18}) in the house mouse showing partial duplications. J. Exp. Zool. *143*, 203–219.

BENNETT, D., and DUNN, L. C. (1964).
Repeated occurrences in the mouse of lethal alleles of the same complementation group. Genetics *49*, 949–958.

BENNETT, D., and DUNN, L. C. (1967).
Studies of effects of *t*-alleles in the house mouse on spermatozoa. I. Male sterility effects. J. Reprod. Fertil. *13*, 421–428

BENNETT, D., and DUNN, L. C. (1969)
Genetical and embryological comparisons of semilethal *t*-alleles from wild mouse populations. Genetics *61*, 411–422.

BENNETT, D., and DUNN, L. C. (1971).
Transmission ratio distorting genes on chromosome IX and their interactions. *In* Proc. Symp. Immunogenetics of the H-2 System, A. Lengerova

and M. Vojteskova, eds. (Basel, Switzerland: Karger), pp. 90–103.

BENNETT, D., DUNN, L. C., and ARTZT, K. (1976).
Genetic change in mutations at the *T/t*-locus. Genetics *83*, 361–372.

BENNETT, D., DUNN, L. C., and BADENHAUSEN, S. (1959a).
A second group of similar lethals in populations of wild house mice. Genetics *44*, 795–802.

BENNETT, D., DUNN, L. C., SPIEGELMAN, M., ARTZT, K., COOKINGHAM, J., and SCHERMERHORN, E. (1975).
Observations on a set of radiation-induced dominant *T*-like mutations in the mouse. Genet. Res. *26*, 95–108.

BENNETT, D., GOLDBERG, E., DUNN, L. C., and BOYSE, E. A. (1972b).
Serological detection of a cell-surface antigen specified by the T (Brachyury) mutant gene in the house mouse. Proc. Nat. Acad. Sci. USA *69*, 2076–2080.

BOYD, W. C. (1966).
Fundamentals of Immunology. Chapter 4. New York: Interscience.

BRADEN, A. W. H. (1958).
Influence of time of mating on the segregation ratio of alleles at the T locus in the house mouse. Nature *181*, 786–787.

BRADEN, A. W. H. (1972).
T-Locus in mice: Segregation distortion and sterility in the male. *In* Proc. Int. Symp. The Genetics of the Spermatozoon, R. A. Beatty and S. Gluecksohn-Waelsch, eds. (Copenhagen: Bogtrykkiert Forum), pp. 289–305.

BRADEN, A. W. H., and
GLUECKSOHN-WAELSCH, S.
(1958).
Further studies of the effect of the T
locus in the house mouse on male fertility. J. Exp. Zool. *138*, 431–452.

BRYSON, V. (1944).
Spermatogenesis and fertility in *Mus
musculus* as affected by factors at the
T locus. J. Morph. *74*, 131–187.

BUC-CARON, M. H., GACHELIN,
G., HOFNUNG, M., and JACOB, F.
(1974).
Presence of a mouse embryonic anti-
gen on human spermatozoa. Proc.
Nat. Acad. Sci. USA *71*, 1730–1733.

CALARCO, P. G., and BROWN,
E. H. (1968).
Cytological and ultrastructural comparisons of t^{12}/t^{12} and normal mouse
morulae. J. Exp. Zool. *168*, 169–186.

CARTER, T. C., and PHILLIPS,
R. S. (1950).
Three recurrences of mutants in the
house mouse. J. Heredity *41*, 252.

CHESLEY, P. (1932).
Lethal action in the short-tailed muta-
tion in the house mouse. Proc. Soc.
Exp. Biol. *29*, 437–438.

CHESLEY, P. (1935).
Development of the short-tailed mu-
tant in the house mouse. J. Exp.
Zool. *70*, 429–459.

CHESLEY, P., and DUNN, L. C.
(1936).
The inheritance of taillessness
(Anury) in the house mouse. Genetics
21, 525–536.

DOBROVOLSKAIA-ZAVADSKAIA,
N. (1927).
Sur la mortification spontanée de la
queue chez le souris nouveau-née et

sur l'éxistence d'un charactère
(facteur) héréditaire "non-viable."
C. R. Soc. Biol. *97*, 114–116.

DOBROVOLSKAIA-ZAVADSKAIA,
N., and KOBOZIEFF, N. (1927).
Sur la reproduction des souris
anoures. C. R. Soc. Biol. *97*, 116–118.

DOBROVOLSKAIA-ZAVADSKAIA,
N., and KOBOZIEFF, N. (1932a).
Terminaison suspelvienne de la
colonne vertébrale au niveau de tronc
chez les souris adultes sans queue.
C. R. Soc. Biol. *104*, 969–971.

DOBROVOLSKAIA-ZAVADSKAIA,
N., and KOBOZIEFF, N. (1932b).
Les souris anoures et la queue fili-
forme qui se reproduisent entre elles
sans disjunction. C. R. Soc. Biol. *110*,
782–784.

DOOHER, G. B., and BENNETT, D.
(1974).
Abnormal microtubular systems in
mouse spermatids associated with a
mutant gene at the T-locus. J. Embryol. Exp. Morph. *32*, 749–761.

DUNN, L. C. (1937).
A third lethal in the T (Brachy) series
in the house mouse. Proc. Nat. Acad.
Sci. USA *23*, 474–477.

DUNN, L. C. (1939).
The inheritance of taillessness (anury)
in the house mouse. III. Taillessness
in the balanced lethal line 19.
Genetics *24*, 728–731.

DUNN, L. C. (1956).
Analysis of a complex gene in the
house mouse. Cold Spring Harbor
Symp. Quant. Biol. *21*, 187–195.

DUNN, L. C. (1957).
Studies of the genetic variability in
populations of wild house mice. II.

Analysis of eight additional alleles at locus *T*. Genetics *42*, 299–311.

DUNN, L. C. (1960)
Variations in the transmission ratios of alleles through egg and sperm in *Mus musculus*. Am. Naturalist *94*, 385–393.

DUNN, L. C. (1964).
Abnormalities associated with a chromosome region in the mouse. Science *144*, 260–263.

DUNN, L. C., and BENNETT, D. (1968).
A new case of transmission ratio distortion in the house mouse. Genetics *61*, 570–573.

DUNN, L. C., and BENNETT, D. (1969).
Studies of effects of t-alleles in the house mouse on spermatozoa. II. Quasi-sterility caused by different combinations of alleles. J. Reprod. Fertil. *20*, 239–246.

DUNN, L. C., and BENNETT, D. (1971a).
Lethal alleles near locus T in house mouse populations on the Jutland Peninsula, Denmark. Evolution *25*, 451–453.

DUNN, L. C., and BENNETT, D. (1971b).
Further studies of a mutation (*low*) which distorts transmission ratios in the house mouse. Genetics *67*, 543–558.

DUNN, L. C., BENNETT, D., and BEASLEY, A. B., (1962).
Mutation and recombination in the vicinity of a complex gene. Genetics *47*, 285–303.

DUNN, L. C., BENNETT, D., and COOKINGHAM, J. (1973).
Polymorphisms for lethal alleles in European populations of *Mus musculus*. J. Mammal. *54*, 822–830.

DUNN, L. C., and CASPARI, E. (1945).
A case of neighboring loci with similar effects. Genetics *30*, 543–568.

DUNN, L. C., and GLUECKSOHN-SCHOENHEIMER, S. (1943).
Tests for recombination amongst three lethal mutations in the house mouse. Genetics *28*, 29–40.

DUNN, L. C., and GLUECKSOHN-SCHOENHEIMER, S. (1950).
Repeated mutations in one area of a mouse chromosome. Proc. Nat. Acad. Sci. USA, *36*, 233–237.

DUNN, L. C., and GLUECKSOHN-WAELSCH, S. (1953a).
Genetic analysis of seven newly discovered mutant alleles at locus T in the house mouse. Genetics *38*, 261–271.

DUNN, L. C., and GLUECKSOHN-WAELSCH, S. (1953b).
The failure of a t-allele (t^3) to suppress crossing over in the mouse. Genetics *38*, 512–517.

DUNN, L. C., and MORGAN, W. C. (1953).
Alleles at a mutable locus found in populations of wild house mouse (*Mus musculus*). Proc. Nat. Acad. Sci. USA *39*, 391–402.

DUNN, L. C., and SUCKLING, J. (1956).
Studies of the genetic variability in wild populations of house mice. 1. Analysis of seven alleles at locus T. Genetics *41*, 344–352.

EDIDIN, M., PATTHEY, H. L., McGUIRE, E. J., and SHEFFIELD, W. L. (1971).
An antiserum to "embryoid body" tumor cells that reacts with normal mouse embryos. *In* Conference and Workshop on Embryonic and Fetal Antigens in Cancer, N. G. Anderson and J. H. Coggin, eds. (Oak Ridge, Tenn.: Oak Ridge National Laboratory), pp. 239–247.

EL-SHERSHABY, A. M. and HINCHLIFFE, J. R. (1974).
Cell redundancy in the zona-intact preimplantation mouse blastocyst: A light and electron microscope study of dead cells and their fate. J. Embryol. Exp. Morph. *31*, 643–654.

EPHRUSSI, B. (1935).
The behavior in vitro of tissues from lethal embryos. J. Exp. Zool. *70*, 197–204.

ERICKSON, R. P. (1973).
Haploid gene expression versus meiotic drive: the relevance of intercellular bridges during spermatogenesis. Nature New Biol. *243*, 210–212.

ERICKSON, R. P., BETLACH C. J., and EPSTEIN, C. J. (1974).
Ribonucleic acid and protein metabolism of t^{12}/t^{12} embryos and T/t^{12} spermatozoa. Differentiation *2*, 203–209.

ERICKSON, R. P., and PEDERSEN, R. A. (1975).
In vitro development of t^6/t^6 embryos. J. Exp. Zool. *193*, 377–384.

FELLOUS, M., and DAUSSET, J. (1973).
Histocompatibility antigens on human spermatozoa. *In* Immunology of Reproduction, K. Bratanov, ed. (Sofia, Bulgaria: Bulgarian Acad. Sci. Press), pp. 332–350.

FELLOUS, M., GACHELIN, G., BUC-CARON, M. H., DUBOIS, P., and JACOB, F. (1974).
Similar location of an early embryonic antigen on mouse and human spermatozoa. Dev. Biol. *41*, 331–337.

FOREJT, J. (1976).
Spermatogenic failure of translocation heterozygotes affected by H-2-linked gene in mouse. Nature *260*, 143–145.

GINSBERG, L., and HILLMAN, N. (1974).
Meiotic drive in t^n-bearing mouse spermatozoa: a relationship between aerobic respiration and transmission frequency. J. Reprod. Fertil. *38*, 157–163.

GINSBERG, L., and HILLMAN, N. (1975).
ATP metabolism in t^n/t^n mouse embryos. J. Embryol. Exp. Morph. *33*, 715–723.

GLUECKSOHN-SCHOENHEIMER, S. (1938a).
The development of two tailless mutants in the house mouse. Genetics *25*, 573–584.

GLUECKSOHN-SCHOENHEIMER, S. (1938b).
Time of death of lethal homozygotes in the *T* (Brachyury) series of the mouse. Proc. Soc. Exp. Biol. Med. *39*, 267–268.

GLUECKSOHN-SCHOENHEIMER, S. (1940).
The effect of an early lethal (t^0) in the house mouse. Genetics *25*, 391–400.

GLUECKSOHN-SCHOENHEIMER, S. (1944).
The development of normal and homozygous brachy (T/T) mouse embryos in the extraembryonic coelom

of the chick. Proc. Nat. Acad. Sci.
USA *30*, 134–140.

GLUECKSOHN-SCHOENHEIMER,
S., SEGAL, R., and FITCH, N.
(1950).
Embryological tests of genetic male
sterility in the house mouse. J. Exp.
Zool. *113*, 621–632.

GLUECKSOHN-WAELSCH, S.
(1963).
Lethal genes and analysis of differentiation. Science *142*, 1269–1276.

GLUECKSOHN-WAELSCH, S.
(1972).
Models for mechanisms of segregation
distortion. *In* Proc. Int. Symp. The
Genetics of the Spermatozoon, R. A.
Beatty and S. Gluecksohn-Waelsch,
eds. (Copenhagen: Bogtrykkiert
Forum), pp. 306–309.

GLUECKSOHN-WAELSCH, S., and
ERICKSON, R. P. (1970).
The *T*-locus of the mouse: implica-
tions for mechanisms of development.
Current Topics Dev. Biol. *5*, 281–316.

GLUECKSOHN-WAELSCH, S., and
ERICKSON, R. P. (1971).
Cellular membranes: a possible link
between H-2 and T-locus effects. *In*
Proc. Symp. Immunogenetics of the
H-2 System, A. Lengerova and M.
Vojteskova, eds. (Basel, Switzerland:
Karger), pp. 120–122.

GOLDBERG, E. H., AOKI, T.,
BOYSE, E. A., and BENNETT, D.
(1970).
Detection of H-2 antigens on mouse
spermatozoa by the cytotoxicity test.
Nature *228*, 570–572.

GOLDBERG, E. H., BOYSE, E. A.,
BENNETT, D., SCHEID, M. and
CARSWELL, E. A. (1973).
Serological demonstration of H-Y
(male) antigen on mouse sperm.
Nature *232*, 478–480.

GRÜNEBERG, H. (1952).
The Genetics of the Mouse. 2nd ed.
The Hague: Martinus Nijhoff.

GRÜNEBERG, H. (1958).
Genetical studies on the skeleton of
the mouse. XXIII. The development
of Brachyury and Anury. J. Embryol.
Exp. Morph. *6*, 424–443.

HÅKANSSON, S., HEYNER, S.,
SUNDQVIST, K. G., and
BERGSTROM, S. (1975).
The presence of paternal H-2 antigens
on hybrid mouse blastocysts during
experimental delay of implantation
and the disappearance of these antigens after onset of implantation. Int.
J. Fertil. *20*, 137–140.

HAMMERBERG, C., and KLEIN, J.
(1975a).
Linkage relationships of markers on
chromosome 17 of the house mouse.
Genet. Res. *26*, 203–211.

HAMMERBERG, C. and KLEIN, J.
(1975b).
Evidence for postmeiotic effect of *t*
factors causing segregation distortion
in mouse. Nature *253*, 137–138.

HILLMAN, N. (1972).
Autoradiographic studies of t^{12}/t^{12}
mouse embryos. Am. J. Anat. *134*,
411–424.

HILLMAN N. (1975).
Studies of the *T*-locus. *In* The Early
Development of Mammals, M. Balls
and A. E. Wild, eds. (London: Cambridge University Press), pp. 189–206.

HILLMAN, N., and HILLMAN, R. (1975).
Ultrastructural studies of t^{w32}/t^{w32} mouse embryos. J. Embryol. Exp. Morph. *33*, 685–695.

HILLMAN, N., HILLMAN, R., and WILEMAN, G. (1970).
Ultrastructural studies of cleavage t^{12}/t^{12} mouse embryos. Am. J. Anat. *128*, 311–340.

HILLMAN, N., and TASCA, R. J. (1973).
Synthesis of RNA in t^{12}/t^{12} mouse embryos. J. Reprod. Fertil. *33*, 501–506.

JOHNSON, D. R. (1974).
Hairpin-tail: a case of post-reductional gene action in the mouse egg? Genetics *76*, 795–805.

KEMLER, R., BABINET, C., CONDAMINE, H., GACHELIN, G., GUENET, J. L., and JACOB, F. (1976).
Embryonal carcinoma antigen and the T/t locus of the mouse. Proc. Nat. Acad. Sci. USA *73*, 4080–4084.

KLEIN, J. (1971).
Cytological identification of the chromosome carrying the IXth linkage group (including H-2) in the house mouse. Proc. Nat. Acad. Sci. USA *68*, 1594–1597.

KLEIN, J. (1975).
Biology of the Mouse Histocompatibility-2 Complex. Berlin: Springer-Verlag.

KLEIN, J., and RASKA, K. (1968).
Deficiency of "ribosomal" DNA in t^{12} mutant mice. Proc. XII Inter. Cong. Genetics *1*, 149.

KLYDE, B. J. (1970).
Selection in mice for change in viability of a semilethal genotype at the T locus. J. Heredity *61*, 39–43.

KOBOZIEFF, N. (1935).
Recherches morphologiques et génétiques sur l'anourie chez la souris. Bull. Biol. *69*, 267–398.

LYON, M. F. (1956).
Hereditary hair loss in the tufted mutant of the house mouse. J. Heredity *47*, 101–103.

LYON, M. F. (1959).
A new dominant T-allele in the house mouse. J. Heredity *50*, 140–142.

LYON, M. F. (1960).
Effect of X-rays on the mutation of t-alleles in the mouse. Heredity *14*, 247–252.

LYON, M. F., GLENISTER, P. H., and HAWKER, S. B. (1972).
Do the H-2 and T-loci of the mouse have a function in the haploid phase of sperm? Nature *240*, 152–153.

LYON, M. F., and MEREDITH, R. (1964a)
Investigations of the nature of t-alleles in the mouse. I. Genetic analysis of a series of mutants derived from a lethal allele. Heredity *19*, 301–312.

LYON, M. F., and MEREDITH, R. (1964b).
Investigations of the nature of t-alleles in the mouse. II. Genetic analysis of an unusual mutant allele and its derivatives. Heredity *19*, 313–325.

LYON, M. F., and MEREDITH, R. (1964c)
Investigations of the nature of t-alleles in the mouse. III. Short tests of some further mutant alleles. Heredity *19*, 327–330.

LYON, M. F., and PHILLIPS, R. J. S. (1959).
Crossing-over in mice heterozygous for t-alleles. Heredity *13*, 23–32.

MERUELO, D., and EDIDIN, M. (1975).
Association of mouse liver adenosine 3′:5′-cyclic monophosphate (cyclic AMP) levels with *Histocompatibility-2* genotype. Proc. Nat. Acad. Sci. USA *72*, 2644–2648.

MICKOVÁ, M., and IVÁNYI, P. (1974).
Sex-dependent and H-2-linked influence on expressivity of the Brachyury gene in mice. J. Heredity *65*, 369–372.

MINTZ, B. (1963).
Growth *in vitro* of t^{12}/t^{12} lethal mutant mouse eggs. Am. Zoologist *3*, 550–551.

MINTZ, B. (1964a).
Gene expression in the morula stage of mouse embryos, as observed during development of t^{12}/t^{12} lethal mutants *in vitro*. J. Exp. Zool. *157*, 267–272.

MINTZ, B. (1964b).
Formation of genetically mosaic mouse embryos, and early development of "lethal (t^{12}/t^{12})-normal" mosaics. J. Exp. Zool. *157*, 273–292.

MINTZ, B. (1971).
Control of embryo implantation and survival. Advan. Biosciences *6*, 317–340.

MOSER, G. C., and GLUECKSOHN-WAELSCH, S. (1967).
Developmental genetics of a recessive allele at the complex T-locus in the mouse. Dev. Biol. *16*, 564–576.

NADIJCKA, M., and HILLMAN, N. (1975a).
Studies of t^6/t^6 mouse embryos. J. Embryol. Exp. Morph. *33*, 687–713.

NADIJCKA, M., and HILLMAN, N. (1975b).
Autoradiographic studies of t^n/t^n mouse embryos. J. Embryol. Exp. Morph. *33*, 725–730.

NICOLAS, J. F., AVNER, P., GAILLARD, J., GUENET, J. L., JAKOB, H., and JACOB, F. (1976).
Cell lines derived from teratocarcinomas. Cancer Res. *36*, 4224–4231.

OLDS, P. J. (1970).
Effect of the T locus on sperm distribution in the house mouse. Biol. Reprod. *2*, 91–97.

OLDS, P. J. (1971).
Effect of the T locus on fertilization in the house mouse. J. Exp. Zool. *177*, 417–433.

SEARLE, A. G. (1966).
Curtailed, a new dominant T-allele in the house mouse. Genet. Res. *7*, 86–95.

SHERMAN, M. I. (1975).
The role of cell-cell interaction during early mouse embryogenesis. *In* The Early Development of Mammals, M. Balls and A. E. Wild, eds. (London: Cambridge University Press), pp. 145–165.

SHERMAN, M. I. (1976).
Generation of cell lines from preimplantation mouse embryos. *In* Tissue Culture Association Manual, V. J. Evans, V. P. Perry, and M. M. Vincent, eds., vol. 1 (Rockville, Md.: TCA, Inc.), pp. 199-201.

SHERMAN, M. I., ATIENZA, S. B.,
SALOMON, D. S., and WUDL, L. R.
(1977).
Progesterone formation and metabo-
lism by blastocysts and trophoblast
cells in vitro. *In* Development in
Mammals, M. H. Johnson, ed., vol. 2
(Amsterdam: Associated Scientific
Publishers), in press.

SHERMAN, M. I., and WUDL, L. R.
(1976).
The implanting mouse blastocyst. *In*
The Cell Surface in Animal Development, G. Poste and G. L. Nicolson,
eds. (Amsterdam: Associated
Scientific Publishers), pp. 81–125.

SILAGI, S. (1962).
A genetical and embryological study
of partial complementation between
lethal alleles at the T locus of the
house mouse. Dev. Biol. *5*, 35–67.

SILAGI, S. (1963).
Some aspects of the relationship of
RNA metabolism to development in
normal and mutant mouse embryos
cultivated *in vitro*. Exp. Cell Res. *32*,
149–152.

SMITH, L. J. (1956).
A morphological and histochemical investigation of a preimplantation lethal
(t^{12}) in the house mouse. J. Exp.
Zool. *132*, 51–84.

SNELL, G. D., and STEVENS, L. C.
(1966).
Early embryology. *In* Biology of the
Laboratory Mouse, E. L. Green, ed.
(New York: McGraw-Hill),
pp. 205–245.

SPIEGELMAN, M. (1976).
Electron microscopy of cell associations in T-locus Mutants. *In* Embryogenesis in Mammals. K. Elliott

and M. O'Connor, eds. (Amsterdam:
Associated Scientific Publishers),
pp. 199–226.

SPIEGELMAN, M., ARTZT, K., and
BENNETT, D., (1976).
Embryological study of a T/t locus
mutation (t^{w73}) affecting trophecto-
derm development. J. Embryol. Exp.
Morph. *36*, 373–381.

SPIEGELMAN, M., and BENNETT,
D. (1974).
Fine structural study of cell migration
in the early mesoderm of normal and
mutant mouse embryos (T-locus:
t^9/t^9). J. Embryol. Exp. Morph. *32*,
723–738.

SUZUKI, D. T. (1970).
Temperature-sensitive mutations in
Drosophila melanogaster. Science
170, 695–706.

THEILER, K. (1972).
The House Mouse. Berlin: Springer-
Verlag.

VAN VALEN, P. (1964).
Genetic studies on t^{w18} and t^9: lack of
complementation in t^{w18}/t^9 embryos.
Ph.D. dissertation, Columbia
University.

VITETTA, E. S., ARTZT, K.,
BENNETT, D., BOYSE, E. A., and
JACOB, F. (1975).
Structural similarities between a product of the T/t-locus isolated from
sperm and teratoma cells, and H-2
antigens isolated from splenocytes.
Proc. Nat. Acad. Sci. USA *72*,
3215–3219.

VOJTISKOVA, M., POLACKOVA,
M., and POKORNA, A. (1969).
Histocompatibility antigens on mouse
spermatozoa. Folia Biol. *15*, 322–332.

WOMACK, J. E., and RODERICK,
T. H. (1974).
T-alleles in the mouse are probably
not inversions. J. Heredity 65,
308–310.

WUDL, L. R., and SHERMAN, M. I.
(1976).
In vitro studies of mouse embryos
bearing mutations at the T locus: t^{w5}
and t^{12}. Cell 9, 523–531.

YANAGISAWA, K., BENNETT, D.,
BOYSE, E. A., DUNN, L. C., and
DIMEO, A. (1974a).
Serological identification of sperm
antigens specified by lethal t-alleles in
the mouse. Immunogenetics 1, 57–67.

YANAGISAWA, K., DUNN, L. C.,
and BENNETT, D. (1961)
On the mechanism of abnormal transmission ratios at T locus in the house
mouse. Genetics 46, 1635–1644.

YANAGISAWA, K. O., and
KITAMURA, K. (1975).
Effects of the Brachyury (T) mutation
on mitotic activity in the neural tube.
Dev. Biol. 47, 433–438.

YANAGISAWA, K., POLLARD,
D. R., BENNETT, D., DUNN, L. C.,
and BOYSE, E. A. (1974b).
Transmission ratio distortion at the T-
locus: serological identification of two
sperm populations in t-heterozygotes.
Immunogenetics 1, 91–96.

5 CELL SURFACE PROPERTIES OF EARLY MAMMALIAN EMBRYOS

Eric J. Jenkinson and
W. David Billington

5.1 Introduction

The cell surface is of special interest in the study of embryogenesis because it forms the interface for interactions between the cell and its environment. In particular, it is now well established that cellular recognition is mediated in whole or in part by the cell membrane and that, as indicated by studies on invertebrates and inframammalian vertebrates, surface phenomena play an important part in developmental processes (Moscona, 1973). However, it is only recently that investigations have been carried out on the surface properties of mammalian embryonic cells during early development and during the initial establishment of the feto-maternal relationship, which involves interactions between cells of two different genotypes. This chapter reviews current knowledge on the surface characteristics of the component cells of the mammalian embryo from fertilization to the early postimplantation stages and considers what insight this knowledge gives into the processes of embryogenesis and the success of the embryo in the potentially hostile allogeneic maternal environment.

5.2 Surface Membrane Structure

To provide a wider context in which to consider the nature of the embryonic cell surface, it is pertinent first to outline some general principles of membrane structure derived from the examination of other cell types.

It is evident that the surface membrane must have sufficient stability to maintain the integrity of the cell and, at the same time, possess the necessary plasticity to undergo the changes associated with transduction of external stimuli and with growth and differentiation. Various hypotheses have been proposed to explain membrane structure, but the "fluid mosaic" model that Singer and Nicolson (1972) described has received general acceptance in recent times. In common with earlier models, this model assumes that the structural framework of the membrane consists of phospholipids arranged in the form of a bilayer. Intercalated into this lipid matrix to varying degrees are the intrinsic membrane proteins, which are considered to be amphipathic with their hydrophobic regions located within the nonpolar interior of the lipid bilayer and their hydrophilic polar regions protruding on one or both sides of the membrane, the outer end usually bearing carbohydrate side chains (see review by Bretscher and Raff, 1975). These oligosaccharide side chains of the plasma membrane glycoproteins (and, to a lesser extent, glycolipids) are primarily responsible for the carbohydrate-rich surface coat, or "glycocalyx," seen at the periphery of most cells (see Leblond and Bennett, 1974).

The intercalated components are envisaged as floating in a sea of lipid and being capable of lateral movement within the plane of the membrane, a view supported by experiments showing that cross-linking of surface proteins by suitable agents (such as specific antibody or lectins) can cause their redistribution (Taylor et al., 1971; de Petris and Raff, 1973; Munro, 1975). Similarly, the intermingling of surface antigens in Sendai-virus-induced heterokaryons of mouse and human cells implies the lateral diffusion of these particular membrane glycoproteins (Frye and Edidin, 1970; Edidin, 1974). At present, the degree to which various intercalated membrane components are free to move under normal conditions is uncertain, although there are indications that their movement is somewhat constrained (e.g., by changes in lipid viscosity (Edidin, 1974). Recent evidence also suggests that some degree of control over membrane protein movement and distribution may be achieved by linkage of intramembranous components to submembranous cytoskeletal elements, such as the microtubular and microfilament systems (Yahara and Edelman, 1972; Edelman et al., 1973), or to components on the outer surface, including the so-called extrinsic proteins, components of the substratum, or even other cells.

Clearly, then, the surface characteristics of cells fall into two broad categories: those concerned with the lipid phase and those concerned with the components intercalated into or associated with it. Although methods for the investigation of the physicochemical properties of the lipid phase are now available (see Edidin, 1974; Johnson, 1975a), to our knowledge, these have not been applied to the early mammalian embryo. This review therefore focuses primarily on the protein and glycoprotein components of the cell surface, which appear to be responsible for many of the antigenic and other biological properties of the membrane (see Winzler, 1970).

5.3 Techniques for the Study of Membrane Glycoproteins

Whereas various methods are available for investigation of the nature of cell surfaces, only those that have already been employed or have clear potential application in the analysis of surface properties of the mammalian embryo are outlined here.

5.3.1 Biochemical Methods

In addition to direct chemical analysis (Winzler, 1970), biochemical investigations of membrane-associated proteins have been carried out by the application of separation techniques, particularly SDS polyacrylamide gel electrophoresis, to preparations of solubilized membrane proteins (see Bretscher, 1974; Bretscher and Raff, 1975). Similarly, the incorporation of labeled sugars into surface material as assessed by autoradiography (Leblond and Bennett, 1974) or gel filtration of material removed from the cell surface by proteolytic enzymes (Buck et al., 1970, 1971) has been used to tag surface glycoproteins and to study their synthesis and turnover. This approach has facilitated the identification of different glycoprotein species associated with a particular cell type (Glossman and Neville, 1971) and of changes in surface glycoproteins associated with malignant transformation (Buck et al., 1970, 1971).

5.3.2 Histochemical Methods

Indications of the general composition of the outer surface coat of the membrane have come to light with the use of histochemical procedures such as the periodic acid Schiff (PAS) technique in the light microscope and techniques of similar specificity in the electron microscope that demonstrate the presence of 1,2-glycol-containing carbohydrates. Acidic surface groups such as sialic acid, on the other hand, are detectable with cationic dyes such as colloidal iron (Benedetti and Emmelot, 1967; Gasic et al., 1968) and thorium (Rambourg and Leblond, 1967), and with ruthenium red (Luft, 1971).

5.3.3 Plant Lectin Probes

Although they provide useful information, the above methods are limited to some extent by their lack of specificity. More recently, the identification and localization of specific saccharide components of the complex oligosaccharides of surface glycoproteins have become possible through the application of various plant lectins (see section 5.4.1) that bind to specific sugar ligands (Sharon and Lis, 1972; Nicolson and Singer, 1971, 1974). Conveniently, these are visible in the light microscope by coupling to fluorescein or in the electron microscope by conjugation with markers such as ferritin or peroxidase, which are electron-dense or which form an electron-dense reaction product (Avrameas, 1969; Nicolson and Singer, 1971, 1974).

5.3.4 Immunological Probes

The cell membrane can also be analyzed in terms of those structures that are recognized as antigens by virtue of their reaction with specific immunological reagents. Although this kind of analysis includes antigen detection by the susceptibility of target cells to lysis by specifically sensitized lymphoid cells, much more precise information on localization can be obtained by determination of the degree of binding of antisera coupled to an appropriate label. For evaluation of the overall level and pattern of antigen expression on monolayer cultures derived from a particular tissue or organ, coupling of the antiserum to red cells is effective (e.g., Hausman and Palm, 1973). For precise quantitation, isotope labeling and radioassay are particularly appropriate (Harder and McKhann, 1968), and spatial distribution on a cell-by-cell basis can be examined using fluorochrome dyes in the light microscope or electron-dense markers such as peroxidase, ferritin, and haemocyanin at the ultrastructural level.

Antigenic structures that yield to these means include those characteristic of a species (xenogeneic antigens), those found in genetically different individuals within a species (alloantigens), and those unique to a particular tissue or to a stage of development (tissue-specific or phase-specific antigens). The alloantigens, especially the major histocompatibility antigens of mouse (H-2) and man (HL-A), are of particular interest, not only because of their immunological importance in tissue transplantation, of which pregnancy is a naturally occurring example, but also because of their value as genetically well-defined glycoprotein surface markers (Nathenson, 1970).

5.4 The Preimplantation Embryo

Biochemical studies on the membrane of the early mammalian embryo so far have been restricted by the difficulties in obtaining adequate quantities of material. However, information has begun to accumulate on the histochemical properties, lectin binding sites, and antigenic status of preimplantation-stage embryos.

5.4.1 Histochemical and Lectin-Binding Properties

The unfertilized egg and zona pellucida
The use of colloidal iron staining in association with neuraminidase treatment and KOH saponification has demonstrated the presence of surface negative sialic acid groups in the N-acetyl-O-diacetyl configuration on the rabbit egg plasma membrane; hence it is possible for such residues to be involved in sperm receptor sites (Cooper and Bedford, 1971). These surface negative groups also increase in concentration in association with fertilization, suggesting that the resultant increase in negative repulsive forces at the egg surface is perhaps an important factor in the block to polyspermy. On the other hand, acidic anionic groups on the membrane of the hamster egg have not been found to undergo any obvious alteration in density or distribution at the time of fertilization (Yanagimachi et al., 1973).

The presence of specific saccharides in the zona pellucida and on the plasma membrane of unfertilized eggs of the hamster, mouse, and rat has been investigated using ferritin-lectin and ferritin-fluorescein conjugates in combination with appropriate inhibitors to indicate the specificity of the lectin binding (Nicolson et al., 1975). In all these species, receptors for Ricinus communis agglutinin (RCA; inhibited by D-galactose or lactose) and wheat germ agglutinin (WGA; inhibited by N-acetyl-D-glucosamine) were present in the zona pellucida with an asymmetric distribution showing greater density in the outer regions. Receptors for concanavalin A (Con A; inhibited by sucrose or α-methyl-D-mannoside), however, were relatively sparse and evenly distributed throughout the zona. Similarly, low pKa anionic groups (possibly sialic acid) seem also to be concentrated in the outer region of the zona (Yanagimachi et al., 1973). At present the significance of the existence and pattern of distribution of these residues is not clear, although their presence is consistent with the proposal by Nicolson et al. (1975) that the zona is a matrix of polymerized glycopeptide-glycoprotein units, which could act like a charged molecular sieve to impede the entry of certain macromolecules. It is unlikely, however, that this matrix excludes immunoglobulins, because the zona of the mouse has recently been shown to be highly permeable, at least to heterologous immunoglobulin molecules (Sellens and Jenkinson, 1975).

Attachment of spermatozoa to the outer surface of the zona represents a preliminary step in the fertilization process. Thus the ability of WGA (but not Con A, RCA, or Dolichos biflorus agglutinin) to block sperm binding to the zona in the hamster (Oikawa et al., 1973, 1974) might imply that the receptor for this agglutinin is structurally related to, or sufficiently close to, the postulated zona sperm receptor for steric interference to occur (Nicolson et al., 1975). In the mouse, however, sperm attachment to the zona is not affected by WGA (Parkening and Chang, 1976), so there might be a species difference in the nature or arrangement of the receptor sites. Similarly, differences between the agglutinability of unfertilized eggs and changes in the light-scattering properties of the zona induced by various lectins led to the speculation that differences in oligosaccharide residues of zona glycoproteins are involved in the species specificity of fertilization (Oikawa et al., 1975).

All the above lectins also bind to the plasma membrane of ovulated, but unfertilized, rat, hamster, and mouse eggs, with a random distribution on aldehyde-fixed eggs or living eggs maintained near 0°C. At room temperature on unfixed eggs, however, Con A and WGA receptors aggregate rapidly into patches or clusters as a result of lectin binding; therefore, the membrane must be sufficiently fluid to allow their lateral displacement (Nicolson et al., 1975). Nicolson et al. have suggested that, in view of the importance of membrane fluidity for cell fusion in other systems, this mobility of certain egg-membrane components might have implications for the fusion of the egg and sperm membranes at fertilization. Recent studies have also shown that the surface of the unfertilized mouse egg is a mosaic in terms of Con A binding sites, the area overlying the metaphase spindle being deficient in these receptors (Johnson et al., 1975). This mosaicism appears to be influenced by the submembranous microfilament system, although it is not clear whether its basis is structural or molecular. As in the study mentioned previously, mobility of these receptors also occurred at 37°C, and patching was induced at greater Con A dilutions on the fertilized compared to the unfertilized egg, again suggesting some form of alteration in membrane characteristics at the time of fertilization.

On the basis of these limited studies, it seems likely that membrane saccharide-containing components are in a position to participate in the fertilization process and that changes in the dynamic properties of the membrane also occur at this time. Thus, in the light of hypotheses postulating a major role for complex carbohydrates in cell recognition and adhesion (Roseman, 1970; 1974) and the known involvement of carbohydrate components in gamete attachment in invertebrates (see Winzler, 1970), the potential role of lectin-binding oligosaccharides in sperm-egg interaction in mammals is worth further investigation. At least in the hamster, however, treatment of zona-free eggs with RCA, WGA, and Con A appears to be without effect on sperm-egg fusion (Yanagimachi and Nicolson; quoted in Nicolson et al., 1975).

Early cleavage-stage embryos
Little information is available on the changes in membrane biochemistry associated with cleavage and the first morphologically detectable differentiation of the embryo into inner cell mass (ICM) and trophectoderm. However, Pinsker and Mintz (1973), by preincubating mouse embryos with labeled carbohydrate precursors and applying separation techniques to the pronase digest of the material initially removed from the surface of the embryos by trypsin, demonstrated a shift to higher-molecular-weight products in the digest during development from the 8-cell to blastocyst stage. Because this shift was comparable to that seen in the elution pattern of pronase-digested trypsinates from the surface of virally transformed cells, they suggested that these changes might be associated with the capacity of the trophoblast to attach to the uterine wall and become invasive. However, these events appear to be associated more with the appearance of the trophectoderm, which does not acquire the ability to attach and invade in vivo or to produce outgrowths in vitro before reaching the late blastocyst stage (Jenkinson and Wilson, 1973).

The lectin binding pattern in cleaving embryos has also received little attention, although Yanagimachi and Nicolson (1974) have shown that in the hamster Con A binds strongly up to the 4- to 8-cell stage but its binding capacity declines at the blastocyst stage. Similarly, RCA binding is strong up to the 4- to 8-cell stage and only moderate later. However, in the case of WGA, which labels the pronuclear egg very strongly, staining diminishes slightly at the 2-cell stage and markedly at the 4- to 8-cell and blastocyst stages. Although still to be extended to other species, these findings seem to indicate alterations in the composition and/or distribution of the oligosaccharide moieties of surface glycoproteins as development proceeds.

The blastocyst

Several workers have investigated the presence of negatively charged groups on the surface of the trophectoderm of the blastocyst. Holmes and Dickson (1973), using Prussian blue staining to assess colloidal iron binding to whole mounts of mouse blastocysts, were unable to detect surface negative groups until the late preimplantation stage when the blastocysts had been "activated" to implant. However, at the ultra-structural level, Nilsson et al. (1973) observed considerable binding of positively charged colloidal iron particles to the trophectoderm of the normal blastocyst. Using an essentially similar technique, we have confirmed this finding and extended it to show a marked reduction in colloidal iron binding following pretreatment with neuraminidase (Figures 5.1 and 5.2) and the proteolytic enzymes pronase and trypsin (Jenkinson and Searle, 1977). The implication is that many of the surface negative sites represent terminal N-acetyl-neuraminic (sialic) acid groups and that these are carried on components linked to peptides at the cell surface.

The presence of acidic glycoproteins on the preimplantation blastocyst surface has also been demonstrated with a number of other stains broadly indicative of these components, including both ruthenium red and colloidal thorium (Enders and Schlafke, 1974; Schlafke and Enders, 1975). Similarly, Con A binding to the trophectoderm of the blastocyst also indicates the presence of surface-located saccharides containing mannose, although, as Schlafke and Enders (1975) have pointed out, this reaction is not wholly specific because Con A binds, with lower affinity, to a series of other sugars.

Future approaches

Although information on the histochemical and biochemical characteristics of the surface of the early mammalian embryo is still fragmentary, a number of potentially useful lines of approach can now be recognized. The application of a battery of lectins might make it possible to assemble a picture of compositional changes in the membrane in terms of the surface molecules recognized by these probes during different phases of development. Changes in the patching or capping ability (the latter refers to the aggregation of determinants at one pole of the cell) of various lectin receptors could provide useful information not only

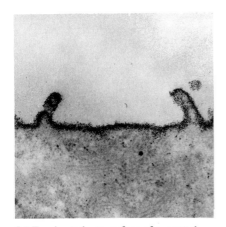

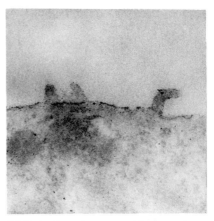

5.1 Trophectoderm surface of a normal 4th-day preimplantation mouse blastocyst showing heavy electron-dense deposits of colloidal iron particles indicative of the presence of negatively charged groups (× 32,500).

5.2 Preimplantation mouse blastocyst incubated in neuraminidase prior to staining with colloidal iron. The significant reduction in binding compared to that seen in figure 5.1 demonstrates that much of the surface negative charge is due to sialic acid residues (× 33,000).

about the relative mobility and degree of association of membrane components, but also about the fluidity of the membrane during successive phases and in different cell populations. As Schlafke and Enders (1975) suggested, the use of other enzymes, in addition to neuraminidase (various glycosidases and endopeptidases, for example), should facilitate analysis of the composition and possible function of the complex surface groups. Similarly, assessment of the incorporation of different labeled sugars into surface materials could be employed to investigate the turnover and appearance of new oligosaccharide-containing components. In addition, the tagging of surface proteins by radiolabeled membrane-impermeable acylating agents such as ^{35}S-formylmethionyl-sulfone methyl phosphate (Bretscher, 1974), in combination with increasingly sophisticated small-scale electrophoretic separation procedures now being applied to the preimplantation embryo, might provide a powerful method for the analysis of changes in membrane proteins associated with primary differentiative events in trophoblast and ICM formation.

5.4.2 Antigenic Determinants

Embryo- and phase-specific antigens
Kirby (1968) was the first to report the existence of embryo-specific antigen(s) on the preimplantation mouse embryo. He found that repeated ectopic transfer of allogeneic blastocysts to recipient mice resulted in inhibition not only of their development, but also of subsequently transferred syngeneic blastocysts. In a further series, this effect was confirmed with sequential transfers using only blastocysts syngeneic with the recipients (Kirby, 1968). More recent studies have attempted to demonstrate the existence of embryo-associated antigens in other ways, largely by the use of xenoantisera absorbed to leave only the specificity desired.

The immunization of guinea pigs with unfertilized mouse eggs to produce antisera with cytoxicity against both the eggs and Simian Virus 40 transformed adult cells, but not against their normal counterparts, has led some workers to suggest the presence at this stage of antigens that in adult tissues become latent or unexpressed until derepressed by viral transformation (Baranska et al., 1970). Similar antisera have been used in cytotoxicity studies to show that the so-called "egg antigen" is present on the 2-cell mouse egg through to the blastocyst stage in vitro, but disappears by the egg-cylinder stage (Moskalewski and Koprowski, 1972).

Immunofluorescent labeling of early embryos with antisera prepared by immunizing rabbits with homogenized whole mouse placentae (which actually consist of a diversity of trophoblastic and other fetal and maternal cell types—a fact apparently not appreciated by many investigators) has yielded variable results. Kometani et al. (1973) reported that fluorescein-labeled antiplacental serum did not stain intact unfertilized and fertilized eggs, but gave positive results from the 2- to 4-cell stage onwards, causing staining to be visible on both the embryonic and trophoblastic components of the eighth-day conceptus. Staining was detected also on the placenta at later stages, although there was no distinction drawn between the trophoblastic and fetal components and it remains unclear whether the antigen(s) involved is confined to the trophoblast in later development. In contrast, an antiplacental serum prepared in rabbits by Wiley and Calarco (1975) reacted with unfertilized eggs, all preimplantation stages, both the ICM and trophoblast of sectioned blastocysts, and later placental trophoblast, but not with other fetal or adult tissues. These authors postulated that the "placental-specific" molecules, which are detected by their antiserum from the earliest stages of development and which subsequently segregate to the placental trophoblast, could represent, in part, surface components that provide a protective immunological barrier at the feto-maternal interface. Clearly, this is entirely speculative.

In addition to their antiplacental serum, Wiley and Calarco (1975) described a rabbit antiserum to mouse blastocysts that, after absorption against adult mouse tissues, gave weak positive reactions in immunofluorescence tests on the 1- to 2-cell stage. The reactions increased to a maximum on the 8- to 12-cell embryo and then declined to undetectable levels on in vitro blastocyst outgrowths. The unfertilized ovum and later fetal and placental tissues were also negative. Unlike the antiplacental serum, the antiblastocyst serum arrested the in vitro development of embryos at the 8- to 12-cell stage; thus the surface moieties that it recognizes might have a functional role with which antibody binding interferes. It was proposed, inter alia, that this phenomenon might be related to processes of cell recognition and adhesion. In this respect, there is an intriguing parallel with the findings of Artzt and her colleagues (1973) that an antiserum prepared in syngeneic mice against embryonal carcinoma cells (F9 cells; see chapter 7) reacts not only with these cells, but also with an antigen on sperm and on preimplantation embryos, and that it reaches maximum reactivity around the 8-cell stage, after which a decline occurs (Artzt and Bennett, 1975).

Subsequent studies (Artzt et al., 1974) have suggested that the antigen present on these teratocarcinoma cells is specified by a wild-type (+) gene of the T complex, whose mutation (t^{12}) in the homozygous condition causes developmental arrest at the morula stage. This developmental block might be due to the absence of stage-specific genetically controlled surface differentiation antigens specified by the $+^{t^{12}}$ gene, which play a role, as yet undefined, in the cellular interactions necessary for development beyond the morula stage (Bennett, 1975). The inability of the blastomeres from t^{12}/t^{12} embryos to adhere to one another (Calarco and Brown, 1968) as well as to those of normal embryos (Mintz, 1964) also indicates that these cells lack the surface properties necessary for normal cellular association. (For a detailed consideration of t mutations, see chapter 4.)

It is clear that much remains to be resolved concerning the identity and function of the various embryo-associated antigens so far detected.

The relatively transient expression of some of these can be compared to the brief appearance of specific bands seen in protein separation studies on embryos at various stages of cleavage, implying a rapid switching-off as well as the switching-on of genetic information as development proceeds (Van Blerkom and Brockway, 1975; Van Blerkom, 1975; see chapter 2). As yet the relationship between the appearance of new antigens on the cell surface and synthetic activity within the embryo has not been elucidated. It is becoming apparent, however, that the period from the 8-cell to the morula stage is a time of considerable activity in the membrane. This conclusion is indicated by the maximal expression of a number of surface antigens, including a recently described hCG-like (human chorionic gonadotrophic hormone) moiety (Wiley, 1974) and by the shift in surface glycoprotein profile discussed earlier. This period is also one of considerable biochemical change with increasing protein and RNA synthesis and evidence of activity on the part of the embryonic genome, as, for example, in the appearance of paternal enzyme variants (see Epstein, 1975, and chapter 3) and in surface histocompatibility antigens (see below).

It is also at the morula stage that the first discernible morphological change of the embryo occurs in the form of a "compaction" as the outer blastomeres, previously rounded, flatten and form tight junctions at their outer margins (Calarco and Brown, 1969; Calarco and Epstein, 1973) as a preliminary step to the formation of trophectoderm, which will enclose the ICM. The appearance or maximal expression of the various antigens described above, which might well indicate considerable membrane glycoprotein changes in the morula, therefore might be involved in the induction of, or at least in some way associated with, this primary differentiation of the embryo. In view of the potentially diverse roles of such surface components in cellular interactions, including adhesion, in substrate or precursor transport systems, and as transducers of external stimuli, further investigation of the identity and distribution of determinants recognizable as embryo- or phase-specific antigens during preimplantation development seems desirable.

An important point to clarify in future investigations of surface determinants on the embryo is the extent to which temporal coincidence of their maximal expression simply reflects recognition of the same components by different antisera, as is possible in the case of antigens identified by the antiblastocyst and antiteratoma antisera. The use of cocapping experiments to map these determinants (Edidin, 1974) and cross-absorption of antisera should resolve this problem. Future immunological analyses of events associated with early differentiation will also require the use of more homogeneous tissue for the preparation and testing of specific antisera. With the advent of methods for the isolation of trophoblast (Snow, 1973; Sherman and Atienza, 1975) and ICMs (Solter and Knowles, 1975) in relatively large numbers and with the development of microsurgical techniques for the early postimplantation embryo (see chapter 1), it should be possible to investigate surface antigenic differences associated with the appearance and subsequent differentiation of the different cell populations in the early mammalian embryo.

Histocompatibility antigens
The availability of a wide range of genetically well-defined inbred strains of mice has made this the animal of choice for studies on the cell surface expression of histocompatibility antigens. These antigens can be classified in two basic groups: those controlled by the minor, or non-H-2, genetic loci and those that appear to play a dominant role in transplantation reactions and are controlled by the H-2, or major histocompatibility, complex (MHC), which is known to consist of at least five different regions (figure 5.3). Almost all the information on the antigenic status of the early mouse embryo relates to certain specified or unspecified non-H-2 antigens and to the classical serologically defined H-2 antigens controlled by the *H-2K* and *H-2D* regions of the MHC.

Studies employing both transplantation and in vitro assays have demonstrated the presence of alloantigens from as early as the 2-cell stage (Simmons and Russell, 1965; Kirby et al., 1966; Olds, 1968; Vandeputte and Sobis, 1972; see review by Billington and Jenkinson, 1975), but the mouse strain combinations used did not allow an evaluation of the relative contribution of the H-2 and non-H-2 determinants. More recently, investigators have attempted a more precise characterization and have identified various non-H-2 antigens on all cleavage-stage embryos and on blastocysts by such techniques as cytotoxicity (Heyner et al., 1969), indirect immunofluorescence (Palm et al., 1971; Muggleton-Harris and Johnson, 1976), and ectopic transplantation (Searle et al., 1974; Hetherington and Humber, 1975). Muggleton-Harris and Johnson (1976) obtained weakly positive immunofluorescent staining of non-H-2 components up to the 6- to 8-cell stage with increasing intensity there-

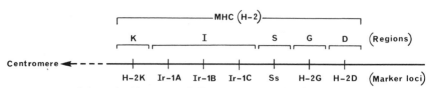

5.3 Structure of the major histocompatibility complex (MHC) of the mouse located on chromosome 17.

after. They suggested that these results probably reflect de novo synthesis from embryonic mRNAs because during the same period paternal antigens in heterozygous embryos make their first appearance and increase to the same level as maternal antigens by the 8- to 12-cell stage. This finding is consistent with other evidence of early activation of the embryonic genome (see chapter 2).

In contrast to their observations on non-H-2 antigens, none of the workers cited above were able to detect antigens determined by the MHC, nor could Gardner et al. (1973) who used mixed antiglobulin and mixed agglutination tests. Consequently, it was believed that there was lack of expression of major histocompatibility antigens during pre-implantation development. However, studies employing radiolabeled antisera (Håkansson et al., 1975) and electron-microscopic visualization of immunoperoxidase label (Searle et al., 1976) have both demonstrated the presence of H-2 on the surface of the blastocyst. The discrepancy between these and the previous findings probably reflects the greater sensitivity of the latter techniques (see figures 5.4–5.6). In the immunoperoxidase study, H-2 antigens were not detectable at the 8-cell stage, although more recent work aimed at identifying the precise time of their appearance indicates that they are expressed on the trophectoderm of the early cavitating blastocyst (R. F. Searle and E. J. Jenkinson, unpublished). There is also a claim that H-2 could be present on the unfertilized ovum (Edidin et al., 1974). Confirmation of this claim would imply that as cleavage proceeds these antigens are either diluted out or lost. The appearance (or reappearance?) of H-2, in common with some of the surface phenomena described above, is likely to be associated with the morula phase when, in addition to the primary morphological differentiation into ICM and trophectoderm cells, there is evidence of increasing activity on the part of the embryonic genome.

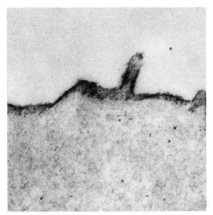 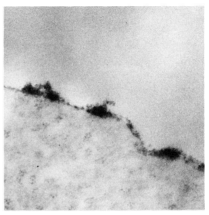

5.4 Preimplantation mouse blastocyst exposed to alloantiserum directed against combined H-2 and non-H-2 antigens followed by peroxidase-conjugated rabbit antimouse immunoglobulin. Note the heavy and continuous deposit of electron-dense peroxidase conjugate (× 33,000).

5.5 Blastocyst surface following immuno-peroxidase treatment with antiserum directed against H-2 only (× 33,500). The labeling is much less regular and less dense than that seen with antiserum against combined H-2 and non-H-2 antigens (cf. figure 5.4).

5.6 Blastocyst surface after experimentally induced estrogen activation and treatment as described in figure 5.4. The absence of labeling (seen also in all specificity controls in the immunoperoxidase series) indicates antigen loss from the trophectoderm when implantation begins (× 33,000).

Evidence to date suggests differential activity of the genes controlling H-2 and non-H-2 antigen expression until at least the 8-cell stage, although it is not yet clear whether the apparent absence of H-2 determinants during the early cleavage stages involves a total repression of their synthesis or whether they are produced but in some way fail to be expressed. It has been suggested recently that certain antigens present both on some forms of teratoma and on preimplantation embryos have a reciprocal relationship with H-2 in their expression on the cell surface. If so, they might be precursors of the complete H-2 molecule (Edidin et al., 1974; Nicolas et el., 1975; Artzt and Bennett, 1975). Thus it is conceivable that the failure to detect H-2 determinants on the early embryo is due to their synthesis in an incomplete form during these stages.

The identification of H-2 antigens on the blastocyst relates so far only to their presence on the trophectoderm because the incubation of intact embryos with specific antisera precludes access to the ICM. Expression of non-H-2, but not of H-2, antigens on ICMs isolated by microsurgery has recently been demonstrated by immunofluorescence (Muggleton-Harris and Johnson, 1976), but whether or not this cell population also expresses H-2 at the low levels detectable by radiolabeling and immunoperoxidase techniques is not yet known. At whatever time these antigens first appear, however, their expression might increase to readily demonstrable levels only with the onset of the rapid processes of differentiation and embryogenesis in the early postimplantation period (see section 5.6).

Clearly, the next step in analyzing the antigenic status of the developing embryo is to examine the relative roles of the various regions of the MHC in specifying cell surface determinants. In this respect, in addition to *H-2K* and *H-2D*, there are two other categories of antigen that will undoubtedly prove to be of considerable interest. These are the Ia antigens, which are serologically detectable on the surface of certain cells, including B-lymphocytes, macrophages, epidermal cells, and spermatozoa, and the strong LD-1 determinants, which lead to stimulation in mixed lymphocyte cultures. Both groups are controlled by the I region of the MHC, which lies between the K and D regions and is known to contain the immune response (Ir) genes (see figure 5.3). Within the I-A subregion, the genes specifying the Ia antigens and the LD determinants have not been genetically separated, and there is some evidence that there is some degree of identity between their

products. It is also known that LD differences are potent elicitors of
graft-versus-host reactions in vivo and that products of the I region,
which could be identical to LD determinants, are involved in graft
rejection. Bach et al. (1976) have recently reviewed these concepts
relating to the MHC in more detail. Investigation of these antigens on
the early embryo should prove valuable in relation both to the ontogeny of antigen expression and to a consideration of the embryo as an
intrauterine allograft.

Finally, there are other antigen systems known to be associated with
certain cell types that might well be represented on the surface of the
early embryo. Among those that might be significant are the LDH-X
antigen detected on spermatozoa (Goldberg, 1974) and the H-Y male-
specific antigen on spermatozoa (Goldberg et al., 1971) and at least
some differentiated somatic cells (Hausman and Palm, 1973).

5.5 Cell Surface Changes Associated with Implantation

To ensure its continued development beyond the blastocyst stage, the
mammalian embryo must implant and establish an intimate relationship
with the uterine tissues leading ultimately to the formation of a placenta. At the time of implantation, the blastocyst becomes increasingly
resistant to removal from the uterus by flushing, implying an increase
in its adhesive properties. Similarly, in vitro, at the late blastocyst
stage the trophectoderm acquires the ability to attach and to produce
outgrowths on a suitable substratum, an ability that suggests a functional change in its surface characteristics (Jenkinson and Wilson,
1973).

Although the ultrastructure of the implanting mouse blastocyst is
well documented, it is only recently that consideration has been given
to molecular events in the membrane during the process of implantation (Nilsson, 1974; Schlafke and Enders, 1975; Sherman and Wudl,
1976). To allow a more detailed analysis of this critical phase of development, blastocysts have been maintained in a state of experimentally
induced delay such that they do not implant until "activated" by the
administration of exogenous estrogen (Yoshinaga and Adams, 1966).
With this model it has been shown that following activation there is a
marked decrease in the ability of the outer membrane of the trophectoderm to bind positively charged colloidal iron particles, indicating
a decrease in surface negative groups (Nilsson et al., 1973, 1975;

Jenkinson and Searle, 1977). In the same system a temporally similar pattern is seen with cell surface antigens recognized by rabbit anti-mouse serum (Håkansson, 1973) and by alloantisera (Håkansson and Sundqvist, 1975; Håkansson et al., 1975; Searle et al., 1976) that undergo a marked decline in their expression, below detectable levels in the case of H-2 antigens, following activation (see figure 5.6). These changes in both colloidal iron binding and alloantigen expression on the trophectoderm also occur when blastocysts produce outgrowths in vitro, a process considered to be analogous to implantation (Jenkinson and Searle, 1977; Searle et al., 1976). The membrane modifications must hence be intrinsic and dependent upon the activity of the embryonic genome and not due simply to changes in the ionic composition of the uterine fluid at implantation (Clemetson et al., 1972) in the case of surface charge or to the activity of uterine proteases such as the recently postulated "implantation initiating factor" (Pinsker et al., 1974), stripping off surface negative or antigenic groups.

The enzyme susceptibility of colloidal iron binding (see section 5.4.1) indicates that at least some of the negatively charged sites on the trophectoderm represent sialic (or, more specifically, N-acetyl-neuraminic) acid groups of surface glycoproteins. Together with the reduction in antigen expression, of which at least the H-2 antigens are known to be glycopeptides (Nathenson, 1970), this finding provides evidence for a considerable modification of the glycoprotein profile of the trophectoderm membrane at the time the blastocyst acquires the ability to attach to the uterine epithelium and invade the maternal tissues. In this context, it is interesting to note that malignant transformation in some cells is accompanied by alterations in surface glycoproteins (Buck et al., 1970).

It is not yet known whether the components detected on the blastocyst surface by colloidal iron are part of the same molecular complex as those detected by either xenoantisera or alloantisera. The nature of the changes in these components also remains speculative and could involve relatively simple modifications, such as the removal of terminal sialic acid groups or of more extensive parts of the oligosaccharide moieties of surface glycoproteins. Apart from modification, or masking (see section 5.6), changes in the detectability of specific glycoprotein membrane components could also be explained in terms of removal either by "internalization" or "shedding," which is known to occur in other situations (see Bretscher and Raff, 1975). It has been postulated that the ability of rabbit antimouse antiserum to induce "patching" or "capping" on the trophectoderm of the late-stage mouse blastocyst, but not the earlier embryo, might reflect an increase in the mobility of membrane components that could facilitate their removal from the surface following antibody binding (Johnson, 1975b). Changes in antigen expression on the trophectoderm in vivo, however, cannot depend entirely upon modulation by maternal antibody in view of the antigen loss seen on blastocyst outgrowths in vitro. The fact that the surface glycoproteins of most cells do not represent a static molecular population but undergo a constant process of turnover and replacement (Leblond and Bennett, 1974) suggests an alternative possibility. In the case of the trophectoderm, there is evidence of an increase in surface activity at implantation, including the acquisition of phagocytic properties (Gardner, 1975) in association with the so-called "giant cell transformation" (see Sherman and Wudl, 1976). Thus a switch-off of the mechanism for replacing specific glycoproteins at a time of increased membrane activity would lead to their effective disappearance from the cell surface.

The various changes in the composition of the outer trophectoderm membrane have several implications for feto-maternal interaction. The lyophobic colloid theory of cell adhesion (Curtis, 1967) proposes that the ability of cells to adhere to each other depends upon the balance of attractive, as opposed to repulsive, negative forces at the cell surface. On this basis, the decrease in negatively charged sialic acid residues on the blastocyst surface at implantation might facilitate its attachment to the uterine epithelium. In addition, decreases in sialic acid content and surface negative charge have been associated with increased membrane deformability (Weiss, 1965) and loss of contact inhibition by transformed malignant cells in vitro (Ohta et al., 1968; Perdue et al., 1972). The acquisition of similar properties by the trophoblast as a result of alterations in membrane sialic acid content might also correlate with its ability to penetrate and invade the maternal tissues.

As an alternative to purely electrostatic effects, recent theories suggest that cell adhesion could result either from enzyme-substrate-type interactions between glycosyl-transferases on one surface and incomplete oligosaccharide chains on another or by the binding of carbohydrates on apposing surfaces via hydrogen bonds (Roseman, 1970, 1974;

Vickers and Edwards, 1972). In terms of these models, the appearance of modified (desialyated?) oligosaccharide chains on the trophectoderm could also provide appropriate conditions for blastocyst attachment.

From an immunological point of view, the disappearance of trophoblast surface antigens concurrently with implantation could be an important factor in the survival of the fetal allograft. At implantation, the zona pellucida, which surrounds the embryo prior to this time, is shed, and following blastocyst attachment and degeneration of the uterine epithelium the trophoblast comes into direct cellular contact with maternal stroma and blood elements. Under these circumstances, the ability of the embryo to present an outer surface deficient in antigens, at a time when antigen expression on the ICM is almost certainly increasing, should reduce the danger of recognition and response by the maternal immunological surveillance system. In this way the fetal allograft might avoid destruction in the initial stages of its association with the mother when the protective mechanisms operative later in pregnancy (see Adinolfi and Billington, 1976) are unlikely to be functional.

5.6 The Early Postimplantation Embryo

A consideration of the surface properties of the various cells comprising the organs and tissues of the definitive embryo is beyond the scope of this review; discussion here is restricted largely to the early postimplantation period. Some mention of subsequent stages is relevant, however, because the detection of particular surface characteristics later in development raises the question of the time of their first appearance.

Shortly after implantation, the embryo embarks upon a rapid process of germ-layer formation and differentiation so that, by the eighth day in the mouse, it consists of an elongated egg cylinder or embryonic sac derived from the ICM, enclosed by a shell of trophoblast with a mesometrial proliferation known as the ectoplacental cone (EPC). This provides a very convenient stage for experimental investigation because relatively pure populations of trophoblastic and embryonic cells can be obtained by microdissection, although little is known of the surface properties of these component cells, with the notable exception of histocompatibility antigen expression.

The precise time course of the expression of H-2 antigens on the embryonic tissues has yet to be established. As indicated earlier, the antigenic status of the ICM in this respect is undefined, and it has so far proven possible to identify these antigens only at a stage equivalent to the eighth day in utero in experiments involving ectopic grafting of postimplantation embryos to preimmunized recipients (Patthey and Edidin, 1973) and examination of blastocyst outgrowth in vitro by immunofluorescence (Heyner, 1973) and by immunoperoxidase labeling (Searle et al., 1976). By the eighth day of gestation, strong staining for H-2 can be demonstrated by the immunoperoxidase technique on the visceral endoderm of the intact egg cylinder (Searle et al., 1976). This evidence suggests an increase in the level of H-2 expression as development proceeds. It is consistent with evidence from other studies that the overall antigenicity of the embryo increases during gestation and that a pattern of different antigenic activities appears in various organs (see Edidin, 1972). Apart from the endoderm, however, the relative degree of antigen expression on the main components of the egg cylinder is unexplored, although the recent development of procedures for separation of the germ layers and the embryonic and extraembryonic ectoderm (see chapter 1) now makes this feasible.

The finding of H-2 on the eighth day endoderm confirms and extends the previous demonstration of these (and non-H-2) antigens on the outer endodermal surface of the definitive yolk sac (Jenkinson and Billington, 1974a). These antigens render this important fetal membrane susceptible to both cell-mediated (Jenkinson and Billington, 1974a) and humoral (Jenkinson et al., 1975) immune attack in vitro. The cells of the yolk-sac endoderm also possess surface Fc receptors at the fifteenth- to eighteenth-day stages (Elson et al., 1975), and these might prove useful as markers for endoderm if present from the earliest stages of differentiation.

In addition to the antigens defined by the above techniques, it seems likely that the egg cylinder also expresses products of the I region or those surface determinants involved in lymphocyte stimulation, because transplanted egg cylinders are immunogenic and capable of inducing sensitization as assessed by accelerated rejection of subsequent skin grafts (Simmons et al., 1971; Searle et al., 1975).

In marked contrast to the observations on the egg cylinder, neither H-2 nor non-H-2 antigens have been detected on the EPC cells by ectopic grafting (Simmons et al., 1971; Searle et al., 1975), susceptibility to lysis by immune lymphoid cells in vitro (Vandeputte and Sobis, 1972; Jenkinson and Billington, 1974b), or immunoperoxidase labeling (Searle et al., 1976). An earlier report that the antigens are present but masked by a neuraminidase-sensitive surface coating material (Currie et al., 1968) is unconfirmed (Simmons et al., 1971; Jenkinson and Billington, 1974b; Searle et al., 1975). Although there are a number of alternative explanations for the apparent absence of antigens on the EPC (Billington, 1976; see below), the data appear to be consistent with the conclusion that this tissue is relatively deficient in at least the serologically defined antigens of the MHC. Nothing is yet known of the possible expression of any of the products of the I region.

The EPC consists of two populations of trophoblast cells, an outer layer of polyploid giant cells and an inner, probably diploid, core, that make a major contribution to the definitive placenta. Preliminary results from our laboratory demonstrate that a later population of cells (diploid core?) do express histocompatibility antigens as detected by a mixed hemagglutination assay on EPC outgrowths in vitro (Sellens, 1976). In addition, this assay has detected antigens on the majority of cells in cultures of thirteenth-day placenta (Sellens, 1976), consistent with the proposal that antigen-bearing trophoblast cells arise earlier in the course of development.

It must be stressed that many different biological forms of trophoblast arise during early differentiation and placental ontogeny. It is not unlikely that there is variation in their surface antigenic and other properties, perhaps even among different strains of mice, as recent in vitro studies analyzing antigen expression on early trophoblast (J. Carter, personal communication; Sellens, 1976) seem to indicate. It should also be recognized that the antigenic properties of trophoblast cells must be considered in terms of their geographical location, especially in relation to the immunological interplay with the maternal organism (see Billington, 1976).

5.7 Conclusion

In recent years various new techniques have led to considerable advances in understanding cell surface structure and function. Many of these are now being applied to the early mammalian embryo, and evidence is beginning to accumulate, particularly from biochemical, lectin-binding, and immunological studies. Although much of this information is still fragmentary, it is becoming clear that whereas the embryo shares certain characteristics with other cells, it displays some unique features in terms of both histocompatibility antigen expression and embryo-specific surface products, especially in association with trophoblast differentiation.

Because the entire period of mammalian embryogenesis takes place within the maternal environment, knowledge of these surface properties is fundamental to an appreciation of interactions both within the embryo and between the mother and her genetically alien fetus, as well as to the identification of potential targets for fertility control procedures. As far as embryogenesis is concerned, the role of surface antigenic determinants in cellular recognition and organization is of particular interest. The presence of H-2 antigens on the early embryo is clearly significant in immunological events, and Johnson (1975b) has argued that this is their primary, if not sole, function. Their involvement in other cellular interactions is not yet clear. It seems reasonable, however, to assume that other surface moieties of a different nature, possibly including the non-H-2 antigens and, in particular, T-complex gene products (Bennett, 1975) and phase-specific or tissue-specific antigens, are important in developmental processes. This view derives additional support from the finding that in vitro reaggregation of dissociated embryonic chick organs can be inhibited by antibody directed against cell surface antigens (Szulman, 1973) and from experimental and clinical studies showing that a continuous and changing phase-specific fetal antigen challenge elicits a maternal antibody response that appears to be "a necessary circumstance for the normal course of embryogenesis" (Volkova and Maysky, 1969).

References

ADINOLFI, M. A., and
BILLINGTON, W. D. (1976).
Ontogeny of acquired immunity and
feto-maternal immunological interac-
tions. *In* Fetal Physiology and Medi-
cine, R. W. Beard and P. W.
Nathanielsz, eds. (London: Saunders
Co.), pp. 17–42.

ARTZT, K., and BENNETT, D.
(1975).
Analogies between embryonic (T/t)
antigens and adult major histocompatibility (H-2) antigens. Nature *226*,
545–547.

ARTZT, K., BENNETT, D., and
JACOB, F. (1974).
Primitive teratocarcinoma cells express a differentiation antigen specified by a gene at the T-locus in the
mouse. Proc. Nat. Acad. Sci. USA
71, 811–814.

ARTZT, K., DUBOIS P.,
BENNETT, D., CONDAMINE, H.,
BABINET, C., and JACOB, F.
(1973).
Surface antigens common to mouse
cleavage embryos and primitive ter-
atocarcinoma cells in culture. Proc.
Nat. Acad. Sci. USA *70*, 2988–2992.

AVRAMEAS, S. (1969).
Coupling of enzymes to proteins with
glutaraldehyde. Use of conjugates for
the detection of antigens and antibodies. Immunochemistry *6*, 43–52.

BACH, F. H., BACH, M. L., and
SONDEL, P. M. (1976).
Differential function of major histocompatibility complex antigens in T-
lymphocyte activation. Nature *259*,
273–281.

BARANSKA, W., KOLDOVSKY,
P., and KOPROWSKI, H. (1970).
Antigenic study of unfertilized mouse
eggs: cross reactivity with SV-40 in-
duced antigens. Proc. Nat. Acad. Sci.
USA *67*, 193–199.

BENEDETTI, E. L., and EMME-
LOT, P. (1967).
Studies on plasma membranes. IV.
The ultrastructural localisation and
content of sialic acid in plasma membranes isolated from rat liver and
hepatoma. J. Cell. Sci. *2*, 499–512.

BENNETT, D. (1975).
T-locus mutants: suggestions for the
control of early embryonic organiza-
tion through cell surface components.
In The Early Development of Mam-
mals, M. Balls and A. E. Wild, eds.
(London: Cambridge University
Press), pp. 207–218.

BILLINGTON, W. D. (1976).
The immunobiology of trophoblast. *In*
Immunology of Human Reproduction,
W. R. Jones and J. S. Scott, eds.
(New York: Academic Press),
pp. 81–102.

BILLINGTON, W. D., and
JENKINSON, E. J. (1975).
Antigen expression during early
mouse development. *In* The Early
Development of Mammals, M. Balls
and A. E. Wild, eds. (London:
Cambridge University Press),
pp. 219–232.

BRETSCHER, M. S. (1974).
Some general principles of membrane
structure. *In* The Cell Surface in
Development, A. A. Moscona, ed.
(New York: Wiley), pp. 17–27.

BRETSCHER, M. S., and RAFF, M. C. (1975).
Mammalian plasma membranes.
Nature *258*, 43–49.

BUCK, C. A., GLICK, M. C., and WARREN, L. (1970).
A comparative study of glycoproteins from the surface of Rous sarcoma virus transformed cells. Biochemistry *9*, 4567–4576.

BUCK, C. A., GLICK, M. C., and WARREN, L. (1971).
Glycopeptides from the surface of control and virus transformed cells.
Science *172*, 169–171.

CALARCO, P. G., and BROWN, E. H. (1968).
Cytological and ultrastructural comparisons of t^{12}/t^{12} and normal mouse morulae. J. Exp. Zool. *168*, 169–186.

CALARCO, P. G., and BROWN, E. H. (1969).
An ultrastructural and cytological study of pre-implantation development of the mouse. J. Exp. Zool. *171*, 253–284.

CALARCO, P. G., and EPSTEIN, C. J. (1973).
Cell surface changes during pre-implantation development in the mouse (SEM). Dev. Biol. *32*, 208–213.

CLEMETSON, C. A. B., KIM, J. K., MALLIKARJUNESWARA, V. R., and WILDS, J. H. (1972).
The sodium and potassium concentrations in the uterine fluid of the rat at the time of implantation. J. Endocrin. *54*, 417–423.

COOPER, G. W., and BEDFORD, J. M. (1971).
Charge density change in the vitelline surface following fertilization of the rabbit egg. J. Reprod. Fertil. *25*, 431–436.

CURRIE, G. A., VAN DOORNINCK, W., and BAGSHAWE, K. D. (1968).
Effect of neuraminidase on the immunogenicity of early mouse trophoblast. Nature *219*, 191–192.

CURTIS, A. S. G. (1967).
The Cell Surface: Its Molecular Role in Morphogenesis. New York: Academic Press.

DE PETRIS, S., and RAFF, M. C. (1973).
Normal distribution, patching and capping of lymphocyte surface immunoglobulins studied by electron microscopy. Nature New Biol. *241*, 257–259.

EDELMAN, G. M., YAHARA, I., and WANG, J. C. (1973).
Receptor mobility and receptor-cytoplasmic interactions in lymphocytes. Proc. Nat. Acad. Sci. USA *70*, 1442–1446.

EDIDIN, M. (1972).
Histocompatibility genes, transplantation antigens and pregnancy. *In* Transplantation Antigens: Markers of Biological Individuality, B. D. Kahan and R. A. Reisfeld, eds. (New York: Academic Press), pp. 75–114.

EDIDIN, M. (1974).
Arrangement and rearrangement of cell surface antigens in a fluid plasma membrane. *In* Cellular Selection and Regulation in the Immune Response, G. M. Edelman, ed. (New York: Raven Press), pp. 121–132.

EDIDIN, M., GOODING, L. R., and JOHNSON, M. H. (1974).
Surface antigens of normal early em-

bryos and a tumour model system useful for their further study. *In* Immunological Approaches to Fertility Control. Karolinska Symposium No. 7, E. Diczfalusy, ed. (Copenhagen: Bogtrykkeriet Forum), pp. 336–356.

ELSON, J., JENKINSON, E. J., and BILLINGTON, W. D. (1975) Fc receptors on mouse placenta and yolk sac cells. Nature 255, 412–414.

ENDERS, A. C., and SCHLAFKE, S. (1974). Surface coats of the mouse blastocyst and uterus during the pre-implantation period. Anat. Rec. 180, 31–46.

EPSTEIN, C. J. (1975). Gene expression and macromolecular synthesis during pre-implantation embryonic development. Biol. Reprod. 12, 82–105.

FRYE, L. D., and EDIDIN, M. (1970). The rapid intermixing of cell surface antigens after formation of mouse-human heterokarons. J. Cell Sci. 7, 319–335.

GARDNER, R. L. (1975). Origins and properties of trophoblast. *In* Immunobiology of Trophoblast, R. G. Edwards, C. W. S. Howe, and M. H. Johnson, eds. (London: Cambridge University Press), pp. 43–65.

GARDNER, R. L., JOHNSON, M. H., and EDWARDS, R. G. (1973). Are H-2 antigens expressed in the pre-implantation blastocyst? *In* Immunology of Reproduction, K. Bratanov, ed. (Sofia, Bulgaria: Bulgarian Academy of Sciences Press), pp. 480–486.

GASIC, G. J., BERWICK, L., and SORRENTINO, M. (1968). Positive and negative colloidal iron as cell surface electron stains. Lab. Invest. 18, 63–71.

GLOSSMAN, H., and NEVILLE, D. M. (1971). Glycoproteins of cell surfaces. A comparative study of three different cell surfaces of the rat. J. Biol. Chem. 246, 6339–6346.

GOLDBERG, E. H. (1974). Effects of immunisation with LDH-X on fertility. *In* Immunological Approaches to Fertility Control. Karolinska Symposium No. 7, E. Diszfalusy, ed. (Copenhagen: Bogtrykkeriet Forum), pp. 202–222.

GOLDBERG, E. H., BOYSE, E. A., BENNETT, D., SCHEID, M., and CARSWELL, E. A. (1971). Serological demonstration of H-Y (male) antigen on mouse sperm. Nature 232, 478–480.

HÅKANSSON, S. (1973). Effects of xenoantiserum on the development in vitro of mouse blastocysts from normal pregnancy and from delay of implantation with and without oestradiol. Contraception 8, 327–342.

HÅKANSSON, S., HEYNER, S., SUNDQVIST, K. G., and BERGSTRÖM, S. (1975). The presence of paternal H-2 antigens on hybrid mouse blastocysts during experimental delay of implantation and the disappearance of these antigens after onset of implantation. Int. J. Fertil. 20, 137–140.

HÅKANSSON, S., and SUNDQVIST, K. G. (1975). Decreased antigenicity of mouse blastocysts after activation for implantation from experimental delay. Transplantation *19*, 479–484.

HARDER, F. H., and MCKHANN, C. F. (1968).
Demonstration of cellular antigens on sarcoma cells by an indirect ^{125}I-labeled antibody technique. J. Nat. Cancer Inst. *40*, 231–241.

HAUSMAN, S. J., and PALM, J. (1973).
Variable expression of Ag-B and non-Ag-B histocompatibility antigens on cultured rate cells of different histological origin. Transplantation *16*, 313–324.

HETHERINGTON, C. M., and HUMBER, D. P. (1975).
The effects of active immunization on the decidual cell reaction and ectopic blastocyst development in mice. J. Reprod. Fertil. *43*, 333–336.

HEYNER, S. (1973).
Detection of H-2 antigens on the cells of the early mouse embryo. Transplantation *16*, 675–678.

HEYNER, S., BRINSTER, R. L., and PALM, J. (1969).
Effect of iso-antibody on pre-implantation mouse embryos. Nature *222*, 783–784.

HOLMES, P. V., and DICKSON, A. D. (1973).
Estrogen induced surface coat and enzyme changes in the implanting mouse blastocyst. J. Embryol. Exp. Morph. *29*, 639–645.

JENKINSON, E. J., and BILLINGTON, W. D. (1974a).
Studies on the immunobiology of mouse fetal membranes: the effect of cell-mediated immunity on yolk sac

cells *in vitro*. J. Reprod. Fertil. *41*, 403–412.

JENKINSON, E. J., and BILLINGTON, W. D. (1974b).
Differential Susceptibility of mouse trophoblast and embryonic tissue to immune cell lysis. Transplantation *18*, 286–289.

JENKINSON, E. J., BILLINGTON, W. D., and ELSON, J. (1975).
The effect of cellular and humoral immunity on the mouse yolk sac. *In* Transmission of Immunoglobulins from Mother to Young, W. A. Hemmings, ed. (London: Cambridge University Press), pp. 225–232.

JENKINSON, E. J., and SEARLE, R. F. (1977).
Cell surface changes on the mouse blastocyst at implantation. Exp. Cell. Res. (in press).

JENKINSON, E. J., and WILSON, I. B. (1973).
In vitro studies on the control of trophoblast outgrowth in the mouse. J. Embryol. Exp. Morph. *30*, 21–30.

JOHNSON, M. H. (1975a).
The macromolecular organisation of membranes and its bearing on events leading up to fertilization. J. Reprod. Fertil. *44*, 167–184.

JOHNSON, M. H. (1975b).
Antigens of the peri-implantation trophoblast. *In* Immunobiology of Trophoblast, R. G. Edwards, C. W. S. Howe, and M. H. Johnson, eds. (London: Cambridge University Press), pp. 87–100.

JOHNSON, M. H., EAGER, D., MUGGLETON-HARRIS, A., and GRAVE, H. M. (1975).
Mosaicism in organisation of concana-

valin A receptors on the surface membrane of the mouse egg. Nature *257*, 321–322.

KIRBY, D. R. S. (1968).
The immunological consequences of extrauterine development of allogeneic mouse blastocysts. Transplantation *6*, 1005–1009.

KIRBY, D. R. S., BILLINGTON, W. D., and JAMES, D. A. (1966).
Transplantation of eggs to the kidney and uterus of immunised mice. Transplantation *4*, 713–718.

KOMETANI, K., PAINE, P., COSSMAN, J., and BEHRMAN, S. J. (1973).
Detection of antigens similar to placental antigens in mouse fertilized eggs by immunofluorescence. J. Obstet. Gynec.. *116*, 351–357.

LEBLOND, C. P., and BENNETT, G. (1974).
Elaboration and turnover of cell coat glycoproteins. *In* The Cell Surface in Development, A. A. Moscona, ed. (New York: Wiley), pp. 29–49.

LUFT, J. H. (1971).
Ruthenium red and violet 11. Fine structural localisation in animal tissues. Anat. Rec. *171*, 369–416.

MINTZ, B. (1964).
Formation of genetically mosaic mouse embryos, and early development of "lethal (t^{12}/t^{12})-normal" mosaics. J. Exp. Zool. *157*, 273–292.

MOSCONA, A. A. (1973).
Cell Aggregation. *In* Cell Biology in Medicine. E. E. Bittar, ed. (New York: Wiley), pp. 571–591.

MOSKALEWSKI, S., and KOPROWSKI, H. (1972).
Presence of egg antigen in immature oocytes and pre-implantation embryos. Nature *237*, 167.

MUGGLETON-HARRIS, A. L., and JOHNSON, M. H. (1976).
The nature and distribution of serologically-detectable alloantigens on the pre-implantation mouse embryo. J. Embryol. Exp. Morph. *35*, 59–72.

MUNRO, A. J. (1975).
Antigenic topography and the interaction of antibodies and immune cells with surface antigens. *In* Immunobiology of Trophoblast, R. G. Edwards, C. W. S. Howe and M. H. Johnson, eds. (London: Cambridge University Press), pp. 5–12.

NATHENSON, S. G. (1970).
Biochemical properties of histocompatibility antigens. Ann. Rev. Genet. *4*, 69–90.

NICOLAS, J. F., DUBOIS, P., JAKOB, H., GAILLARD, J., and JACOB, F. (1975).
Teratocarcinome de la souris: Differenciation en culture d'une lignée de cellules primitives à potentialités multiples. Ann. Microbiol. (Inst. Pasteur) *126A*, 3–22.

NICOLSON, G. L., and SINGER, S. J. (1971).
Ferritin-conjugated plant agglutinins as specific saccharide stains for electron microscopy: application to saccharides bound to cell membranes. Proc. Nat. Acad. Sci. USA *68*, 942–945.

NICOLSON, G. L., and SINGER, S. J. (1974).
The distribution and assymetry of mammalian cell surface saccharides utilising ferritin-conjugated plant

agglutinins as specific saccharide stains. J. Cell Biol. *60*, 236–247.

NICOLSON, G. L., YANAGIMACHI, R., and YANAGIMACHI, H. (1975).
Ultrastructural localization of lectin-binding sites on the zonae pellucidae and plasma membranes of mammalian eggs. J. Cell Biol. *66*, 263–274.

NILSSON, O. (1974).
The morphology of blastocyst implantation. J. Reprod. Fertil. *39*, 187–194.

NILSSON, O., LINDQVIST, I., and RONQUIST, G. (1973).
Decreased surface charge of mouse blastocysts at implantation. Exp. Cell Res. *83*, 421–423.

NILSSON, O., LINDQVIST, I., and RONQUIST, G. (1975).
Blastocyst surface charge and implantation in the mouse. Contraception *11*, 441–450.

OHTA, N., PARDEE, A. B., MCAUSLAN, B. R., and BURGER, M. M. (1968).
Sialic acid contents and controls of normal and malignant cells. Biochim. Biophys. Acta. *158*, 98–102.

OIKAWA, T., NICOLSON, G. L., and YANAGIMACHI, R. (1974)
Inhibition of hamster fertilization by phytoagglutinins. Exp. Cell Res. *83*, 239–246.

OIKAWA, T., YANAGIMACHI, R., and NICOLSON, G. L. (1973).
Wheat germ agglutinin blocks mammalian fertilization. Nature *241*, 256–259.

OIKAWA, T., YANAGIMACHI, R., and NICOLSON, G. L. (1975).
Species differences in the lectin-binding sites on the zona pellucida of rodent eggs. J. Reprod. Fertil. *43*, 137–140.

OLDS, P. J. (1968).
An attempt to detect H-2 antigens on mouse eggs. Transplantation *6*, 478–479.

PALM, J., HEYNER, S., and BRINSTER, R. L. (1971).
Differential immunofluorescence of fertilized mouse eggs with H-2 and non-H-2 antibody. J. Exp. Med. *133*, 1282–1293.

PARKENING, T. A., and CHANG, M. C. (1976).
Effects of wheat germ agglutinin on fertilization of mouse ova *in vivo* and *in vitro*. J. Exp. Zool. *195*, 215–222.

PATTHEY, H. L., and EDIDIN, M. (1973).
Evidence for the time of appearance of H-2 antigens in mouse development. Transplantation *15*, 211–214.

PERDUE, J. F., KLETZIEN, R., and WRAY, V. L. (1972).
The isolation and characterization of plasma membrane from cultured cells. IV. The carbohydrate composition of membranes isolated from oncogenic RNA virus-converted chick embryo fibroblasts. Biochim. Biophys. Acta. *266*, 505–510.

PINSKER, M. C., and MINTZ, B. M. (1973).
Changes in cell surface glycoproteins of mouse embryos before implantation. Proc. Nat. Acad. Sci. USA *70*, 1645–1648.

PINSKER, M. C., SACCO, A. G., and MINTZ, B. M. (1974).
Implantation associated protease in mouse uterine fluid. Dev. biol. *38*, 285–290.

RAMBOURG, A., and LEBLOND, C. P. (1967).
Electron microscopic observations on the carbohydrate-rich cell coat present at the surface of cells in the rat. J. Cell. Biol. *32*, 27–53.

ROSEMAN, S. (1970).
The synthesis of complex carbohydrates by multiglycosyl transferase systems and their potential function in cell adhesion. Chem. Phys. Lipids. *5*, 270–297.

ROSEMAN, S. (1974).
Complex carbohydrates and intercellular adhesion. *In* The Cell Surface in Development, A. A. Moscona, ed. (New York: Wiley), pp. 255–271.

SCHLAFKE, S., and ENDERS, A. C. (1975).
Cellular basis of interaction between trophoblast and uterus at implantation. Biol. Reprod. *12* 41–65.

SEARLE, R. F., JENKINSON, E. J., and JOHNSON, M. H. (1975).
Immunogenicity of mouse trophoblast and embryonic sac. Nature *255*, 719–720.

SEARLE, R. F., JOHNSON, M. H., BILLINGTON, W. D., ELSON, J., and CLUTTERBUCK-JACKSON, S. (1974).
Investigation of H-2 and non-H-2 antigens on the mouse blastocyst. Transplantation *18*, 136–141.

SEARLE, R. F., SELLENS, M. H., ELSON, J., JENKINSON, E. J., and BILLINGTON, W. D. (1976).
Detection of alloantigens during preimplantation development and early trophoblast differentiation in the mouse by immunoperoxidase labelling. J. Exp. Med. *143*, 348–359.

SELLENS, M. H. (1976).
Studies on the antigenic status of the early mouse embryo. Ph.D. dissertation, University of Bristol.

SELLENS, M. H. and JENKINSON, E. J. (1975).
Permeability of the mouse zona pellucida to immunoglobulin. J. Reprod. Fertil. *42*, 153–157.

SHARON, N., and LIS, H. (1972).
Lectins: cell-agglutinating and sugar-specific proteins. Science *177*, 949–959.

SHERMAN, M. I., and ATIENZA, S. B. (1975).
Effects of bromodeoxyuridine, cytosine arabinoside and Colcemid upon *in vitro* development of mouse blastocysts. J. Embryol. Exp. Morph. *34*, 467–484.

SHERMAN, M. I., and WUDL, L. R. (1976).
The implanting mouse blastocyst. *In* The Cell Surface in Animal Development, G. Poste and G. L. Nicolson, eds. (Amsterdam: Associated Scientific Publishers), pp. 81–125.

SIMMONS, R. L. LIPSCHULTZ, M. L. RIOS, A., and RAY, P. K. (1971).
Failure of neuraminidase to unmask histocompatibility antigens on trophoblast. Nature New Biol. *231*, 111–112.

SIMMONS, R. L., and RUSSELL, P. S. (1965).
Histocompatibility antigens in transplanted mouse eggs. Nature *208*, 698–699.

SINGER, S. J., and NICOLSON, G. L. (1972).
The fluid mosaic model of the struc-

ture of cell membranes. Science *175*, 720–731.

SNOW, M. H. L. (1973).
The differential effect of H³-thymidine upon two populations of cells in pre-implantation mouse embryos. *In* The Cell Cycle in Development and Differentiation, M. Balls and F. S. Billett, eds. (London: Cambridge University Press), pp. 311–324.

SOLTER, D., and KNOWLES, B. B. (1975).
Immunosurgery of mouse blastocyst. Proc. Nat. Acad. Sci. USA *72*, 5099–5102.

SZULMAN, A. E. (1973).
ABO incompatibility in foetal wastage. Personal communication. Res. Reprod. *5*(1), 3–4.

TAYLOR, R. B., DUFFUS, W. P. H., RAFF, M. C., and DE PETRIS, S. (1971).
Redistribution and pinocytosis of lymphocyte surface immunoglobulin molecules induced by anti-immunoglobulin antibody. Nature New Biol. *233*, 225–229.

VAN BLERKOM, J. (1975).
Qualitative aspects of protein synthesis during the pre-implantation stages of pregnancy. Personal communication. Res. Reprod. *7*(5), 2–3.

VAN BLERKOM, J., and BROCKWAY, G. O. (1975).
Qualitative patterns of protein synthesis in the pre-implantation mouse embryo. 1. Normal pregnancy. Dev. Biol. *44*, 148–157.

VANDEPUTTE, M., and SOBIS, H. (1972).
Histocompatibility antigens on mouse

blastocysts and ectoplacental cones. Transplantation *14* 331–338.

VICKERS, M. G., and EDWARDS, J. G. (1972).
The effect of neuraminidase on the aggregation of BHK21 cells transformed by polyoma virus. J. Cell. Sci. *10*, 759–768.

VOLKOVA, L. S., and MAYSKY, I. N. (1969).
Immunological interaction between mother and embryo. *In* Immunology and Reproduction, R. G. Edwards, ed. (London: International Planned Parenthood Federation), pp. 211–230.

WEISS, L. (1965).
Studies on cell deformability. I. Effect of surface charge. J. Cell. Biol. *26*, 735–739.

WILEY, L. D., (1974).
Presence of gonadotrophin on the surface of pre-implanted mouse embryos. Nature *252*, 715–716.

WILEY, L. D., and CALARCO, P. G. (1975).
The effects of anti-embryo sera and their localization on the cell surface during mouse pre-implantation development. Dev. Biol. *47*, 407–418.

WINZLER, R. (1970).
Carbohydrates in cell surfaces. Int. Rev. Cytol. *29*, 77–125.

YAHARA, I., and EDELMAN, G. M. (1972).
Restriction of the mobility of lymphocyte immunoglobulin receptors by concanavalin A. Proc. Nat. Acad. Sci. USA *69*, 608–612.

YANAGIMACHI, R., and NICOLSON, G. L. (1974).
Changes in lectin binding to plasma membranes of hamster eggs during

maturation and pre-implantation development. J. Cell. Biol. *63*, 381.

YANAGIMACHI, R., NICOLSON, G. L., NODA, Y. D., and FUJIMOTO, M. (1973). Electron microscopic observations on the distribution of acidic anionic residues on hamster spermatozoa and eggs before and during fertilization. J. Ultrastruct. Res. *43*, 344–353.

YOSHINAGA, K., and ADAMS, C. E. (1966). Delayed implantation in the spayed, progesterone treated adult mouse. J. Reprod. Fertil. *12*, 593–595.

6 TUMOR VIRUS EXPRESSION DURING MAMMALIAN EMBRYOGENESIS

Rudolf Jaenisch and
Anton Berns

6.1 Introduction

Over the last decade tumor viruses have contributed to an increased understanding of the fundamental processes of cellular growth control in a great variety of systems (Tooze, 1973). Tumor viruses having either DNA or RNA as genetic material can transform cells in vitro and can cause tumors in animals. RNA tumor viruses can be divided into two groups: exogenous viruses and endogenous viruses. The genetic information of endogenous viruses is present in all the somatic and germ cells of all individuals of a species, and these virus-related genes are transmitted genetically according to Mendelian expectations (Payne and Chubb, 1968; Bentvelzen, 1972; Rowe, 1972). The most extensively researched viruses have been murine C-type viruses. As a result of this research a number of distinct classes of endogenous viruses have been defined. Ecotropic viruses, by definition, are able to infect the cells of the organism that carries them. Murine ecotropic viruses might cause leukemia or other diseases in mice (Gross, 1970). In contrast, xeno-tropic viruses are not able to infect cells of species that carry them, but can readily infect and replicate in the cells of other species (Levy, 1973; Stephenson et al., 1975). The chromosomal loci of the structural genes of some endogenous viruses have been mapped genetically (Rowe et al., 1972; Stockert et al., 1972, 1976) or biochemically (Chattopadhya et al., 1975). It is apparent that we are just beginning to understand the very complex biology of RNA tumor viruses, and it is likely that the mouse genome harbors the genetic information for many more still undefined C-type viruses that might or might not give rise to infectious particles. Several reviews on RNA tumor viruses have been published recently (Hirsch and Black, 1975; Vogt, 1977).

In contrast to endogenous viruses, exogenous or horizontally infect-ing viruses are not transmitted genetically from parents to offspring and thus are not part of the normal genetic complement of an animal (Essex, 1975). Whereas it is reasonable to assume that exogenous C-type viruses evolve from endogenous viruses by recombination or mutation events (Weiss et al., 1973; Stephenson et al., 1974), the converse event, namely, the conversion of exogenous viruses into endogenous viruses, seems to be very rare. However, recent evidence suggests that the transfer of exogenous viruses into the germ line of another species has occurred during evolution (Todaro et al., 1975; Benveniste and Todaro, 1975) and can be demonstrated under defined laboratory conditions (Jaenisch, 1976). In contrast, no endogenous DNA-containing viruses have been described so far, and all known DNA tumor viruses are exogenous or horizontally infecting viruses.

The genetic information of an RNA tumor virus is sufficient to code for proteins of a total molecular weight of only 300,000 daltons (Vogt, 1977). Isolation of most, if not all, gene products of C-type leukemia viruses is complete, and sensitive assays have been developed to test for their cellular expression (Strand and August, 1974). The major proteins of murine leukemia viruses are the envelope protein (gp70), the core proteins (p30, p15, p12, and p10), and the reverse transcriptase (Tooze, 1973). The so-called src gene is an integral part of avian and murine sarcoma viruses (Stehelin et al., 1976; Scolnick et al., 1976) and is thought to be responsible for the maintenance of fibroblast transformation (Martin, 1970; Wang et al., 1976). The product of this gene is not known, but it has been shown that the viral src gene is homologous to cellular sequences present in the normal untransformed cell (Stehelin et al., 1976; Scolnick et al., 1976). The expression of these sequences is not restricted to tumor cells or to embryonic cells. It appears to occur in most tissues or tissue culture cells tested and is possibly related to the process of cell division (Varmus et al., 1976).

As discussed above, endogenous RNA tumor viruses are an integral part of the genetic makeup of an animal and can be considered essentially normal cellular genes. Because these virus-related genes can be isolated in purified form as particles, they offer a unique opportunity to study the regulation and controls of a discrete set of mammalian genes on a molecular basis.

A number of hypotheses postulate a role for tumor viruses in embryogenesis (Huebner et al., 1970), cellular differentiation, and the evolution of species (Temin, 1971). Indeed, a close relationship appears to exist between the extent of differentiation of a cell and the way it responds to tumor virus infection or expression:

1. RNA tumor viruses integrate into the genome of infected cells (Varmus et al., 1973), and it is possible that the site of integration is specific for a given virus (Rowe et al., 1972; Stockert et al., 1972; Shoyab et al., 1976). When injected into animals, different viruses cause very different diseases. For example, AKR-mouse leukemia virus (AKR-MuLV) and Moloney-mouse leukemia virus (M-MuLV) cause predominantly a thymus-derived leukemia (Gross, 1970), but not an erythroblastic leukemia, which Friend virus induces, or a myloid

leukemia, which Abelson virus induces (Lilly and Pincus, 1972; Scher and Siegler, 1975; Rosenberg et al., 1975). Mouse mammary tumor virus, on the other hand, induces specifically mammary carcinoma (Bentvelzen, 1972). Thus it appears that each virus characteristically transforms specific "target" cells. Similarly, for avian RNA tumor viruses, agents have been isolated that cause specifically nephroblastomas, osteopetrosis, myeloblastosis, or lymphoblastosis (Ogura et al., 1974; Moscovici et al., 1975; Smith et al., 1976; Graf et al., 1976). That this tissue specificity, or "organtropism," holds true for exogenously infecting agents could simply be explained by assuming that a given virus can infect only its specific target cells, but it is also observed with endogenous viruses, the genetic information of which resides in every "target" and "nontarget" cell of an animal. These observations suggest that the state of differentiation of a cell determines its susceptibility to infection and transformation by a given tumor virus.

2. The expression of specific tumor virus genes responsible for the maintenance of the transformed state can directly influence the potential of a cell to differentiate. This fact has been demonstrated recently in elegant experiments using chick myoblasts infected with temperature-sensitive mutants of Rous sarcoma virus (Fiszman and Fuchs, 1975; Holtzer et al., 1975).

3. The transformation of differentiated cells with tumor viruses can activate specific genes that are normally expressed only during embryogenesis and are repressed in the healthy adult. This activation has been demonstrated by the induction of fetal cell surface antigens (Ambrose et al., 1971; Girardi et al., 1973) and of fetal globin (Groudine and Weintraub, 1975) following transformation by tumor viruses. These experiments raise the possibility that a close linkage exists between cellular transformation by tumor viruses and activation of genes involved in early development.

That cancer cells and embryonic cells have much in common is an old idea supported by the discovery of fetal antigens (Gold and Friedman, 1965; Brawn, 1971) and fetal isozymes (Shapira et al., 1970) in many tumors of the adult. These and other considerations have led to the notion that "oncogeny is blocked ontogeny" (Anderson and Coggin, 1971). That certain tumor cells (teratomas) can be induced to develop into healthy mice when placed in the right environment (e.g., Mintz and Illmensee, 1975) underlines the close similarity between tumor and embryo.

The oncogene hypothesis tries to bridge this relation by postulating that RNA-tumor-virus-related genes are not only involved in the tumor formation of adult animals, but also play a decisive role during early embryogenesis (Huebner and Todaro, 1969; Huebner et al., 1970). It is proposed that these "oncogenes" are expressed during early development and become repressed during the course of cellular differentiation. Tumor formation might result if accidental derepression occurs during later life.

This review considers some of the experimental evidence that relates the expression of tumor viruses to embryonic development and cellular differentiation. Because most experiments have been performed in the mouse system, we refer to other systems only peripherally. The first part considers the interaction of exogenous tumor viruses with developing mouse embryos; subsequently we discuss the expression of endogenous tumor viruses during embryogenesis.

6.2 Effect of Exogenous Tumor Viruses on Embryonic Development

The effect of infection by a variety of different viruses on cellular differentiation has been studied in vivo and in vitro in a number of systems. In this section, we discuss the interactions of DNA and RNA tumor viruses with developing mouse embryos. Embryos can be infected before implantation or during uterine development after implantation. Numerous experiments have been conducted to investigate the effect of tumor viruses on implanted embryos by infecting females before or during pregnancy, but the success of such an infection is difficult to assess (see below). Preimplantation embryos, on the other hand, can be infected reproducibly under well-defined conditions in vitro.

Preimplantation mouse embryos can be isolated from the fallopian tubes or the uterus of pregnant females at specific times after fertilization (Mintz, 1971). Isolated embryos at different stages of development can be cultured successfully up to the blastocyst stage in a simple, chemically defined medium (Whitten, 1971) and subsequently transferred to the uteri of pseudopregnant foster mothers to allow further development in vivo (Mintz, 1971). Thus, preimplantation embryos can be exposed to various viruses in vitro and the effect of virus infection on further development can be studied in vitro and in vivo.

Preimplantation embryos are surrounded by a protective protein layer, the zona pellucida, which can be removed with proteolytic enzymes (Mintz, 1971). Some lytic viruses, such as Mengo virus, can penetrate through the intact zona and and replicate in the embryos (Gwatkin, 1971), whereas other viruses, such as vesicular stomatitis virus, can replicate only after removal of the zona with pronase (R. Jaenisch, unpublished). Mouse embryos exposed to human adenovirus were arrested in development after one cell cycle (Chase et al., 1972), but it is not clear whether the zona has to be removed for successful penetration of this virus. The DNA tumor viruses, simian virus 40 (SV40) and polyoma, and the RNA Moloney leukemia virus appear to infect successfully only after digestion of the zona (Baranska et al., 1971; Sawiciki et al., 1971). Successful infection of mouse blastocysts without removal of the zona has been accomplished by microinjection of SV40 DNA into the blastocoel (Jaenisch and Mintz, 1974).

The following principal questions deserve attention here:

1. Do the conditions that allow successful infection of embryos in vitro exist under in vivo conditions?

2. What is the effect of infection on further cellular development and differentiation and what is the ultimate fate of infected mouse embryos?

3. Is permissiveness for viral replication and virus gene expression related to the stage of cellular differentiation?

We shall discuss the effect of two classes of oncogenic viruses on the development of the mouse, the DNA tumor viruses, SV40 and polyoma, and the RNA-containing Moloney murine leukemia virus.

6.2.1 Infection of Mouse Embryos with SV40 and Polyoma

When mouse fibroblasts are infected with SV40 or polyoma, two entirely different virus-cell interactions occur (Tooze, 1973). Polyoma is a lytic mouse virus, and early and late viral functions are expressed after infection. Virus DNA is synthesized, and the degenerating cells release infectious progeny particles. In contrast, SV40 infection of mouse cells can result in cellular transformation, but no lytic interactions occur. Thus, most cells are nonpermissive for virus DNA replication and no synthesis of virus particles takes place. However, early viral genes (T antigen), which are responsible for cellular transformation, are expressed. It appears that integration of polyoma and SV40 into the cellular genome is a prerequisite for successful transformation (Sambrook et al., 1968).

The evidence that preimplantation embryos are susceptible to polyoma and SV40 is based on the electron-microscopic detection of virus particles in the cytoplasm after infection (Biczysko et al., 1973), on rescue of infectivity by permissive indicator cells (Baranska et al., 1971), and on the experimental recovery of SV40-specific DNA sequences in adult mice derived from microinjected blastocysts (Jaenisch and Mintz, 1974). Preimplantation mouse embryos have been infected as early as the 1-cell stage and as late as the blastocyst stage. The normal in vitro development of infected 2- to 4-cell stage embryos (up to the blastocyst stage) was not detectably affected (Sawicki et al., 1971), and live mice have been obtained from SV40- and polyoma-infected blastocysts (Jaenisch and Mintz, 1974; unpublished observations). This result was surprising in view of the fact, mentioned above, that polyoma is a lytic virus for mouse cells. Therefore, experiments were performed to study whether these viruses can replicate in early mouse embryos.

Permissiveness of mouse embryos for replication of SV40 and polyoma
The possibility that mouse embryos at preimplantation stage are permissive for replication of SV40 and polyoma was raised in a study by Biczysko et al. (1973) and reviewed by Solter et al. (1974). Cellular uptake of SV40, but not of polyoma, was found by electron microscopy when 2-cell embryos were exposed to virus. When blastocysts were infected with both SV40 and polyoma, penetration of virus particles into nuclei was observed. The authors claimed that virus replication and synthesis of progeny particles took place in a number of different cells when explanted blastocysts or egg cylinders were exposed to SV40 and polyoma. These findings are surprising because, as pointed out above, SV40 does not replicate in mouse somatic cells.

However, there are serious objections to the interpretations of these data. The only evidence to support the notion that SV40 and polyoma can replicate in embryos consisted of electron-microscopic detection of viruslike particles in the nuclei of embryo cells several days after infection (Biczysko et al., 1973a). A distinction between input and progeny virus cannot be made using this method. Because both SV40 and polyoma are relatively stable viruses, it is possible that virus particles adsorbed to the cells and penetrated the cells only days after the original infection or that particles penetrated the cells at the time of infection and persisted in the cells without being uncoated. Therefore, we reinvestigated this question with different techniques (Jaenisch, 1974).

Mouse embryos were isolated on the third day of development (4- to 8-cell stage) and treated with pronase to remove the zona pellucida (Mintz, 1971). We placed the denuded embryos in Whitten's (1971) medium containing 1×10^7 plaque-forming units/ml of SV40. Four hours later we washed the embryos and incubated them for another 24 hr at 37°C. At this time, approximately 70 percent had developed to the blastocyst stage. They were labeled with ^{3}H-thymidine for an additional 30 hr, after which we extracted the DNA and centrifuged it to equilibrium in an ethidium bromide (EtBr)-CsCl gradient. Because SV40 DNA is a covalently closed circular molecule, newly synthesized viral DNA would band at a heavy position in such a gradient (Jaenisch et al., 1971). Figure 6.1 shows that most of the replicated cellular DNA banded at the density position of form II SV40 DNA (which was added as a marker). A small shoulder (seen on the expanded scale) appeared at the position of supercoiled SV40 DNA. When uninfected embryos were labeled and analyzed in a similar gradient, an almost identical profile was obtained, suggesting that the small shoulder at the position of supercoiled SV40 DNA represents mitochondrial DNA.

In order to detect SV40-specific DNA synthesized during the labeling period, the fractions indicated by bars in figure 6.1 were pooled and hybridized to membrane filters containing SV40 DNA or mouse DNA or to blank filters (table 6.1). Approximately 12–14 percent of the DNA extracted from uninfected or infected embryos hybridized to filters containing mouse DNA. No significant hybridization to filters containing SV40 DNA was observed. Under the same conditions, about 70 percent of SV40 DNA was bound to filters containing viral DNA.

These results suggest that preimplantation mouse embryos are not permissive for SV40 DNA replication. The limit of detection of virus replication in these experiments was 0.2 percent of the total cellular DNA replicated; we cannot exclude the possibility that virus replication took place to an extent that was not detected in these experiments.

Similarly, when polyoma-infected mouse embryos were labeled with ^{3}H-thymidine and DNA banded in EtBr-CsCl gradients, no virus DNA replication was detected (R. Jaenisch, unpublished).

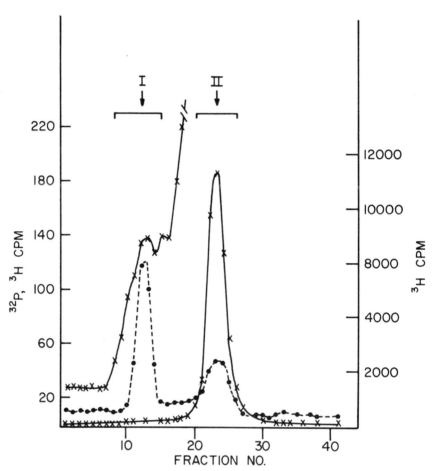

6.1 Density equilibrium centrifugation in EtBr-CsCl of ^{3}H-labeled DNA extracted from mouse blastocysts. Approximately 450 mouse embryos were isolated at the 4- to 8- cell stage, treated with pronase, infected with SV40 and labeled with ^{3}H-thymidine (5 mCi/ml. 55 Ci/mmole). DNA was isolated and analyzed in EtBr-CsCl gradients as described in the text. ●––●, ^{32}P-labeled SV40 DNA marker, showing positions of supercoils (I) and relaxed circles, also density position of cellular DNA (II). From Jaenisch (1974); reprinted, with permission, from *Cold Spring Harbor Symposium of Quantitative Biology*.

Table 6.1. Hybridization of Mouse Embryo DNA to Filters Containing SV40 or Mouse DNA

DNA in Solution			Input cpm	DNA Hybridized to Filters Containing					
				Mouse DNA		SV40 DNA		Blank	
				cpm	%	cpm	%	cpm	%
Mouse Embryo DNA	Infected with SV40	I[a]	750	110	14	22	—	12	—
		II[a]	69,000	7880	11.5	88	0.13	125	0.18
	Uninfected	II[a]	20,000	2580	13	54	0.27	40	0.2
SV40 DNA			400	8	—	280	70	10	—

[a] Denotes heavy (I) or light (II) positions of the EtBr-CsCl gradient.
Tritium-labeled DNA was extracted from mouse blastocysts and handed in EtBr-CsCl gradients, as described in figure 6.1. The fractions indicated by bars in figure 6.1, corresponding to the positions of SV40 components I and II, were pooled, treated with isopropanol and precipitated with ethanol. The DNA was dissolved and hybridized to membrane filters (6 mm diameter) previously loaded with 5μg denatured SV40 DNA or 50μg denatured mouse liver DNA. The incubation volume was 0.25 ml and hybridization proceeded for 18 hr at 68°C. The filters were extensively washed with 4 × SSC at 68°C. dried and counted. From Jaenisch (1974); reprinted, with permission, from *Cold Spring Harbor Symposium of Quantitative Biology*.

Our results suggest that preimplantation mouse embryos are nonpermissive for replication of SV40 and polyoma, in contrast to the experiments published by Biczysko et al. (1973a). Because polyoma can interact lytically with mouse somatic cells and replicate its DNA (see above), our results point to an interesting relation between the differentiated state of a cell and its permissiveness to oncogenic viruses. However, the experiments discussed above do not exclude a limited expression of early virus functions such as T antigen, whose expression might be necessary for viral integration. A similar relation has been found with Moloney leukemia virus, which can replicate efficiently in mouse fibroblasts but cannot express any functions in preimplantation mouse embryos, although the virus can integrate into the genome of the embryo (see below). In agreement with these interpretations are our observations that infection of embryos with SV40, polyoma, or Moloney leukemia virus does not detectably interfere with normal in vitro or in vivo development (Jaenisch and Mintz, 1974; Jaenisch et al., 1975). Biczysko et al. (1973a), on the other hand, have observed degenerated cells in polyoma and SV40-infected embryos after prolonged cultivation in vitro. It is not clear whether these differences are due to differences in toxicity of the virus stocks used or to other experimental variations.

A similar relation between permissiveness for tumor virus replication and extent of cellular differentiation has been observed in the teratoma system (see chapter 7). When undifferentiated teratoma stem cells were infected with SV40 or polyoma, no expression of T antigen (SV40) or virus replication (polyoma) was observed (Swartzendruber and Lehman, 1975). Differentiated cells derived from the embryonal carcinoma stem cells, on the other hand, could efficiently replicate polyoma and express SV40 T antigen upon infection. This appears not to be due to a block of virus uptake by the cells (Lehman et al., 1975). Likewise, Moloney leukemia virus replicates in differentiated teratoma cells but is unable to grow in the undifferentiated stem cell (G. Martin and N. Teich, personal communication). So far it has not been excluded that the block in M-MuLV virus replication in the undifferentiated stem cells is not due to lack of receptors on those cells.

Persistence of SV40-specific sequences throughout mouse development
The observation that infection with virus does not interfere with normal in vitro development of preimplantation embryos could be due to the trivial possibility that SV40 or polyoma cannot stably infect mouse embryos and that infection is abortive. Direct biochemical analysis to demonstrate virus integration into the genome of infected embryos is not possible due to the limited number of embryos one can isolate.

To decide whether integration can take place and to study long-term effects of SV40 infection of mouse embryos on embryogenesis and adult life, we transplanted infected blastocysts into the uterine horns of pseudopregnant foster mothers to ensure further in vivo development (Jaenisch and Mintz, 1974; Jaenisch, 1974). In these experiments, we microinjected high concentrations of purified SV40 DNA into the blastocoel of blastocysts. The same fraction of injected and control embryos developed to birth and to healthy adult animals, indicating that injection with SV40 DNA does not interfere with embryonic development. The crucial question was whether the adult animals infected at the preimplantation stage carry any SV40-specific sequences in their somatic cells.

To decide this question, we killed the animals at one year of age and

Table 6.2. Detection of SV40 DNA in DNA Extracted from Various Mouse Organs

| Mouse No. | No. of SV40 Copies per Diploid Mouse DNA Value | |
	Liver and Kidneys	Brain
1	0	13
3	0.62	0.7
7	1.5	0
8	0.72	n.t.
11	1.0	n.t.
15	0	8.5
19	0	1.95
20	1.6	0
24	3.6	n.t.
29	0.5	0

DNA was extracted from mice derived from SV40 DNA-infected blastocysts and the number of SV40 copies per diploid mouse genome was determined by molecular hybridization. n.t. = not tested. From Jaenisch and Mintz (1974).

performed molecular hybridization experiments with DNA extracted from different organs to detect SV40-specific sequences. Table 6.2 summarizes our results. Between 5 and 13 SV40 genome equivalents were found in some organs of 40 percent of the adult survivors. This finding indicates that SV40 sequences can, in effect, survive through at least a substantial part of the animal's lifespan in the cells of mice injected at the blastocyst stage. The results suggest that the virus DNA could have become stably integrated into the mouse genome and thus was replicated with the cellular DNA. This replication of the integrated SV40 DNA is thought to occur by a cellular replication mechanism without expression of virus-specific functions. However, direct evidence for integration has not yet been obtained.

We did not observe any signs of expression of viral genetic functions, i.e., tumor formation, up to one year of age, and in breeding experiments with a limited number of animals, we have not found genetic transmission of the SV40 gene. We observed both virus expression and genetic transmission with preimplantation embryos infected with M-MuLV.

6.2.2 Interaction of Mouse Embryos with RNA Tumor Viruses

A common feature of RNA tumor viruses is their remarkable specificity of transformation, i.e., the restricted tissue or cell type a given virus can transform. This section is a review of the interaction of exogenous RNA tumor viruses with cells at different stages of embryonic development and differentiation. Many of these studies were undertaken to understand the basis of the organtropism of RNA tumor viruses.

Infection of newborn animals

Newborn mice infected with murine leukemia viruses such as M-MuLV develop a specific thymus-derived leukemia after a latency of several months (Gross, 1970). Two aspects of this process are important in this discussion. First, the development of leukemia follows the appearance of high titers of infectious leukemia virus in the blood (Lilly et al., 1975). Second, leukemic males do not transmit the virus genetically to the offspring (Gross, 1961; Law, 1966). Leukemic females, on the other hand, do transmit the disease efficiently to their offspring through the milk (Ida et al., 1966), but true genetic transmission has never been observed. These observations allow two conclusions. Injection of newborn animals with the virus leads to infection of certain cells, such as the thymus-derived lymphocytes, which probably produce the virus found in the blood, or to infection of the mammary epithelial cells, which possibly secrete high titers of infectious virus into the milk (Jenson et al., 1976). However, other tissues, such as the germ line cells, never seem to become infected. Similar conclusions arise from studies with chickens congenitally infected with avian leukosis virus (Rubin et al., 1961, 1962).

In experiments to determine which tissues of the newborn mouse can successfully be infected with M-MuLV, we extracted DNA from different organs of leukemic mice infected with virus after birth. Molecular hybridization experiments to determine the presence or absence of virus-specific DNA sequences in the different tissues demonstrated clearly that of seven different tissues tested, only the target organs of M-MuLV (i.e., the thymus, the spleen, and tumor tissues) contained virus-specific sequences (Jaenisch et al., 1975; Jaenisch, 1976). We detected no virus-specific sequences in other organs, such as kidneys, liver, brain, testes, muscle, and lungs. The parenchymal cells of these

nontarget organs were not available for infection even in the advanced
stages of the disease, when high titers of infectious virus are present in
the serum. Similar observations have been made with chickens infected
with avian myeloblastosis virus (Shoyab and Baluda, 1975). Dougherty
and DiStefano (1967), on the other hand, reported on the basis of
electron-microscopic detection of C-type particles that avian leukosis
virus replicated in many organs of congenitally infected chickens.
These experiments should be repeated with the more quantitative molecular hybridization techniques now available, because these methods
allow the identification of the virus and a clear distinction between
infecting and endogenous virus.

The results discussed above might illustrate the observed tissue
specificity of exogenous RNA tumor viruses (Jaenisch et al., 1976b).
They suggest that infection with these agents is restricted to only a few
cell types (i.e., target cells) of the differentiated organism. There are
possible exceptions to this rule. The mammary epithelial cells appear to
be susceptible to infection, as mentioned above. These cells are nontarget cells in the sense that they cannot be transformed by leukemia
virus, but they seem to become "productively" infected, i.e., they
produce virus (Jenson et al., 1976). Similarly, the uterus of AKR mice
carrying the endogenous AKR-MuLV has been demonstrated to
produce high titers of infectious virus (Rowe and Pincus, 1972). It is
interesting that cellular proliferation in both of these nontarget organs
producing leukemia virus is under hormonal control (see below).

A newborn mouse can be considered as an essentially fully differentiated animal. Which tissues are susceptible to RNA tumor viruses when
animals are infected at earlier stages of development, i.e., in utero, or
at preimplantation stages? Are embryos at different stages of development permissive for virus expression?

Infection of preimplantation embryos

When denuded early mouse embryos were infected with 10^4 plaque-
forming units/ml of Moloney sarcoma virus for one hour and subsequently cultured in vitro, no harmful effects on further development
were noticed (Sawicki et al., 1971; Baranska et al., 1971). These authors claimed to have rescued infectious virus from embryos four days
after infection, suggesting that preimplantation embryos are permissive
for virus replication. We reinvestigated this question, using for infections much higher concentrations of virus for longer periods of time.

We infected pronase-treated 4- to 8-cell stage mouse embryos with 10^8 xc plaque-forming units (Rowe et al., 1970) of cloned M-MuLV for six hours (Jaenisch et al., 1975). We washed the embryos and cultured them in vitro for 30 hr until they reached the blastocyst stage. As reported by Sawicki et al. (1971), no adverse effects on development were noticed. But, in contrast to their results, we did not detect the production of infectious virus, although we used highly sensitive techniques that would have detected a single infectious particle. By electron microscopy we could not demonstrate the presence of C-type particles in infected blastocysts, nor could we detect the expression of virus proteins p30 or gp70 with immunofluorescence (R. Jaenisch, unpublished). These experiments indicate that preimplantation mouse embryos are nonpermissive for leukemia virus expression. The results are similar to observations for the DNA tumor viruses, SV40 and polyoma (Jaenisch, 1974).

To study the further development and the ultimate fate of the infected embryos, we transplanted treated blastocysts to pseudopregnant foster mothers (Jaenisch et al., 1975). Survival to birth was equal for infected and uninfected embryos, suggesting that there are no detrimental effects of the virus on further in vivo development to term. That infectious competent virus did indeed integrate into some of the blastomeres of the 4- to 8-cell stage embryos was demonstrated by the observation that some animals developed an M-MuLV-induced leukemia at later times. Molecular hybridization experiments revealed that these animals carried M-MuLV-specific sequences in all organs tested. This discovery contrasts sharply with the results obtained with leukemic animals infected after birth (see above). The results indicate that infection of embryos at the preimplantation stage, before any differentiation into target and nontarget tissues has taken place, circumvents the organtropism of M-MuLV, so that the affected animals can carry Moloney-specific sequences in all cells. So far, however, we have obtained only one animal that carried the maximum of two to three M-MuLV copies per haploid mouse genome equivalent integrated into every cell (Jaenisch et al., 1975). The majority of animals obtained after infection of preimplantation embryos appear to be mosaics that can carry M-MuLV integrated in any fraction of cells of a given organ (Jaenisch, 1976; Jaenisch et al., 1976a).

As mentioned above, viremic mice infected after birth with leukemia virus never transmit the virus genetically. We tested whether the

nontarget organ germ line carried M-MuLV in animals derived from
M-MuLV-infected embryos. A viremic male was bred with uninfected
females to yield the first backcross generation (i.e., N-1 animals), and
the progeny were tested for infectious virus (Jaenisch, 1976). Figure 6.2
shows that the animal infected at the 4- to 8-cell stage (No. 339) transmitted the virus to his offspring in a non-Mendelian manner at decreas-
ing frequency with increasing age. That genetic transmission did occur
was demonstrated by breeding his viremic sons with uninfected females
to obtain the N-2 generation. Table 6.3 indicates that viremic N-1, N-2,
and N-3 animals derived from Male No. 339 transmitted the disease in
a strictly Mendelian manner to 50 percent of their offspring. The implication is strong that these animals were heterozygous for a single locus
responsible for M-MuLV synthesis. Heterozygosity of backcrossed
animals is supported by molecular hybridization experiments demonstrating the presence of one-half copy of M-MuLV per haploid mouse
genome (Jaenisch, 1976). Recently, animals homozygous for this locus
were obtained from offspring of brother-sister matings (R. Jaenisch,
unpublished). During the process of leukemia development, somatic
amplification of M-MuLV-specific sequences occurred in the target
cells only, not in nontarget tissues (Jaenisch, 1976; Berns and Jaenisch,
1976). The non-Mendelian transmission (figure 6.2) of virus from the
original male (No. 339) infected at the preimplantation stage can be
explained by assuming that this animal was a "germ-line" mosaic, as
discussed elsewhere (Jaenisch, 1976).

Based on our experimental results, we propose that preimplantation
mouse embryo cells are not permissive for leukemia virus replication
and expression, although DNA transcribed from the infecting virus can
integrate into the mouse genome. The integrated virus DNA appears
not to express any virus-specific functions, but is replicated by a cellular mechanism as part of the cellular genome. This nonpermissivity of
early multipotential embryonic cells for tumor virus replication changes
to permissivity during subsequent differentiation in certain tissues (target tissues), but not in other tissues. Several observations support this
hypothesis. First, the occurrence of mosaic animals after infection of
preimplantation embryos with M-MuLV supports the idea that early
mouse embryos are nonpermissive for virus replication. If embryonic
cells were permissive for virus replication at the time of infection, virus
should have spread throughout the embryo and given rise to a majority
of animals carrying equal amounts of M-MuLV sequences in all cells.

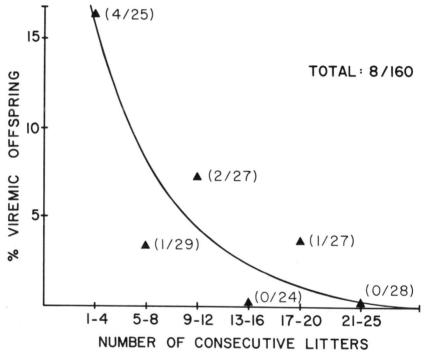

GENETIC TRANSMISSION OF M-MuLV FROM
MALE INFECTED AT PRE-IMPLANTATION
STAGE (#339) TO NEXT GENERATION

6.2 Transmission of M-MuLV to N-1 generation from viremic male infected at the pre-implantation stage. Male No. 339 was originally infected with M-MuLV at the 4- to 8-cell stage and mated with uninfected BALB/c females when he was 10 weeks old. The progeny were tested for M-MuLV in the serum. The last five litters tested were fathered when the animal was 10–12 months old. From Jaenisch (1976).

Table 6.3. Genetic Transmission of M-MuLV

Male Mouse No.	Mode of Infection with M-MuLV	Transmission of M-MuLV to Offspring after Mating with Uninfected Females	
		No. of Viremic Offspring	Percent
339	4- to 8-cell preimplantation stage	8/160	5
921	N−1 of No. 339	21/39	54
984		16/30	53
901−3	N−2 of No. 339	14/31	45
901−10		28/59	48
901−10−5	N−3 of No. 339	10/22	45
		Total 89/181	49
1	Infected as	0/35	0
2	newborns with	0/29	0
4	M-MuLV	0/32	0
9		0/25	0
11		0/46	0
		Total 0/167	

Male No. 339, infected at the 4- to 8-cell stage with M-MuLV, was bred with uninfected BALB/c females to give the first backcross generation (N-1 generation; figure 6.2). Virus-positive N-1 animals were identified as described in the text and bred with uninfected females to give the second backcross (N-2 generation). N-3 animals were obtained similarly. Viremic males infected as newborns with M-MuLV were bred with BALB/c females as controls. From Jaenisch (1976).

This is not the case. Most animals obtained from infected preimplantation embryos carry no or different concentrations (due to somatic
mosaicism) of M-MuLV copies in their respective tissues (Jaenisch et
al., 1976a).

Second, somatic amplification is detectable in target tissues only, and
this fact suggests that only these cells can replicate virus and integrate
additional copies (Jaenisch, 1976; Berns and Jaenisch, 1976). Although
the mechanism of the somatic amplification of virus DNA is not
known, the available evidence suggests that superinfection of the target
cells with virus might be involved in this process [see Berns and Jaenisch (1976) for discussion of this point]. In agreement with this concept
is the fact that only target cells are susceptible to infection when animals are exposed to virus after birth (Jaenisch, 1976).

Third, hybridization experiments to quantitate the concentration of
virus-specific RNA in target and nontarget tissues of M-MuLV-carrying
animals indicated that nontarget tissues contained 50- to 100-fold less
virus-specific RNA as compared with tumor tissues, indicating repression of the virus genes in nontarget tissues (Jaenisch et al., 1975,
1976b).

Fourth, hybridization experiments with RNA extracted from AKR
embryos showed little or no evidence for AKR-virus-specific transcription except in fetal thymus at late gestation (A. Berns and R. Jaenisch,
unpublished). And finally, radioimmunoassays with monospecific antisera detected no or little virus protein expression in AKR embryos or
adults except in embryonic thymus or adult target organs (Hilgers et
al., 1974; Strand et al., 1977). All these experiments suggest that the
expression of leukemia virus genes integrated into cells of the developing mouse is repressed and becomes activated in specific cells only at a
later stage of development.

In agreement with this concept is our observation that infection with
M-MuLV has no detectable adverse effects on normal fetal development. This observation has been made also in chicks heavily infected
with avian leukosis virus through the ovum (Rubin et al., 1961). However, other workers claim to have seen inhibition of fetal development
after infection with RNA tumor viruses. Chapman et al. (1974) reported
that infection of hamster blastocysts with feline leukemia virus caused
death of the embryos during early gestation. Their conclusions are
clearly equivocal because essential controls are missing. The authors
did not remove the zona pellucida prior to infection. They reported no

evidence that successful infection with feline leukemia virus occurred, because they failed to characterize the virus they detected by electron microscopy in implanted embryos. Furthermore, they were unable to distinguish between dead and live embryos, and it seems quite possible that the in vitro manipulations of the blastocysts caused unspecific cell death after implantation. Therefore, the C-type particles they detected could represent activated endogenous virus in dying cells and could have nothing to do with the exogenous virus used for infection.

Congenital infection
RNA tumor viruses can be passed from parents to offspring by vertical transmission. Two modes of vertical transmission are recognized: congenital infection and genetic transmission (Rubin et al., 1961, 1962; Tooze, 1973). During genetic transmission, the viral genome is vertically transmitted from one generation to the next as a DNA provirus according to Mendelian expectations (Rowe, 1972; Payne and Chubb, 1968). Congenital infection, on the other hand, occurs when infectious virus particles released by the mother infect the offspring. The virus may infect the ovum directly (Rubin et al., 1961) or be transmitted via the placenta or milk. It must be released from a maternal cell and then enter a cell of the progeny where provirus is transcribed and integrated into the cell genome. The subject of this section is congenital infection of mice with leukemia virus and the effect on embryonal development.

Theoretically, a viremic mother can infect embryos during three different periods of development: prior to implantation, at or after implantation, or after birth via the milk. The clearest evidence is for milk transmission of leukemia virus. A number of investigators have demonstrated that up to 100 percent of normal newborn mice become infected with leukemia virus by foster nursing on viremic or leukemic mothers (Gross, 1961; Law, 1962; Ida et al., 1966; Buffet et al., 1969). The observation that milk of viremic females contains higher concentrations of infectious virus than serum (Jenson et al., 1976) underlines the important role of mother's milk in congenital infection of the offspring.

To demonstrate true transplacental infection of embryos with virus by testing for development of viremia or leukemia it is obviously important to prevent milk transmission. A variable percentage of newborn mice that were removed immediately after birth and foster-nursed on a normal, nonviremic female subsequently came down with leukemia. Law (1962) observed that 7 of 51, and Ida et al. (1966) found that 19 of

156, foster-nursed babies became viremic and leukemic. These authors
concluded that transplacental infection of embryos can take place, although with significantly lower frequency compared to milk infection.
However, the design of these experiments did not rigorously exclude
the possibilities that a few mice suckled before removal from their
mother and that infection of the embryos took place during birth, in the
birth canal. The latter possibility deserves serious consideration because the uterus of the mouse produces large amounts of virus (Rowe
and Pincus, 1972). These complications can be avoided by delivering
embryos by cesarean section and subsequent foster-nursing on normal
females. We have performed cesarean sections of embryos from highly
viremic mothers that were infected after birth with M-MuLV. To inactivate exogenous maternal virus, the embryos were dipped in ether
prior to being freed from fetal membranes and subsequently given to
nonviremic foster mothers. Our preliminary observations indicate that
none of 45 mice born and nurtured in this way became infected with
Moloney virus (R. Jaenisch, unpublished). Although the number of
mice is small, these observations raise the possibility that transplacental infection of embryos occurs at a lower frequency than has been
assumed (Buffet et al., 1969).

   Umar et al. (1976), on the other hand, reported evidence that
Rauscher leukemia virus can cross the placenta under certain circumstances. They infected females with 4 mg of purified virus shortly before or after mating. The effect of infection with such a high concentration of virus was severe. Many females were unable to mate and only a
few became pregnant. However, even in those that became pregnant,
embryos did not survive past midgestation. At earlier developmental
stages embryos were found to have been infected: infectious Rauscher
virus was recovered from the embryos, and positive immunofluorescence for virus-specific proteins was observed in embryonic cells. It is
not clear in these experiments whether fetal death occurred as a consequence of virus infection of the embryos or as a consequence of the
severe effect of virus infection on the mother. These experiments
should be repeated under less severe conditions, e.g., with congenitally
infected mice during the preleukemic phase.

   There is no convincing evidence that preimplantation embryos can be
infected in vivo with RNA tumor viruses. In vitro experiments
(Sawicki et al., 1971; Jaenisch et al., 1975) indicate that removal of the
zona pellucida is essential for successful infection by viruses such as

SV40, polyoma, vesicular stomatitis virus, and M-MuLV. Because the zona is not lost until just prior to implantation, preimplantation embryos should be resistant to tumor viruses under in vivo conditions. The following observations support this conclusion. It is clear from our data (Jaenisch, 1976) that preimplantation embryos can be infected success-fully in vitro with M-MuLV after removal of the zona pellucida and that genetic transmission of the virus can occur as one consequence of such an infection. Genetic transmission of exogenous virus has never been observed in congenitally infected animals, even when preimplan-tation embryos are exposed to high titers of virus in the genital tract (Rubin et al., 1961, 1962; Rowe and Pincus, 1972; B. Croker, personal communication). Hence congenital infection of embryos with RNA tu-mor viruses must occur at a later stage in development than the pre-implantation stage.

6.2.3 In Vitro Infection of Differentiating Cells with Tumor Viruses

Teratocarcinoma cells closely resemble some cells of the embryo (see chapter 7), as is most elegantly demonstrated by their potential to contribute to the development of normal mice (e.g., Mintz and Illmensee, 1975). Therefore, it is not surprising that infection with tumor viruses has the same effect on both teratoma stem cells and normal preimplantation embryos. As mentioned above, undifferentiated teratoma stem cells do not support the replication or gene expression of the DNA tumor viruses, SV40 and polyoma (Swartzendruber and Lehman, 1975), or the RNA leukemia virus, M-MuLV (G. Martin and N. Teich, personal communication). After differentiation of the stem cells, however, all three viruses are able to replicate or to express specific functions. These experiments support our concept that em-bryonic cells are nonpermissive for tumor virus replication but that these cells can become permissive during subsequent differentiation.

Virus infection of embryonic neuroretinal cells or fibroblasts appears to give different results. Rous sarcoma virus was able to transform neuroretinal cells from seven-day-old embryos, but cells derived from ten-day-old chick embryos were not transformed (Pessac and Calothy, 1974). Similarly, a host range mutant of polyoma appears to be able to replicate better in fibroblasts derived from younger mouse embryos than in cells derived from older embryos or adults (Goldman and Benjamin, 1975; T. Benjamin, personal communication). These experiments suggest that permissiveness for virus expression in embryonic cells changes to nonpermissiveness during futher differentiation. However, it should be kept in mind that the cells used in these experiments were derived from already highly differentiated embryos. Furthermore, the virus infections were performed in vitro while the cells were adapting to tissue culture conditons. Adaptation to tissue culture conditions can profoundly change the susceptibility of cells to virus infection and their ability to replicate virus. For example, secondary fibroblasts or established fibroblast lines can efficiently replicate M-MuLV and are used to prepare virus stocks. However, it appears that M-MuLV cannot infect fibroblasts in vivo or replicate when integrated into the genome of fibroblasts in the animal (R. Jaenisch, unpublished). Similarly explanted fetal liver cells are susceptible to infection with M-MuLV in vitro and can efficiently replicate the virus (R. Jaenisch and H. Leffert, preliminary observation). In vivo, however, liver cells do not become infected with virus, even when exposed to high titers of M-MuLV in viremic animals (Jaenisch et al., 1975; Jaenisch, 1976). Thus one should be cautious about extending observations made in vitro to the in vivo situation.

6.3 Expression of Endogenous Viruses during Embryogenesis

The expression of endogenous viruses during embryogenesis has been measured by the observation of virus particles or "viruslike structures" in the electron microscope and by the detection of virus gene products using biochemical or immunological methods. In this section, we describe and summarize such experiments with the intent of discussing the available evidence for a possible role of endogenous viruses in mammalian embryogenesis. It becomes clear that different investigators using similar and different experimental approaches have often reported contradictory results. These contradictions appear to be due to a number of circumstances, such as the complexity of the system under investigation, the difficulty in isolating sufficient amounts of defined embryonic tissues, the nonspecificity of certain techniques used to identify viral gene products, and poor quantitation. Thus, it is difficult, if not impossible, to arrive at a safe conclusion concerning the role of endogenous viruses during embryonic development.

In the first part of this section, we discuss the evidence for viral expression provided by the electron microscope. Subsequent parts deal with the evidence for expression of viral-specific RNA and of viral-specific proteins. The interpretation of the available data is somewhat biased toward a model that will be proposed in section 6.4.

6.3.1 Electron-Microscopic Identification of Viral Particles

During their assembly, RNA tumor viruses go through characteristic stages that can be recognized with the electron microscope. Viral particles of different shapes can be observed. Particles with an annular centrally placed nucleoid, termed A-type particles (see review by Sarkar et al., 1972), are seen intracellularly. Two types are distinguished, one located intracisternally, the other intracytoplasmically. The intracytoplasmic particles reportedly form the nucleoid of the B-particles of the mammary tumor agent (Dalton and Potter, 1968; Sarkar et al., 1972). The intracisternal A-particle seems not to be a precursor of any of the mature RNA tumor viruses (Kuff et al., 1972). Although it has a high-molecular-weight RNA (Yang and Wivel, 1973) and reverse transcriptase, no evidence of biological activity exists (Kuff et al., 1972, Wivel et al., 1973). C-type particles have a centrally placed spherical nucleoid and can be found extracellularly. Budding C-particles show a rather characteristic structure at the cell membrane and can be recognized easily with the electron microscope (Sarkar et al., 1972).

Intracisternal A-particles are abundant in a variety of mouse tumors (Bernard, 1960; Kakefuda et al., 1970; Wivel and Smith, 1971), in normal mouse tissue (Wivel and Smith, 1971), in guinea-pig oogonia (Andersen and Jeppesen, 1972), spermatogonia (Black, 1971), immature mouse oocytes (Calarco and Szollosi, 1973), mouse embryos (Enders and Schlafke, 1965), and rat tumors (Novikoff and Biempica, 1966). Although they are present in mouse oocytes, they are not seen in unfertilized eggs (Biczysko et al., 1973b). However, they are seen in 2-cell mouse embryos (Biczysko et al., 1973b; see also chapter 2) and in parthenogenically stimulated eggs at the 3- to 4-cell stage (Biczysko et al., 1974). The disappearance of these particles in eggs and their reappearance in the 2-cell embryo coincides with the cessation and resumption of ribosomal RNA synthesis (Calarco and Szollosi, 1973; Hillman and Tasca, 1969; Knowland and Graham, 1972). The highest frequency of occurrence is at the 4- to 8-cell stage (Calarco and Szollosi, 1973; Biczysko et al., 1973b; 1974). In blastocysts, A-particles are seen rarely (Biczysko et al., 1973b), and there are no reports of their presence in later-stage embryos. The significance of the observation of A-particles is far from clear. It is unknown how they originate and their viral nature is not established.

In contrast to A-type particles, C-type particles are easier to identify by their typical budding characteristics (Sarkar et al., 1972). Furthermore, the coincidence between electron-microscopic observations of C-particles on the one hand, and infectivity assays (Hartley et al., 1969) and the demonstration of viral RNA (Nowinski et al., 1970), reverse transcriptase (Mayer et al., 1974), and viral antigens (Nowinski et al., 1970) on the other, has made the electron-microscopic observation of C-particles acceptable evidence for the expression of RNA tumor virus genes. However, the reports describing C-type particles in mouse embryos are rather controversial. Dmochowsky et al. (1963) and Feldman and Gross (1964) found immature doughnutlike particles in thymus, spleen, and liver of AKR embryos, whereas Dirksen and Caileau (1967) found no such particles. Similarly, Chase and Pikó (1973) found C-particles in blastocysts of outbred Swiss albino mice, but Biczysko et al. (1973b) did not detect them in C57BL and ICR strains. Vernon et al. (1973) performed a more quantitative assay with embryos from several mouse strains in the second half of gestation. They found a close association between budding particles and hematopoietic cell types. The number of particles in twelfth- to fifteenth-day livers was considerably larger than in livers from embryos at later stages of gestation. This effect was seen with AKR, C3H, BALB/c, C57BL, and NIH strains. The number of particles in spleen and thymus varied significantly, and no consistent pattern emerged. Muscle tissue was uniformly negative for C-particles. Unpublished electron-microscopic examinations performed by B. Croker (personal communication) showed that significant numbers of particles could be found in the gall bladder of NZW embryos. C-type particles have also been observed in placentas from baboon (Kalter et al., 1973a), rhesus monkey (Schidlovsky and Ahmed, 1973), man (Kalter et al., 1973b; Vernon et al., 1974), and other primates (Seman et al., 1975). The presence of virus particles in placenta might seriously complicate the detection of viral gene expression by radioimmunoassay and complement fixation (see below) in early stages of development, as it might be difficult to dissect the embryos completely free of maternal tissue (Hilgers et al., 1974).

So far, the electron-microscopic observations have given rather equivocal results, the main reason being the quantitations that have been used. In our judgment, the best approach is that of B. Croker (personal communication), who determined the number of particles per 100 nuclei. In a similar type of quantitation, Vernon et al. (1973) calculated the number of particles per surface area. However, using the time of scanning as a measurement seems to be inadequate when positive virus expression is concluded from the detection of at least one particle per 30 min (Kalter et al., 1973a) and no virus expression is indicated by the failure to observe a particle in 2.5 hr (Kalter et al., 1973b). The lack of adequate quantitation is compounded by the difficulty in recognizing specific structure by electron microscopy. Except for C-type particles, there are apparently no definitive criteria for ascertaining the viral nature of an observed "viruslike" structure. This seems to be especially true for the identification of A-particles. Furthermore, a poor relationship exists between the occurrence of C-particles and infectivity assays. Although a highly sensitive in vitro plaque assay has been developed by Rowe and Pincus (1972), they were unable to detect virus in seventeenth- to eighteenth-day AKR mouse embryos and only trace amounts in eighteenth- or nineteenth-day embryos using this assay, even though the AKR strain has a high incidence of C-type particles (Vernon et al., 1973). This failure could mean either that the C-particles seen represent not the ecotropic but the xenotropic virus, which cannot be detected in the plaque assay, or that these particles are noninfectious for an unknown reason.

Summarizing briefly the data obtained with the electron microscope, A-type particles seem to be present rather uniformly in preimplantation embryos. A certain regularity is also seen for the frequency with which C-particles are detected in the hematopoietic liver. However, particles do not appear in every liver, as should be the case if this were an essential function for development. Furthermore, a very irregular pattern in the spleen and thymus suggests that the manifestation of particles is more accidental than functional. The observation that in many instances particles can be seen in normal fetal and adult tissues with an unpredictable pattern also argues against a specific role of these genes for cellular differentiation. This is shown clearly for epithelial tissues such as gall bladder, vas deferens, and lactating breast (B. Croker, personal communication), where the occurrence of viral particles differs by more than two orders of magnitude in different strains of mice.

The available evidence does not indicate that the occurrence of C-type particles is more frequent in embryos than in adults.

6.3.2 Identification of Viral-Specific RNA

Few experiments have been done so far to measure the presence of viral-specific RNA in developing embryos. Mukerjee et al. (1975) examined embryos from BALB/c mice for the expression of the endogenous N-tropic virus (Benveniste et al., 1974). Using a complementary DNA made in the endogenous reverse transcriptase reaction to detect viral-specific sequences, they found that the extent of hybridization of 5 to 7 percent of their probe with RNA from embryos was similar to or less than that found with RNA from the adult intestine and uterus and about 2 percent higher than that observed with adult liver and kidney RNA. Their conclusion that viral expression in embryos is greater than in adult tissues seems to be an overinterpretation of their results. The hybridization observed could represent cellular sequences, which often comprise a few percent of the viral genome (Garapin et al., 1973) rather than the viral sequences. Thus it is not clear whether the small extent of hybridization represents normal cellular gene expression, partial expression of the N-tropic endogenous virus, or the expression of another endogenous virus that has partial homology with the N-tropic virus.

Preliminary experiments using a specific AKR probe (A. Berns and R. Jaenisch, in preparation) to monitor AKR RNA expression in AKR midgestation embryos showed that extremely little AKR-virus-specific RNA was present, in agreement with the absence of infectious virus in this period (Rowe and Pincus, 1972). Later in gestation, a rapid increase in AKR-virus-specific RNA occurred in spleen and thymus, which are target organs of AKR-MuLV. Apparently the expression of AKR endogenous N-tropic virus is strictly controlled during embryogenesis. This factor does not mean that no viral expression occurs during earlier stages of embryogenesis. It might well be that the C-type particles seen in AKR embryos during this period of gestation (Vernon et al., 1973) represent one of the xenotropic endogenous viruses whose expression is differently controlled.

Nucleic acid hybridization offers a unique possibility for distinguishing between the expression of different endogenous viruses. By preparing probes that specifically recognize only one of the endogenous viruses, one should be able to follow the differential expression of several viruses at the same time. Furthermore, the technique allows an accurate and quantitative determination of the amount of viral-specific RNA present.

6.3.3 Detection of Viral Proteins

The two major protein components found in RNA tumor viruses are the internal structural proteins, which contain the group specific (gs) determinants, and the envelope glycoprotein gp 69/71. Strand et al. (1974) have shown that the expression of the internal structural protein p30 and the glycoprotein are not coordinate. Antibodies against both proteins have been raised and have been used in a series of immunological tests. The radioimmunoassay (Hunter, 1973), complement fixation test (Huebner et al., 1970), and immunofluorescence absorption test (Hilgers et al., 1972) are the main techniques utilized to detect viral proteins. In addition, immunofluorescence has been used (Hilgers et al., 1972; Lerner et al., 1976) to demonstrate the localization of expression of the RNA tumor virus antigens. The antisera used in earlier studies were often not well characterized and were raised by transplanting virus-induced sarcomas into rats (Huebner et al., 1970). The use of ill-defined antisera and the poor quantitation with complement fixation tests used in the earlier studies are probably the basis for the inability to reproduce various results using the highly sensitive radioimmunoassay. Huebner et al. (1971), using the complement fixation technique, detected gs antigens in a variety of mouse embryos, such as NIH, BALB/c, C3H, C57BL, and wild mice, with high frequency. However, the incidence varied; not all embryos of a given strain were positive or negative, e.g., NIH eleventh-day embryos were invariably positive, whereas feral mice were positive in only 10 percent of the cases. On the eleventh day after insemination, BALB/c embryos were uniformly positive, whereas they were completely negative on the fourteenth day. However, in an earlier paper, this group (Huebner et al., 1970) reported that eleventh-day embryos were negative.

In studies using the highly sensitive radioimmunoassay to detect p30 and gp70, no virus expression was detected during the second half of gestation in BALB/c embryos (figure 6.3; Strand et al., 1977). Similarly, AKR embryos of the same age showed little or no expression of both antigens, whereas C3H embryos expressed high amounts of gp70 only (figure 6.3). A similar picture has been described for NZB and

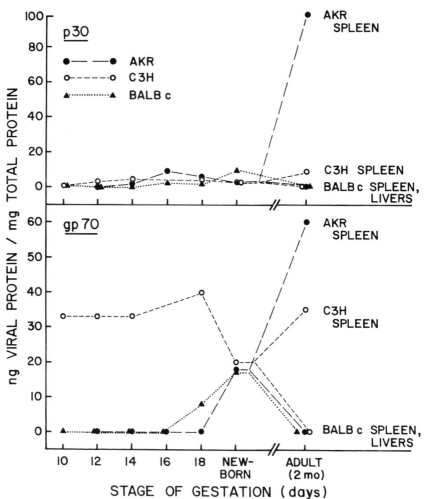

6.3 Expression of virus-related antigens during embryogenesis of AKR, C3H, and
BALB/c mice. Embryos at specific stages of development were isolated after timed
matings and virus-related proteins determined by radioimmunoassay using monospecific
antisera (Strand and August, 1974). In this figure, 10 days gestation age refers to the 11th
day of pregnancy, etc. From Strand et al. (1977).

NZW mice by Lerner et al. (1976), who found high levels of gp70 and moderate levels of p30 expressed during embryogenesis. In nineteenth-day embryos, the concentration of viral protein was significantly greater in tissues such as spleen and thymus than in liver, brain, and kidney (figure 6.3; Strand et al., 1977). Hilgers et al. (1974) had obtained similar results for the gs antigen. Both observations are in good agreement with the lack of significant amounts of infectious virus (Rowe and Pincus, 1972) or viral RNA (A. Berns and R. Jaenisch, in preparation) at this stage of development. The presence of gs antigens has also been reported for earlier-stage embryos. Huebner et al. (1970) detected high titers of gs antigens in the seventh-day BALB/c embryos, with their complement fixation test, and Hilgers et al. (1974) detected gs antigens in seventh- to ninth-day AKR embryos by the immuno-fluorescence absorption test. The latter authors, however, mentioned that they were not able to free the embryos from placental tissues, which have been shown to express viral antigens and consequently might have contaminated the embryo proper. Therefore, no conclusive data on viral antigen expression are available at these early stages of development. The experiments suggest that expression of viral antigens occurs frequently, although not regularly, during embryonic development. However, evidence has yet to be provided that correlates viral gene expression to a certain developmental stage of the mouse.

With immunofluorsecence, Lerner et al. (1976) demonstrated convincingly the localization of virus expression to certain tissues. They examined a variety of mouse strains for the sites of gp70 expression. Little staining was found in eleventh-day NZB, NZW, and C57BL fetuses, whereas positive staining was observed in all three strains on the fifteenth day. On the nineteenth day, the sites of expression were similar to the tissues that express gp70 in the adult mouse, and fluorescence was evident in epithelial, lymphoid, and hematopoietic tissues. Croker et al. (1976), on the other hand, detected 100-fold more budding C-type particles by electron microscopy in embryonic and adult epithelial tissues (vas deferens and gall bladder) than in spleens and lactating breast tissue of NZB and NZW mice.

In summarizing the experimental data on expression of virus-related proteins during early embryogenesis, we would like to emphasize the following points.

1. The available data show large quantitative differences in viral gene expression. Viral protein expression can vary by more than two orders

of magnitude in different mouse strains or between specific tissues of a given strain (Strand et al., 1974, 1977).

2. Viral antigens are never found in all cells of an individual animal, only in certain tissues. The pattern of virus expression is often characteristic for a given strain; e.g., NZB and NZW express gp 69/71 (and C-particles) predominantly in tissues such as gall bladder, vas deferens, epididymis, and significantly less virus proteins appear in breast, spleen and lymphoid tissue (Lerner et al., 1976; Croker et al., 1976). In contrast, AKR mice show exactly the opposite pattern: high particle expression in lactating breast, spleen, and lymphoid tissue and low expression in gall bladder and vas deferens (Croker et al., 1976). On the other hand, BALB/c or C57BL mice show little or no expression in any of these organs. However, explanted BALB/c cells can easily be induced to produce large amounts of viral proteins in vitro (Todaro, 1972; Lieber et al., 1973; Benveniste et al., 1974).

3. Endogenous virus genes can be activated by hormones (Parks et al., 1974). Great variations in inducibility exist among different strains. C57BL mouse tissues do not show any detectable antigen or reverse transcriptase activity following estrogen treatment (Fowler et al., 1973), whereas NIH and BALB/c strains start producing viral antigens and reverse transcriptase after treatment with this hormone (Fowler et al., 1972, 1973). In uteri of AKR mice, which already contain viral antigens, treatment does not increase production of antigens and reverse transcriptase (Fowler et al., 1973). Similarly, one can imagine that cells in developing embryos, inducible by hormones, show the same inconsistent response seen in the adult mouse. Therefore, it is possible that the variations seen during embryonic development is influenced by external factors such as the change of hormonal balances during pregnancy and that these variations might be strain-specific. Thus it appears that the irregularities observed in the expression of viral antigens in embryonic tissues and in adult animals differ greatly among different strains and no pattern of expression consistent with a functional role of these genes in tissue differentiation can be deduced.

4. The use of monospecific antisera has shown either coordinate expression of p30 and gp70 in strains such as AKR and C58/J (Strand et al., 1974) or noncoordiante expression, as in C3H (figure 6.3; Strand et al., 1977), DBA/2 (Strand et al., 1974), and 129 (Lerner et al., 1976). Expression of only gp70 is observed more frequently than the expres-

sion of p30 alone. Coordinate expression of both p30 and gp70 generally accompanies synthesis of virus particles.

6.4 Conclusions and Speculations

The aim of this article was to review the involvement of tumor viruses in cellular differentiation and embryonic development and the effect of differentiation on virus replication and expression. One of the central and most controversial problems of modern tumor virology is the question about the biological role, if any, that endogenous RNA tumor viruses might play in oncogenesis and in embryogenesis. Hypotheses have been proposed for specific and essential functions of tumor viruses in cellular differentiation (Huebner et al., 1970), evolution of the species (Temin, 1971), and oncogenesis (Huebner et al., 1971; Todaro and and Huebner, 1972). Most relevant experiments related to these questions have been performed in the mouse system, so we restrict our discussion to murine tumor viruses.

Several cautions should be exercised in the interpretation of experimental results from RNA tumor virus systems. The biology of RNA tumor viruses is extremely complex and not well understood. So far two major classes, the ecotropic and the xenotropic classes of endogenous murine viruses, have been defined, but it appears likely that the mouse genome harbors many additional virus-related genes, which might or might not give rise to virus particles or virus-related proteins. Detection of viruslike particles by electron microscopy, for example, does not permit any conclusions concerning the relatedness of these particles to the known and characterized viruses. The viral nature of many structures detected by electron microscopists is doubtful, and, as in the case of A-type particles, nothing can be said about their biological function. Furthermore, many results obtained by one group of investigators have not been confirmed or have directly contradicted the observations made by others, so firm conclusions are exceedingly difficult to draw. And finally, one should consider that the complicated biology of RNA tumor viruses appears simple and clear when compared with the extreme complexity of mammalian embryogenesis. With these reservations in mind, we shall attempt to summarize the relevant observations related to the expression of RNA tumor viruses during the embryonic development of the mouse.

Experimental evidence to support the hypothesis that RNA tumor
viruses might be important for embryonic development comes from the
electron-microscopic observation of viruslike structures in embryos and
from the detection of virus-related proteins by immunological procedures. Because the biological relevance of A-type particles is completely unknown, we consider the detection of only C-type particles an
indication of virus gene expression. As discussed above, C-type particles are visible in a variety of mouse strains at different stages of
development, but a satisfying quantitation of the particles seen has
been a problem in most experiments. Although older reports describing
the presence of gs antigen in embryos of various mouse strains by
complement fixation (Huebner et al., 1970) were not confirmed (Strand
et al., 1977) or were only partly confirmed (Hilgers et al., 1974) by
more recent experiments using highly sensitive techniques, it appears
safe to suggest that antigens related to p30 of murine leukemia viruses
are expressed occasionally during embryogenesis in certain mouse
strains. The evidence of expression of proteins related to the viral
glycoprotein gp70 in embryos and a variety of tissues is more convincing (Lerner et al., 1976). Do these types of observations indeed justify
the suggestion that viruses play an important or crucial role (Huebner
et al., 1970; Todaro and Huebner, 1972) during mammalian embryogenesis or even the proposal of a "viral etiology of differentiation" (Kalter
et al., 1973)?

It is obvious that the mere detection of virus-related gene products
or of viruslike structures does not indicate at all that these genes have
any function for cellular differentiation. A causal relationship between
virus gene expression and embryogenesis might be impossible to prove
or disprove experimentally. But answers to the following questions
should yield at least circumstantial evidence in favor of or against the
validity of this concept.

1. If the gene product of an endogenous virus is essential for embryogenesis, it should be needed for the embryonic development of *all*
strains of mice and therefore should be expressed at similar stages of
development and at comparable concentrations in every strain. Does
this, in fact, occur?
2. Does a gene product needed for early embryogenesis show "phase-
specific" expression (Coggin and Anderson, 1974), i.e., expression at
early stages of development but not at later stages and in the adult?

3. Does the infection of embryos with C-type tumor viruses interfere severely with further development, as one might expect if gene expression of endogenous C-type viruses were crucial for development?

The experiments that permitted quantitative determination of viral proteins revealed very different patterns of virus gene expression in the different mouse strains examined. Some strains, such as BALB/c, express very little or no detectable virus proteins during embryogenesis or adult life, but other strains, such as C3H and 129, express high levels of gp70 but little or no p30 (figure 6.3; Strand et al., 1977; Lerner et al. al., 1976). Strains such as NZB or NZW express extremely high levels of gp70 (Lerner et al., 1976) and moderate levels of p30 during embryonic and adult life. On the other hand, AKR mice express little or no detectable virus proteins during embryogenesis (figure 6.3; Rowe et al., 1972; Hilgers et al., 1974) but produce high titers of infectious virus later in life (Rowe and Pincus, 1972; Lilly et al., 1975). Of importance here is that the concentrations of virus proteins detected in the various strains differed by more than two orders of magnitude. If virus gene products played a crucial role for cellular differentiation in the mouse, one would expect a more uniform expression of these proteins in all mouse strains.

As mentioned above, 129 G IX[+] mice express high levels of G IX antigen (i.e., gp70) in certain tissues (Lerner et al., 1976). A congenic strain has been isolated that has lost the capacity to synthesize this virus protein. That these 129 G IX[-] mice develop as normally as their G IX[+] counterparts again suggests that expression of this protein is not crucial for embryogenesis.

The few studies that investigated the expression of viral proteins (Hilgers et al., 1974; Lerner et al., 1976; Strand et al., 1977) or virus-specific RNA quantitatively at different stages of development did not reveal any evidence for phase-specific expression. On the contrary, in the few mouse strains in which significant amounts of virus-related proteins were found, the expression was either constant throughout embryogenesis, as in NZB, NZW (Lerner et al., 1976), and C3H (figure 6.3; Strand et al., 1977) mice, or increased toward the end of gestation in specific target organs, as observed in AKR mice (Strand et al., 1977).

We realize that a major objection can be raised against our interpretation of the data discussed above. This is the argument that the methods used to measure virus gene expression are not sensitive enough to detect minimal but biologically crucial expression or expression in only a few specific cells in *all* developing embryos. There seems to be no way to counter this argument. Furthermore, we would like to emphasize that the expression of two important genes, the reverse transcriptase and the src gene, which are found in RNA tumor viruses, has not been investigated yet. Temin (1971) has proposed a specific role of reverse transcriptase for embryonic development. Because the question as to whether the src gene is a true viral gene or represents normal cellular sequences is currently unsettled (Stehelin et al., 1976; Scolnick et al., 1976), src expression, even if found during embryogenesis, would not necessarily indicate virus gene expression.

That embryonic cells gain rather than lose the potential to express viral functions during the course of cellular differentiation is indicated by our experiments with preimplantation embryos. We have demonstrated that preimplantation mouse embryos cannot express any virus-specific functions after infection with M-MuLV (Jaenisch et al., 1975; Jaenisch, 1976). This nonpermissivity of embryonic multipotential cells for tumor virus replication and expression not only is observed with M-MuLV, but also is seen with the DNA tumor viruses, polyoma and SV40 (Jaenisch, 1974). A similar relation between tumor virus expression and the differentiated state of a cell is observed in the teratoma system. The nonpermissivity for tumor virus expression changes to permissivity during cellular differentiation in both systems, in the normal developing mouse embryo as well as in the differentiating teratoma stem cell. Thus, mice derived from M-MuLV-infected preimplantation embryos develop a Moloney-induced leukemia later in life (Jaenisch et al., 1975), and differentiated teratoma cells can support virus replication in contrast to the undifferentiated stem cell (Swartzendruber and Lehman, 1975; G. Martin and N. Teich, personal communication). Moloney virus-specific DNA was found to be integrated into the genome of all cells of animals infected at the 4- to 8-cell stage, but virus gene expression was restricted to the target cells of this virus, i.e., cells of lymphatic origin (Jaenisch et al., 1975). Similarly, in AKR mice that carry the endogenous Gross virus in all cells, virus expression in adult animals is restricted to the target cells (A. Berns and R. Jaenisch, in preparation). Quantitation of virus-specific RNA and proteins in individual organs of developing AKR embryos suggest that virus gene expression commences in lymphoid cells or their progenitors during embryonic development and remains restricted to these target cells or their descendants throughout life (Strand et al., 1977; A. Berns and R. Jaenisch, in preparation). Consistent with this target-cell-specific expression of M-MuLV and AKR-MuLV is the observation that somatic amplification of the integrated virus DNA is detected in lymphoma cells only (A. Berns and R. Jaenisch, 1976; Jaenisch et al., 1976a). Our results are not consistent with the hypothesis that C-type virus expression is important for early development because this concept predicts permissivity for virus expression during early stages of development rather than during later stages. And finally, the observation that infection of preimplantation mouse embryos with tumor viruses does not interfere detectably with the normal development of the mouse argues against a functional relationship between early embryogenesis and C-type virus expression.

Recent evidence suggests that embryogenesis and cellular differentiation involve sequential activation and repression of different sets of cellular genes at specific stages of embryonic development (Galau et al., 1976). One way to explain the patterns of viral protein expression in mice is to assume that different endogenous viruses are integrated at different, but still virus-specific, chromosomal loci and consequently are under different cellular controls. Thus, the genetic expression or repression of the genes of different viruses might depend on different cellular control elements that might be involved in normal tissue differentiation. The ecotropic Moloney and AKR viruses might be integrated at specific chromosomal loci of the mouse genome that are not expressed during early mouse development, i.e., in preimplantation embryos. During the course of normal tissue differentiation, as these loci become activated in cells of the lymphatic-erythropoetic lineages, the integrated virus genes could be expressed concomitantly. They would remain silent in most other organs of the developing and adult mouse. This model predicts that the remarkable organ specificity for cellular transformation by different RNA tumor viruses might be the consequence of different integration sites determining in which tissues expression and transformation by a given virus occurs. Thus, the specificity of mouse mammary tumor virus for transformation of breast epithelial cells might be due to integration at a chromosomal locus that becomes activated during differentiation only in cells that give rise to breast tissue. In agreement with this concept is the observation that mouse mammary tumor virus expression is found in mammary-gland cells, but not in spleen or thymus cells or in cells of a number of other organs (Varmus et al., 1973).

The different patterns of virus gene expression seen in the various mouse strains can be explained similarly by the simple assumption that in each instance a virus gene is integrated at a different chromosomal region, which is active at specific stages of development or in certain tissues but not at other stages of development or in other tissues. Therefore, the virus expression detected in many epithelial tissues of strains such as NZB, NZW, and 129 might be due to the integration of an "epithelial virus" (B. Croker, personal communication) at a specific chromosomal site that is different from the integration site(s) of the BALB/c and C57BL endogenous viruses, which in turn are not expressed in epithelial cells. In the few instances in which endogenous viruses have been mapped genetically, a different chromosomal locus of the virus indeed revealed a completely different pattern of virus expression (Stockert et al., 1972, 1976; Rowe et al., 1972).

The different patterns of virus expression that follow hormonal induction of endogenous viruses (Parks et al., 1974) might also be a consequence of different viral integration sites in the respective mouse strains. It is interesting that many organs that are known to be susceptible to hormone action, including the placenta (Kalter et al., 1973), uterus (Rowe and Pincus, 1972), vas deferens (Lerner et al., 1976), and breast epithelium (Jensen et al., 1976), are found to produce virus particles or virus-specific proteins. However, this phenomenon is not seen in all mice. Strains such as NZB, 129, and AKR show it, but strains such as BALB/c, C57BL, and NIH do not. Thus, it is possible that the endogenous virus genes in various mouse strains are integrated at specific but different chromosomal loci which are activated by hormonal action in certain organs (e.g., Harris et al., 1975). Virus gene expression might occur concomitantly with activation of their respective chromosomal region and therefore might not play any functional role.

It is clear that our model of virus expression and cellular differentiation is highly speculative and represents an oversimplification of a very complex issue. Furthermore, we realize that the experimental evidence supporting our hypothesis is based mainly on results obtained with the ecotropic Moloney and AKR viruses and therefore does not necessarily apply to xenotropic viruses. But our model makes a number of precise predictions that can be tested experimentally. Techniques that have been developed to introduce well-characterized exogenous tumor viruses into the germ line of mice (Jaenisch, 1976) should make it possible to map the chromosomal integration sites of different viruses by genetic means and to correlate these loci with the tissue-specific expression and transformation of a given virus.

Endogenous viruses have been isolated from many different species. The selective pressures for the persistence of functional viruses in most species are not recognized. But it is possible that these viruses serve as vehicles for gene transfer between different species (Temin, 1971; Todaro et al., 1974) or that they confer resistance to infection by harmful viruses (Todaro et al., 1974). Both these phenomena are known to exist. Cellular genes can be transferred to cells of other species by RNA tumor viruses (Stehelin et al., 1976), and it has been demonstrated that expression of viral functions can confer resistance to superinfection (Vogt and Ishizaki, 1965; Sarma et al., 1969; Weiss et al., 1974; Hunsmann et al., 1975). Alternatively, endogenous viruses might play no functional role in the life cycle of mammals (Baltimore, 1976).

Acknowledgment   We thank all members of the Tumor Virology Laboratory for critical and helpful suggestions, especially W. Eckhart, H. Fan, T. Hunter, B. Sefton, M. Vogt, and G. Walter. We also thank F. Jensen of the Scripps Clinic and Research Foundation for his helpful suggestions.

References

AMBROSE, K., ANDERSON, N., and COGGIN, J. (1971).
Interruption of SV40 oncogenesis with human foetal antigen. Nature 233, 194–195.

ANDERSEN, H. K., and JEPPESEN, T. (1972).
Virus-like particles in guinea pig oogonia and oocytes. J. Nat. Cancer Inst. 49, 1403–1410.

ANDERSON, N. G., and COGGIN, J. H. (1971).
Models of differentiation, retrogression and cancer. In Proc. 1st Conf. Embryonic and Fetal Antigens, N. G. Anderson and J. H. Coggin, eds. (Oak Ridge, Tenn: Oak Ridge National Laboratory), pp. 7–39.

BALTIMORE, D. (1976).
Viruses, polymerases and cancer. Science 192, 632–636.

BARANSKA, W., SAWICKI, W., and KOPROWSKI, J. (1971).
Infection of mammalian unfertilized and fertilized ova with oncogenic viruses. Nature 230, 591–592.

BENTVELZEN, P. (1972).
Hereditary infections with mammary tumor viruses in mice. In RNA Viruses and Host Genome in Oncogenesis, P. Emmelot and P. Bentvelzen, eds. (Amsterdam: North-Holland), pp. 309–337.

BENVENISTE, R. E., LIEBER, M. M., and TODARO, G. J. (1974).
A distinct class of inducible murine type-C viruses that replicates in the rabbit SIRC cell line. Proc. Nat. Acad. Sci. USA 71, 602–606.

BENVENISTE, R. E., and TODARO, G. (1975).
Evolution of type-C viral genes: preservation of ancestral murine type C viral sequences in pig cellular DNA. Proc. Nat. Acad. Sci. USA 72, 4090–4094.

BERNARD, W. (1960).
The detection and study of tumor viruses with the electron microscope. Cancer Res. 20, 712–727.

BERNS, A., and JAENISCH, R. (1976).
Increase of AKR-specific sequences in tumor tissues of leukemic AKR mice. Proc. Nat. Acad. Sci. USA 73, 2448–2452.

BICZYSKO, W., PIENKOWSKI, M., SOLTER, D., and KOPROWSKI, H. (1973b).
Virus particles in early mouse embryos. J. Nat. Cancer Inst. 51, 1041–1050.

BICZYSKO, W., SOLTER, D., GRAHAM, C., and KOPROWSKI, H. (1974).
Synthesis of endogenous type-A virus particles in parthenogenetically stimulated mouse eggs. J. Nat. Cancer Inst. 52, 483–489.

BICZYSKO, W., SOLTER, D., PIENKOWSKI, M., and KOPROWSKI, H. (1973a).
Interactions of early mouse embryos with oncogenic viruses—SV40 and polyoma. I. Ultrastructural studies. J. Nat. Cancer Inst. 51, 1945–1954.

BLACK, V. H. (1971).
Gonocytes in fetal guinea pig testes: phagocytosis of degenerating gonocytes by sertoli cells. Am. J. Anat. 131, 415–426.

BRAWN, R. (1971).
Evidence for association of embryonal antigens with several 3-methylcholantren induced murine sarcomas. In Proc. 1st Conf. Embryonic and Fetal Antigens, N. G. Anderson and J. H. Coggin, eds. (Oak Ridge, Tenn.: Oak Ridge National Laboratory), pp. 143–149.

BUFFET, R., GRACE, J., DIBERARDINO, C., and MIRAND, E. (1969).
Vertical transmission of murine leukemia virus. Cancer Res. 29, 588–595.

CALARCO, P. G., and SZOLLOSI, D. (1973).
Intracisternal A particles in ova and preimplantation stages of the mouse. Nature New Biol. 243, 91–93.

CHAPMAN, A., WEITLAUF, H., and BOPP, W. (1974).
Effect of feline leukemia virus on transferred hamster fetuses. J. Nat. Cancer Inst. 52, 583–586.

CHASE, D. G., and PIKÓ, L. (1973).
Expression of A- and C-type particles in early mouse embryos. J. Nat. Cancer Inst. 51, 1971–1975.

CHASE, D. G., WINTERS, W., and PIKÓ, L. (1972).
Human adenovirus uptake by preimplantation embryos. 30th Ann. Proc. Electron Microscopy Soc. Am. 30, 268–269.

CHATTOPADHYAY, S. K., ROWE, W. P., TEICH, N. M. and LOWRY, D. R. (1975).
Definitive evidence that the murine C-type virus inducing locus Akv-1 is viral genetic material. Proc. Nat. Acad. Sci. USA 72, 906–910.

COGGIN, J. H., and ANDERSON, N. G. (1974).
Cancer, differentiation and embryonic antigens: some central problems. Advan. Cancer Res. 19, 105–165.

CROKER, B., MCCONAHEY, P., and DIXON, F. (1976).
Quantitation of oncorna virus expression in normal, lymphomatous and immunopathologic mice. Fed. Proc., Abstract No. 1025.

DALTON, A. J., and POTTER, M. (1968).
Electron microscopic study of the mammary tumor agent in plasma cell tumors. J. Nat. Cancer Inst. 40, 1375–1385.

DIRKSEN, E. R., and CAILEAU, R. (1967).
An electron microscopic study of the leukemia virus in AKR and hybrid mice inoculated with ascites passage or tissue-cultured leukemic cells. Cancer Res. 27, 568–577.

DMOCHOWSKI, L., GREY, C. R., PADGETT, F., and SYKES, J. A. (1963).
Studies on the structure of the mammary tumor-inducing virus (Bittner) and of leukemia virus (Gross). In Viruses, Nucleic Acids, and Cancer (Baltimore: Williams & Wilkins), pp. 85–121.

DOUGHERTY, R., and DI STEFANO, H. (1967).
Sites of avian leukosis virus multiplication in congenitally infected chickens. Cancer Res. 27, 322–332.

ENDERS, A. C., and SCHLAFKE, S. J. (1965).
The fine structure of the blastocyst: some comparative studies. In Pre-

implantation Stages of Pregnancy, G. E. Wolstenholme and M. O'Conner eds. (Boston: Little, Brown), pp. 29–54.

ESSEX, M. (1975). Horizontally and vertically transmitted oncorna viruses of cats. Advan. Cancer Res. *21*, 175.

FELDMAN, D. G., and GROSS, L. (1964). Electron-microscopic study of the mouse leukemia virus (Gross), and tissues from mice with virus-induced leukemia. Cancer Res. *24*, 1760–1783.

FISZMAN, M., and FUCHS, P. (1975). Temperature-sensitive expression of differentiation in transformed myoblasts. Nature *254*, 429–431.

FOWLER, A. K., MCCONAHEY, P. J., and HELLMAN, A. (1973). Strain dependency of hormonally activated C-type RNA tumor virus markers in mice. J. Nat. Cancer Inst. *50*, 1057–1059.

FOWLER, A. K., REED, C. D., TODARO, G. J., and HELLMAN, A. (1972). Activation of C-type RNA virus markers in mouse uterine tissue. Proc. Nat. Acad. Sci. USA *69*, 2254–2257.

GALAU, G., KLEIN, W., DAVIS, M., WOLD, B., BRITTEN, R., and DAVIDSON, E. (1976). Structural gene sets active in embryos and adult tissues of the sea urchin. Cell *7*, 487–505.

GARAPIN, A., VARMUS, H., FARAS, A., LEVINSON, W., and BISHOP, M. (1973). RNA directed synthesis by virions of

Rous sarcoma virus: further characterization of the templates and extent of their transcription. Virology *52*, 264–274.

GIRARDI, A., REPUCCI, P., DIERLAM, P., RUTALA, W., and COGGIN, J. (1973). Prevention of simian virus 40 tumors by hamster fetal tissue: influence of parity status of donor females on immunogenicity of fetal tissue and on immune cell toxicity. Proc. Nat. Acad. Sci. USA *70*, 183–186.

GOLD, P., and FREEDMAN, S. (1965). Demonstration of tumor-specific antigens in human colonic carcinomata by immunological tolerance and absorption techniques. J. Exp. Med. *121*, 439.

GOLDMAN, E., and BENJAMIN, T. (1975). Analysis of host range of nontransforming polyoma virus mutants. Virology *66*, 372–384.

GRAF, T., ROYER-POKORA, B., and BENG, H. (1976). In vitro transformation of specific target cells by avian leukemia viruses. *In* ICN-UCLA Winter Conference on Animal Viruses, D. Baltimore, A. S. Huang, and T. Fox, eds. (New York: Academic Press), pp. 321–338.

GROSS, L. (1961). Vertical transmission of passage A leukemic virus from inoculated C3H mice to their untreated offspring. Proc. Soc. Exp. Biol. Med. *107*, 90–93.

GROSS, L. (1970). Oncogenic Viruses. 2nd ed. Oxford: Pergamon Press.

GROUDINE, M., and
WEINTRAUB, H. (1975).
Rous sarcoma virus activates embryonic globin genes in chicken fibroblasts. Proc. Nat. Acad. Sci. USA 72,
4464–4468.

GWATKIN, R. (1971).
Studying the effect of viruses on eggs.
In Methods in Mammalian Embryol-
ogy, J. Daniel, ed. (San Francisco:
W. H. Freeman), pp. 228–237.

HARRIS, E., ROSEN, J., MEANS,
A., and O'MALLEY, B. (1975).
Use of a specific probe for ovalbumin
messenger RNA to quantitate
estrogen-induced gene transcription.
Biochemistry 14, 2072–2081.

HARTLEY, J. W., ROWE, W. P.,
CAPPS, W. I., and HUEBNER, R. J.
(1969).
Isolation of naturally occurring
viruses of the murine leukemia virus
group in tissue culture. J. Virol. 3,
126–132.

HILGERS, J., DECLÈVE, A.,
GALESLOOT, J., and KAPLAN,
H. S. (1974).
Murine leukemia virus group-specific
antigen expression in AKR mice.
Cancer Res. 34, 2553–2561.

HILGERS, J., NOWINSKI, R. C.,
GEERING, G., and HARDY, W.
(1972).
Detection of avian and mammalian
oncogenic RNA viruses (oncornaviruses) by immunofluorescence.
Cancer Res. 32, 98–106.

HILLMAN, N., and TASCA, R. J.
(1969).
Ultrastructural and autoradiographic
studies of mouse cleavage stages.
Am. J. Anat. 126, 151–173.

HIRSCH, M., and BLACK, P.
(1974).
Activation of mammalian leukemia
viruses. Advan. Virus Res. 19,
265–313.

HOLTZER, H., BIEHL, J., YEOH,
G., MEGANATHAN, R., and KAJI,
A. (1975).
Effect of oncogenic viruses on muscle
differentiation. Proc. Nat. Acad. Sci.
USA 72, 4051–4055.

HUEBNER, R. J., KELLOFF, G. J.,
SARMA, P. S., LANE, W. T.,
TURNER, A. C., GILDEN, R. V.,
OROSZLAN, S., MERER, H.,
MYERS, D. B., and PETERS, R. L.
(1970).
Group specific antigen expression
during embryogenesis of the genome
of the C-type RNA tumor virus: implications for ontogenesis and oncogenesis. Proc. Nat. Acad. Sci. USA
67, 366–376.

HUEBNER, R. J., SARMA, P. S.,
KELLOFF, G. J., GILDEN, R. V.,
MEIER, H., MYERS, D. D., and
PETERS, R. L. (1971).
Immunological tolerance to RNA
tumor virus genome expressions: significance of tolerance and prenatal expressions in embryogenesis and
tumorigenesis. Ann. N. Y. Acad. Sci.
181, 246–271.

HUEBNER, R. J., and TODARO,
G. J. (1969).
Oncogenes of RNA tumor viruses as
determinants of cancer. Proc. Nat.
Acad. Sci. USA 64, 1087–1094.

HUNSMANN, G., MOENNIG, V.,
and SCHAFER, W. (1975).
Properties of mouse leukemia virus:
IX. Active and passive immunization
of mice against Friend leukemia with

isolated viral gp71 glycoprotein and its corresponding antiserum. Virology 66, 327–329.

HUNTER, W. M. (1973).
Radioimmunoassay. In Handbook of Experimental Immunology, D. M. Weir, ed. (Oxford: Blackwell), pp. 17.1–17.36.

IDA, N., FUKUHARA, A., and OHBA, Y. (1966).
Several aspects of vertical transmission of Moloney virus. Nat. Cancer Inst. Monograph 22, 287–311.

JAENISCH, R. (1974).
Infection of mouse blastocysts with SV40 DNA: normal development of the infected embryos and persistence of SV40 specific DNA sequences in the adult animals. Cold Spring Harbor Symp. Quant. Biol. 39, 375–380.

JAENISCH, R. (1976).
Germ line integration and Mendelian transmission of the exogenous Moloney leukemia virus. Proc. Nat. Acad. Sci. USA 73, 1260–1264.

JAENISCH, R., BERNS, A., DAUSMAN, J., and COX, V. (1976a).
Germ line integration and leukemogenesis of exogenous and endogenous murine leukemia viruses. In ICN-UCLA Winter Conference on Animal Virology, D. Baltimore, A. S. Huang, and T. Fox, eds. (New York: Academic Press), pp. 283–310.

JAENISCH, R., DAUSMAN, J., COX, V., FAN, H., and CROKER, B. (1976b).
Infection of developing mouse embryos with murine leukemia virus: tissue specificity and genetic transmission of the virus. In Modern

Trends in Human Leukemia, II, vol. 19, R. Neth, R. C. Gallo, K. Mannweiler, and W. C. Maloney, eds. (Munich: J. F. Lehmanns Verlag), pp. 341–356.

JAENISCH, R., FAN, H., and CROKER, B. (1975).
Infection of preimplantation mouse embryos and of newborn mice with leukemia virus: tissue distribution of viral DNA and RNA and leukemogenesis in the adult animal. Proc. Nat. Acad. Sci. USA 72, 4008–4012.

JAENISCH, R., MAYER, A., and LEVINE, A. (1971).
Replicating SV40 molecules containing closed circular template DNA strands. Nature New Biol. 233, 72–75.

JAENISCH, R., and MINTZ, B. (1974).
Simian virus 40 DNA sequences in DNA of healthy adult mice derived from preimplantation blastocysts injected with viral DNA. Proc. Nat. Acad. Sci. USA 71, 1250–1254.

JENSON, B., GROFF, D., MCCONAHEY, P., and DIXON, F. (1976).
Transmission of murine leukemia virus from parent to progeny mice as determined by p30 antigenemia. Cancer Res. 36, 1228–1232.

KAKEFUDA, T., ROBERTS, E., and SUNTZEFF, V. (1970).
Electron microscopic study of methylcholanthrene-induced epidermal carcinogenesis in mice: mitochondrial dense bodies and intracisternal A-particles. Cancer Res. 30, 1011–1019.

KALTER, S. S., HELMKE, R. J., HEBERLING, R. L., PANIGEL, M., FOWLER, A. K.; STRICKLAND, J. E., and HELLMAN, A. (1973b). C-type particles in normal human placentas. J. Nat. Cancer Inst. *50*, 1081–1084.

KALTER, S. S., HELMKE, R. J., PANIGEL, M., HEBERLING, R. L., FELSBURG, P. J., and AXELROD, L. R. (1973a). Observations of apparent C-type particles in baboon (Papio cynocephalus) placentas. Science *179*, 1332–1333.

KNOWLAND, J., and GRAHAM, C. F. (1972). RNA synthesis in the two-cell stage of mouse development. J. Embryol. Exp. Morph. *27*, 167–176.

KUFF, E. L., LUEDERS, K. K., OZER, H. L., and WIVEL, N. A. (1972). Some structural and antigenic properties of intracisternal A particles occurring in mouse tumors. Proc. Nat. Acad. Sci. USA *69*, 218–222.

LAW, L. (1962). Influence of foster-nursing on virus-induced and spontaneous leukemia in mice. Proc. Soc. Exp. Biol. Med. *111*, 615–623.

LAW, L. (1966). Transmission studies of a leukemogenic virus, MLV, in mice. Nat. Cancer Inst. Monograph *22*, 267–282.

LEHMAN, J. M., KLEIN, I. B., and HACKENBERG, R. M. (1975). The response of murine teratocarcinoma cells to infection with DNA and RNA viruses. *In* Teratomas and Differentiation, M. I. Sherman and D. Solter, eds. (New York: Academic Press), pp. 289–301.

LERNER, R. A., WILSON, C. B., DEL VILLANO, B. C., MCCONAHEY, P. J., and DIXON, F. J. (1976). Endogenous oncornaviral gene expression in adult and fetal mice: quantitative, histologic, and physiologic studies of the major viral glycoprotein, gp70. J. Exp. Med. *143*, 151–166.

LEVY, J. A. (1973). Xenotropic viruses: murine leukemia viruses associated with NIH Swiss, NZB and other mouse strains. Science *182*, 1151–1153.

LIEBER, M. M. BENVENISTE, R. E., LIVINGSTON, D. M., and TODARO, G. J. (1973). Mammalian cells in culture frequently release type C viruses. Science *182*, 56–59.

LILLY, F., DURAN-REYNALS, M., and ROWE, W. (1975). Correlation of early murine leukemia virus titer and H-2 type with spontaneous leukemia in mice of the Balb/c × AKR cross: a genetic analysis. J. Exp. Med. *141*, 882–889.

LILLY, F., and PINCUS, T. (1973). Genetic control of murine viral leukemogenesis. Advan. Cancer Res. *17*, 231–277.

MARTIN, S. (1970). Rous sarcoma virus: a function required for maintenance of the transformed state. Nature *227*, 1021–1023.

MAYER, R. J., SMITH, R. G., and GALLO, R. C. (1974). Reverse transcriptase in normal Rhesus monkey placenta. Science *185*, 864–867.

MINTZ, B. (1971).
Allophenic mice of multi-embryonic origin. In Methods in Mammalian Embryology, J. Daniel, ed. (San Francisco: W. H. Freeman), pp. 186–214.

MINTZ, B., and ILLMENSEE, K. (1975).
Normal genetically mosaic mice produced from malignant teratocarcinoma cells. Proc. Nat. Acad. Sci. USA 72, 3585–3589.

MOSCOVICI, C., GAZZOLO, L., and MOSCOVICI, G. (1975).
Focus assay and defectiveness of avian myeloblastosis virus. Virology 68, 173–181.

MUKHERJEE, B. B., and MOBRY, P. M. (1975).
Variations in hybridization of RNA from different mouse tissues and embryos to endogenous C-type virus DNA transcripts. J. Gen Virol. 28, 129–135.

NOVIKOFF, A. B., and BIEMPICA, L. (1966).
Cytochemical and electron microscopic examination of Morris 5123 and Reuber H-35 hepatomas after several years of transplantation. GANN Monograph 1, 65–87.

NOWINSKI, R. C., OLD, L. J., SARKAR, N. H., and MOORE, D. H. (1970).
Common properties of the oncogenic RNA viruses (oncornaviruses). Virol. 42, 1152–1157.

OGURA, H., GELDERBLOM, H., and BAUER, H. (1974).
Isolation of avian nephroblastoma virus from avian myeloblastosis virus by the infectious DNA technique. Intervirology 4, 69–76.

PARKS, W., SCOLNICK, F., and RANSOM, J. (1974).
Glucocorticoid induction of murine mammary tumor virus in vitro. Cold Spring Harbor Symp. Quant. Biol. 39, 1151–1158.

PAYNE, L. N., and CHUBB, R. (1968).
Studies on the nature and genetic control of an antigen in normal chick embryos which reacts in the COFAL test. J. Gen. Virol. 3, 379–391.

PESSAC, B., and CALOTHY, G. (1974).
Transformation of chick embryo neuroretinal cells by Rous sarcoma virus in vitro: induction of cell proliferation. Science 185, 709–710.

ROSENBERG, N., BALTIMORE, D., and SCHER, C. (1975).
In vitro transformation of lymphoid cells by Abelson murine leukemia virus. Proc. Nat. Acad. Sci. USA 72, 1932–1936.

ROWE, W. (1972).
Studies of genetic transmission of murine leukemia virus by AKR mice. I. Crosses with Fv-1^n strains of mice. J. Exp. Med. 136, 1272–1285.

ROWE, W. P., HARTLEY, J. W., and BRENNER, T. (1972).
Genetic mapping of a murine leukemia virus-inducing locus of AKR mice. Science 178, 860–862.

ROWE, W. P., and PINCUS, T. (1972).
Quantative studies of naturally occurring murine leukemia virus infection of AKR mice. J. Exp. Med. 135, 429–436.

ROWE, W., PUGH, W., and
HARTLEY, J. (1970).
Plaque assay techniques for murine
leukemia viruses. Virology 42,
1136–1139.

RUBIN, H., CORNELIUS, A., and
FANSHIER, L. (1961).
The pattern of congenital transmission
of an avian leukosis virus. Proc. Nat.
Acad. Sci. USA 47, 1058–1069.

RUBIN, H., FANSHIER, L.,
CORNELIUS, A., and HUGHES, W.
(1962).
Tolerance and immunity in chickens
after congenital and contact infection
with an avian leukosis virus. Virology
17, 143–156.

SAMBROOK, J., WESTPHAL, H.,
SRINIVASAN, P., and DULBECCO,
R. (1968).
The integrated state of DNA in SV40-
transformed cells. Proc. Nat. Acad.
Sci. USA 60, 1288–1295.

SARKAR, N. H., MOORE, D. H.,
and NOWINSKI, R. C. (1972).
Symmetry of the nucleocapsid of the
oncornaviruses. In RNA Viruses and
Host Genome in Oncogenesis,
P. Emmelot and P. Bentvelzen, eds.
(Amsterdam: North Holland),
pp. 71–79.

SARMA, P., CHEANG, M.,
HARTLEY, J., and HUEBNER, R.
(1967).
A viral interference test for mouse
leukemia viruses. Virology 33,
180–188.

SAWICKI, W., BARANSKA, W.,
and KOPROWSKI, H. (1971).
Susceptibility of unfertilized and
fertilized mouse eggs to SV40 and
Moloney sarcoma virus. J. Nat.
Cancer Inst. 47, 1045–1051.

SCHER, C., and SIEGLER, R.
(1975).
Direct transformation of 3T3 cells by
Abelson murine leukemia virus.
Nature 253, 729–731.

SCHIDLOVSKY, G., and AHMED,
M. (1973).
C-type virus particles in placentas and
fetal tissues of Rhesus monkeys. J.
Nat. Cancer Inst. 51, 225–233.

SCOLNICK, E., GOLDBERG, R.,
and WILLIAMS, D. (1976).
Characterization of rat genetic
sequences in Virsten sarcoma virus:
distinct class of endogenous rat type
C viral sequences. J. Virol. 18,
559–566.

SEMAN, G., LEVY, B. M.,
PANIGEL, M., and
DMOCHOWSKI, L. (1975).
Type-C virus particles in placenta of
the cottontop marmoset (Saguinus oe-
dipus). J. Nat. Cancer Inst. 54,
251–252.

SHAPIRA, F., REUBER, M., and
HATZFELD, A. (1970).
Resurgence of two fetal-type aldolases
(A and C) in some fast-growing hepa-
tomas. Biochem. Biophys. Res.
Comm. 40, 321–327.

SHOYAB, M., and BALUDA, M.
(1975).
Acquisition of viral DNA sequences
in target organs of chickens infected
with avian myeloblastosis virus. J.
Virol. 16, 783–789.

SHOYAB, M., DASTOOR, M., and
BALUDA, M. (1976).
Evidence for tandem integration of
avian myeloblastosis virus DNA with
endogenous provirus on leukemic
chicken cells. Proc. Nat. Acad. Sci.
USA 73, 1749–1753.

SMITH, R., DAVIDS, L., and
NEIMAN, P. (1976).
Comparison of avian osteopetrosis
virus with avian lymphomatosis virus
by RNA-DNA hybridization. J. Virol.
17, 160–167.

SOLTER, D., BICZYSKO, W., and
KOPROWSKI, H. (1974).
Host-virus relationship at the
embryonic level. In Viruses, Evo-
lution and Cancer, E. Kuslak and K.
Maramosch, eds. (New York:
Academic Press), pp. 3–30.

STEHELIN, D., GUNTAKA, R.,
VARMUS, H., and BISHOP, M.
(1976).
Purification of DNA complementary
to nucleotide sequences required for
neoplastic transformation by avian
sarcoma viruses. J. Mol. Biol. 101,
349–365.

STEPHENSON, J., ANDERSON,
G., TRONICK, S., and
AARONSON, S. (1974).
Evidence for genetic recombination
between endogenous and exogenous
mouse RNA type C viruses. Cell 2,
87–94.

STEPHENSON, J., REYNOLDS, R.,
TRONICK, S., and AARONSON, S.
(1975).
Distribution of three classes of
endogenous type C RNA viruses
among different strains of mice.
Virology 67, 404–414.

STOCKERT, E., BOYSE, E., SATO,
H., and ITAKURA, K. (1976).
Heredity of the G1X thymocyte antigen associated' with murine leukemia
virus: segregation data simulating
genetic linkage. Proc. Nat. Acad. Sci.
USA 73, 2077–2081.

STOCKERT, E., SATO, H.,
ITAKURA, K., BOYSE, E. A.,
OLD, L., and HUTTON, J. J. (1972).
Location of the second gene required
for expression of the leukemia-
associated antigen G_{IX}. Science 178,
862–863.

STRAND, M., and AUGUST, J. T.
(1974).
Structural proteins of RNA tumor viruses as probes for viral gene expression. Cold Spring Harbor Symp.
Quant. Biol. 39, 1109–1116.

STRAND, M., AUGUST, T., and
JAENISCH, R., (1977).
Oncornavirus gene expression during
embryonal development of the mouse.
Virology. 76, 886–890.

STRAND, M., LILLY, F., and
AUGUST, J. T. (1974).
Host control of endogenous murine
leukemia virus gene expression: concentrations of viral proteins in high
and low leukemia mouse strains.
Proc. Nat. Acad. Sci. USA 71,
3682–3686.

SWARTZENDRUBER, D., and
LEHMAN, J. (1975).
Neoplastic differentiation: interaction
of SV40 and polyoma virus with mur-
ine teratocarcinoma cells in vitro. J.
Cell. Physiol. 89, 179–187.

TEMIN, H. (1971).
The protovirus hypothesis: speculations on the significance of RNA-
directed DNA synthesis for normal
development and for carcinogenesis.
J. Nat. Cancer Inst. 46, III–VII.

TODARO, G. J. (1972).
"Spontaneous" release of type C
viruses from clonal lines of "spontaneously" transformed BALB 3T3

Cells. Nature New Biol. *240*,
157–160.

TODARO, G., BENVENISTE, R.,
CALLAHAN, R., LIEBER, M., and
SHERR, C. (1974).
Endogenous primate and feline type C
viruses. Cold Spring Harbor Symp.
Quant. Biol. *39*, 1159–1168.

TODARO, G., and HUEBNER, R.
(1972).
The viral oncogene hypothesis: new
evidence. Proc. Nat. Acad. Sci. USA
69, 1009–1015.

TOOZE, J. (1973), ed.
The Molecular Biology of Tumour
Viruses. Cold Spring Harbor, New
York: Cold Spring Harbor
Laboratory.

UMAR, M., and VAN GRIENSVEN,
L. (1976).
Effect of Rauscher leukemia virus
infection on Balb/c mouse embryos.
J. Nat. Cancer Inst. *56*, 375–380.

VARMUS, H., QUINTRELL, N.,
MEDEIROS, E., BISHOP, M.,
NOWINSKI, R., and SARKAR, N.
(1973).
Transcription of mouse mammary
tumor virus genes in tissues from high
and low tumor incidence mouse
strains. J. Mol. Biol. *79*, 663–679.

VARMUS, H., STEHELIN, D.,
SPECTOR, D., TAL, J., FUJITA,
D., PADGETT, T., ROULLAND-
DUSSOIX, D., KUNG, H., and
BISHOP, M. (1976).
Distribution and function of defined
regions of avian tumor virus genomes
in viruses and uninfected cells. *In*
ICN-UCLA Winter Conference on
Animal Viruses, D. Baltimore, A. S.
Huang, and T. Fox, eds. (New York:
Academic Press), pp. 339–358.

VARMUS, H., VOGT, P., and
BISHOP, M. (1973).
Integration of deoxyribonucleic acid
specific for Rous sarcoma virus after
infection of permissive and nonpermissive hosts. Proc. Nat. Acad. Sci.
USA *70*, 3067–3071.

VERNON, M. L., LANE, W. T., and
HUEBNER, R. J. (1973).
Prevalence of type-C particles in visceral tissues of embryonic and newborn mice. J. Nat. Cancer Inst. *51*,
1171–1175.

VERNON, M. L., MCMAHON,
J. M., and HACKETT, J. J. (1974).
Additional evidence of type-C parti-
cles in human placentas. J. Nat.
Cancer Inst. *52*, 987–989.

VOGT, P. (1977).
The genetics of RNA tumor viruses.
In Comprehensive Virology 9. In
press.

VOGT, P., and ISHIZAKI, R. (1965).
Reciprocal patterns of genetic resistance to avian tumor viruses in two
lines of chickens. Virology *26*,
664–672.

WANG, L., DUESBERG, P.,
KAWAI, S., and HANAFUSA, H.
(1976).
Location of envelope-specific and sar-
coma specific oligonucleotides on
RNA of Schmidt-Ruppin Rous sar-
coma virus. Proc. Nat. Acad. Sci.
USA *73*, 447–451.

WEISS, R., BOETTIGER, D., and
LOVE, D. (1974).
Phenotypic mixing between vesicular
stomatitis virus and avian RNA tumor
viruses. Cold Spring Harbor Symp.
Quant. Biol. *39*, 913–918.

WEISS, R., MASON, W., and
VOGT, P. (1973).
Genetic recombinants and hetero-
zygotes derived from endogenous and
exogenous RNA tumor viruses.
Virology 52, 535–552.

WHITTEN, W. (1971).
Nutrient requirements for the culture
of preimplantation embryos in vitro.
Advan. Biosciences 6, 129–138.

WIVEL, N. A., LUEDERS, K. K.,
and KUFF, E. L. (1973).
Structural organization of murine
intracisternal A particles. J. Virol. 11,
329–334.

WIVEL, N. A., and SMITH, G. H.
(1971).
Distribution of intracisternal
A-particles in a variety of normal and
neoplastic mouse tissues. Int. J.
Cancer 7, 167–175.

YANG, S. S., and WIVEL, N. A.
(1973).
Analysis of high-molecular-weight
ribonucleic acid associated with
intracisternal A particles. J. Virol. 11,
287–298.

7 TERATOCARCINOMA CELLS AND NORMAL MOUSE EMBRYOGENESIS

C. F. Graham

7.1 Introduction

Teratocarcinomas in mice are tumors that develop spontaneously in the gonads and can be produced experimentally from embryos and primordial germ (PG) cells (section 7.2). Characteristically they contain a rapidly dividing undifferentiated stem cell population of embryonal carcinoma (EC) cells, which can each differentiate into a wide variety of tissues, such as respiratory and alimentary epithelia, muscle, nerve, cartilage, and bone (figure 7.1; Kleinsmith and Pierce, 1964). Teratocarcinomas are considered to be malignant because they kill their hosts; like many other mouse tumors, they rarely metastasize. Tumors that contain this variety of differentiated tissues but lack the EC cells are benign and are called teratomas.

Some of the tumors can be maintained by transplantation. After several transplantation generations they often become chromosomally abnormal and form fewer differentiated cell types (section 7.3). Some solid teratocarcinomas form embryoid bodies when injected into the peritoneal cavity; embryoid bodies are an ascitic form of the tumor that contain a core of aggregated EC cells surrounded by a rind of primitive endoderm (simple embryoid bodies). Embryoid bodies can become cystic, in which case they are vesicular and may also contain hemopoietic yolk sac, muscle, and nerve.

The EC cells can be grown in culture, and under the appropriate conditions, they differentiate into an extensive range of cell types (section 7.4). And so it is possible to take petri dishes full of cells through a developmental sequence that resembles normal embryogenesis; such cultures provide bulk material for biochemical studies on the processes of determination and differentiation that usually occur in small cell populations inside the developing mammalian embryo (see chapter 1). Here we shall concentrate on this aspect of the biology of teratocarcinomas and pay particular attention to the similarities and differences between the development of these tumors and normal embryos (sections 7.4 and 7.5).

EC cells are malignant and they kill their host, but they also closely resemble other embryonic stem cells (section 7.6). No carcinogens or viruses are obviously involved in their induction (section 7.7), and their cells become benign as they differentiate (section 7.8).

The development and differentiation of teratocarcinomas has been extensively reviewed in the past. The accumulation of information

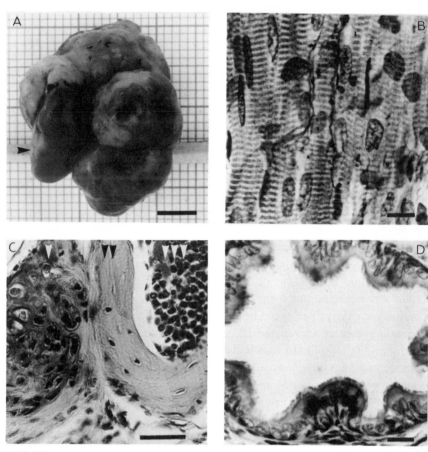

7.1 Solid teratocarcinomas. (a) Teratocarcinoma growing on the kidney (arrow). An F_1 (129/J × C3H) 7th-day egg cylinder had been transferred beneath the kidney capsule 1.5 months before this photograph was taken. The tumor weighed 3.8 g and was transplantable and multipotential. Scale bar = 0.5 cm. (b) Striated muscle with nerve fibers in a solid teratocarcinoma. Scale bar = 20 μm. (c) Cartilage (one arrow), bone (two arrows), and nucleated erythrocytes (three arrows) in a solid teratocarcinoma. Scale bar = 50 μm. (d) Hollow tube with ciliated epithelium inside a solid teratocarcinoma. Scale bar = 50 μm.

about these tumors and about the development of the mouse embryo is so rapid, however, that it is useful to assess the data regularly. The following reviews emphasize other aspects of mouse teratocarcinomas: Stevens (1974), the origin of teratocarcinomas; Pierce (1974), the change from malignant to benign growth during teratocarcinoma differentiation; Damjanov and Solter (1974b), the pathology of teratocarcinomas; Solter et al. (1975a,b), the effect of the host on teratocarcinoma growth and development; Martin (1975) and Evans (1976b), the development of EC cells in culture. Many of the more recent experimental studies are included in the report of the Roche Institute Symposium *Teratomas and Differentiation* (Sherman and Solter, 1975) and in *The Early Development of Mammals* (Balls and Wild, 1975).

7.2 Origin of Embryonal Carcinoma Cells

Teratocarcinomas form when either embryos or genital ridges containing PG cells are transferred to extrauterine sites; the ease with which these tumors can be produced experimentally has allowed a detailed analysis of their origins.

7.2.1 Embryonic Origin of Teratocarcinomas

Embryos and parts of embryos usually survive and grow when taken from the uterus and transplanted to extrauterine sites in syngeneic hosts. The whole embryo can form many of the tissues of the fetus ectopically; these are usually squeezed into a disorderly array, but in certain sites, such as the scrotal testis, morphologically normal twelfth day embryos can develop in rare instances (Kirby, 1963). The disorderly tissue arrangement in most embryos developing outside the uterus is thought to be one of the necessary conditions for subsequent teratocarcinoma formation.

Tissue of origin of embryonal carcinoma cells

The direct way to discover the origin of EC cells is to transplant isolated parts of the embryo to extrauterine sites and then to determine which parts can form teratocarcinomas. The interpretation of these experiments is based on two assumptions: first, it is assumed that the isolated tissue forms EC cells directly without passing through an intermediate stage of differentiation; second, it is supposed that the isolation procedure does not restrict EC cell formation either by immediate damage or by preventing cell interactions. These isolation experiments have been conducted on seventh- and eighth-day egg cylinders because they are multilayered and regularly form teratocarcinomas.

Teratocarcinomas are formed only by transplants that contain the embryonic ectoderm of the egg cylinder (see chapter 1). Diwan and Stevens (1976) divided the seventh-day embryo into three parts: the primitive endoderm, the embryonic ectoderm, and the extraembryonic ectoderm [F_1(129/Sv – Sl^J/CP × A/He) embryos]. Thirty days later the embryonic ectoderm had formed a wide range of differentiated tissues and large areas of undifferentiated cells; these were probably EC cells both because the whole embryonic part of the egg cylinder (endoderm and ectoderm) had previously been shown to form transplantable EC cells [Solter et al. (1970b); Damjanov et al. (1971a) with eighth-day embryos of strain C3H/H] and because the endoderm by itself formed only parietal endoderm.

The other parts of the egg cylinder are developmentally restricted to the formation of extraembryonic ectoderm and extraembryonic endoderm. The extraembryonic ectoderm is a source of trophoblast cells and part of the chorion (see chapter 1), and when it and the extraembryonic endoderm are transplanted together to the kidney, they also form hyaline material that resembles the products of parietal endoderm (Reichert's membrane). In a series of twenty grafts, only one failed to produce hyaline material; two grafts formed a little squamous epithelium, and in another, cartilage was found. Despite these additional tissues, the complete absence of EC cells suggests that the extraembryonic region of the egg cylinder is never a source of pluripotential cells [Solter and Damjanov (1973), eighth-day embryos of strain C3H/He]. However, after the extraembryonic mesoderm has migrated into the extraembryonic region from the embryonic part, then visceral yolk sac grafts can form teratomas (Pierce et al., 1970; Sobis and Vandeputte, 1975).

Developmental stage of the transplanted embryo

The embryo forms teratocarcinomas more frequently as it develops up to the late egg-cylinder stage on the eighth day of pregnancy (figure 7.2). After the eighth day of development; grafted embryos form teratomas, but not teratocarcinomas. When genital ridges of twelfth- and thirteenth-day embryos are grafted to the testis and other scrotal sites

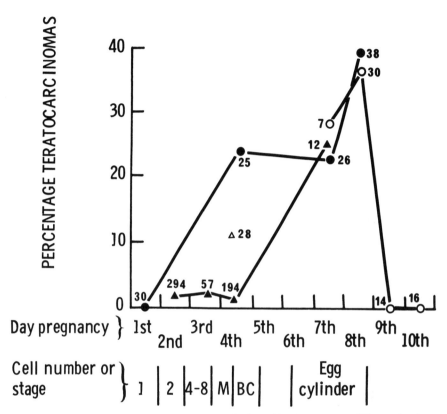

7.2 Teratocarcinoma formation by embryos grafted outside the uterus at different developmental stages. Teratocarcinomas were recognized by the presence of EC cells in histological sections of the tumors. This figure contains information only about grafts that were examined 50 to 100 days after transfer so that the data from different experiments can be compared. The number beside each symbol shows how many embryos were transferred. M = morula; BC = blastocyst.•: C3H embryos to the testis (Iles et al., 1975; Iles, 1976).○: C3H embryos to the kidney (Damjanov et al., 1971b).▲: 129/J embryos to the testis (Stevens, 1968, 1970c). Δ:129/J embryos to the testis (Iles, 1976).

in some strains, teratocarcinomas develop from the PG cells (Stevens, 1970a). The ability of the genital ridges to form teratocarcinomas is lost during the next four days of development. The biphasic ability of the cells of the embryo to form teratocarcinomas is curious and not yet understood.

The increasing frequency of teratocarcinoma formation as the embryo develops up to the eighth day can probably be explained by the increase in the ability of the pluripotential cells in the embryo to survive and form embryonic ectoderm in extrauterine sites as they grow larger. It is known that the growth of the preimplantation embryo is limited in extrauterine sites; the majority of preimplantation embryos form only a few trophoblast giant cells with some extraembryonic membranes after the transplants have developed for one to two weeks (e.g., Kirby, 1965; Billington et al., 1968; Stevens, 1968). The EC cells must develop from the few growths that contain embryonic derivatives because neither of these extraembryonic tissues is able to form teratocarcinomas (mouse: see above; Simmons and Russell, 1962; Simmons and Weintraub, 1965; Billington, 1965; Avery and Hunt, 1968; Pierce et al., 1970; rat: Payne and Payne, 1961; Sobis and Vandeputte, 1975). Embryonic derivatives are found more often as progressively more advanced embryos are transplanted: no 1-cell eggs form embryonic derivatives (Stevens, 1968), about one-third of transferred blastocysts do so (Billington et al., 1968), and all egg cylinders form embryonic structures (Damjanov et al., 1971a). It is not surprising, therefore, that no teratocarcinomas form as a result of the transfer of 1-cell eggs, whereas about half of eighth-day egg cylinders do so (figure 7.2).

It is more difficult to explain the sudden loss of ability to form teratocarcinomas after the eighth day of development. One idea is that the formation of teratocarcinomas depends on the disruption of the normal cellular interactions of embryogenesis (see Stevens, 1974). It might be the case, therefore, that the transplantation of embryos after the eighth day does not sufficiently disturb the embryo (Solter et al., 1975a). Another possibility is that after the eighth day all embryonic cells have experienced the cell interactions that restrict developmental potential, i.e., they have become determined (Solter et al., 1975b). We know that multipotential stem cells are present in the inner cell mass (ICM) of the blastocyst and in the embryonic ectoderm of the egg cylinder (see chapter 1), but we do not know when these multipotential cells disappear. It is hard to believe that such cells are absent from the embryo because PG cells form shortly thereafter in embryonic development; there must therefore be progenitors of the multipotential PG cells in the embryo that either cannot form teratocarcinomas on the ninth and tenth days of development or that are so infrequent that the chance of them doing so is very low. Perhaps multipotential stem cells are particularly sensitive to damage by transplantation over this period; support for this idea is provided by the observation that tenth-day PG cells degenerate soon after grafting (Ozdzenski, 1969), and this evidence has led to the view that the EC cells of teratocarcinomas can form directly from the embryonic ectoderm without proceeding through a developmental stage equivalent to a PG cell (Damjanov et al., 1971a). To establish this view, it is necessary to find out where the PG cells or their precursors are supposed to be on the eighth day of development.

Location of the primordial germ cells and teratocarcinoma formation
The round shape and characteristic staining of PG cells led to their identification in the early part of this century. They were first observed in the yolk sac and gut endoderm, from which they subsequently appeared to migrate to the genital ridges (reviewed by Jenkinson, 1913). In the mouse, the evidence that these cells were PG cells came from two types of study. First, it was shown that transplants of genital ridges form gonads with mature gametes only after PG cells have migrated into the transplants (Everett, 1943). Second, these cells proved to have high alkaline phosphatase activity (Chiquoine, 1954), and only cells with this activity were depleted in mutant embryos that developed into sterile adults (Bennett, 1956; Mintz, 1957; Mintz and Russell, 1957; McCoshan and McCallion, 1975). Although size, shape, and high alkaline phosphatase activity are rather widely distributed characteristics by which to identify one particular cell type, I will assume that such cells are PG cells.

Cells with these characteristics are detectable first early on the eighth day of development in an extension of the embryonic ectoderm or mesoderm (the two layers are not distinct at this time) that pushes down into the extraembryonic mesoderm of the allantois (Odzenski, 1967). During the next day, they appear to migrate into the endoderm overlying this mesoderm, and on the ninth day they are found in the visceral extraembryonic endoderm and the endoderm overlying the primitive streak; the latter tissue forms part of the gut endoderm, and they move through this to the genital ridge (Jeon and Kennedy, 1973; Spiegelman and Bennett, 1973; Zamboni and Merchant, 1973; Clark and Eddy, 1975). The conclusion is that the embryonic ectoderm near the base of the allantois is probably the original source of PG cells on the eighth day of development and probably continues to produce PG cells during the next two days (Odzenski, 1967). It is conceivable, therefore, that the embryonic ectoderm can form teratocarcinomas because it is a source of PG cells, but there are reasons to believe that this is not the case (Stevens, 1974). First, PG-cell-derived teratocarcinomas always have male karyotypes, whereas embryo-derived teratocarcinomas have male and female karyotypes. This observation suggests that teratocarcinoma formation from the embryo differs in kind from that of PG-cell-derived teratocarcinomas. Second, some strains, including C3H, form embryo-derived teratocarcinomas but fail to form PG cell teratomas. It is therefore difficult to believe that the embryos form teratocarcinomas by passing through a stage of PG cell formation, and the conclusion is that EC cells can be derived directly from the embryonic ectoderm of the egg cylinder (Damjanov et al., 1971a).

Later in development there is no doubt that teratomas can form from PG cells (reviewed by Stevens, 1967b, 1974). Genital ridges grafted to the testis can form teratomas between the twelfth and sixteenth days of development (Stevens, 1964, 1966, 1967c, 1970a). Half the genital ridges form ovaries in this site and are therefore believed to derive from female embryos, whereas the remaining grafts form testis tissue and subsequently teratomas, which are believed to derive from male genital ridges. The grafting procedure induces teratoma formation in some strains that do not have sponteneous testicular teratocarcinomas. It is not known why female PG cells fail to form teratomas or why all PG cells should eventually lose the capacity to form these tumors. One possibility is that the events that precede meiosis inhibit teratoma formation. The fact that female PG cells enter meiosis on the fourteenth day of development but male PG cells do not do so until eight to ten days after birth is consistent with this explanation (reviewed by Peters, 1970).

7.2.2 Spontaneous Teratocarcinomas

Testicular teratocarcinomas

The development of spontaneous testicular teratocarcinomas (Stevens, 1959, 1962) is traceable in strains with a high incidence of these tumors; these are 129/Sv–Sl^{J}CP and 129/ter Sv (see section 7.7.2). The teratomas are seen as small groups of undifferentiated cells within the seminiferous tubules of the testis on the fifteenth day of development. They remain in the undifferentiated state until two or three days after birth, at which time they are composed of disorganized ectodermal and endodermal epithelia with surrounding mesodermal cells; on the fourth day after birth they form tubes of neuroepithelial cells that resemble the embryonic neural tube. These tumors regularly become well differentiated and are rarely transplantable.

A conspicuous feature of their embryology is that the undifferentiated cells never form trophectoderm or trophoblast giant cells during their early differentiation (see chapter 1 and section 7.5.2). It is also clear that they have XY near-diploid karyotypes and therefore develop directly from PG cells without a normal meiotic division.

Ovarian teratocarcinomas

In contrast to testicular teratocarcinomas, spontaneous ovarian teratocarcinomas appear to derive from parthenogenetically activated 1-cell eggs, which often appear to undergo abnormal meiotic divisions (see section 7.7.1). Like spontaneous testicular teratocarcinomas, ovarian teratocarcinomas are very rare in mice (Slye et al., 1920), and an analysis of their embryology has depended on the discovery of a mouse strain (LT) with a high incidence of these tumors (Stevens and Varnum, 1974). In the LT strain, all 30-day-old females contain cleaving eggs in the ovary; histologically these appear to develop through stages of blastocyst and egg-cylinder formation before becoming disorganized and forming teratomas in about half the females. Similarly, these eggs develop parthenogenetically up to the eighth day of development inside the uterus after they have been ovulated. During their development these tumors proceed through the normal sequence of embryogenesis of the fertilized mouse egg; they form trophectoderm and trophoblast giant cells and have an "embryology" that differs from the spontaneous testicular teratomas.

7.3 Chromosomal Abnormalities and Their Effect on Development

The majority of transplantable teratocarcinomas and cultured EC cells are chromosomally abnormal. However, the number of their chromosomes is usually close to the mouse diploid count of 40; other mouse tumors often contain cells with much higher chromosome numbers. It is impossible to compare the behavior of EC cells and the cells of karyotypically normal embryos before we have established what effect these abnormalities have on development.

7.3.1 Teratocarcinomas

Limited differentiation
Chromosomal abnormalities often develop in teratocarcinomas as they are progressively transplanted (Stevens and Bunker, 1964; Bunker, 1966). The changes in the karyotype are usually associated with a reduction in the cell types that are obvious in sections of the tumor, and this correlation has led to the view that chromosomal changes might restrict the developmental potential of the EC cells. It is well known that repeated transplantation results in tumors that either do not differentiate at all and consist solely of EC cells (nullipotential) or form predominantly neural tissue from the EC cells and are called neuroteratomas (Stevens, 1970c; Bernstine et al., 1973; Damjanov et al., 1973; Iles, 1977). The latter tumors are quite distinct from neuroblastomas, which do not contain a population of EC stem cells. Likewise, endoderm carcinomas (often called parietal yolk-sac carcinomas), which are formed by teratocarcinomas (e.g., Pierce et al., 1962), do not contain EC cells.

Chromosomal abnormalities
The apparent limitation of differentiation in the tumors just discussed is accompanied by changes in the karyotype. A detailed study was made of the cell types and karyotypes of four transplantable teratocarcinomas using modern chromosome-banding techniques (Iles, 1977; Iles and Evans, 1977). One tumor formed a wide variety of differentiated tissues for ten transplantation generations over two and a half years and retained a normal karyotype throughout; this observation demonstrates that the EC cells do not necessarily develop from the 2.1 percent of cells with spontaneous chromosomal abnormalities in the embryo

(Ford, 1975). During transplantation of this teratocarcinoma, fragments of the solid tumor that were injected into the peritoneum formed embryoid bodies, which subsequently metamorphosed into endoderm carcinoma. After this transition, the modal chromosome number rose from 40 to 41, all karyotypes had trisomy for chromosome 17 and monosomy for 19 and the X, and other chromosomes had banding abnormalities that varied among the karyotypes. Similar changes could also occur inside solid tumors, for in another case, a well-differentiated tumor contained chromosome counts of 40 and 41 in similar proportions at the first and second generations. Eventually there developed a subline in which the EC cells formed mainly neural tissue: in this subline the modal chromosome number was 41 with trisomy for chromosome 11, a 20 percent addition to the distal end of 1 and a 30 percent deletion from the distal end of 14. In this study it was found that on the five occasions in which there were changes in the pattern of differentiation of the tumor, there were also changes in the karyotype; this relationship is probably general, because one neuroteratoma is known to be hypotetraploid and other EC cells with altered differentiation are also chromosomally abnormal (Dominis et al., 1975; Guenet et al., 1974).

Although abnormal karyotypes are characteristic of tumors whose properties have changed, it is clear that an abnormal karyotype is not inconsistent with the pluripotentiality of EC cells. For instance, between 42 and 90 percent of the karyotypes of OTT 6050 embryoid bodies are abnormal (Jakob et al., 1973; Mintz and Illmensee, 1975), and contrary to the report of Guenet et al. (1974), the karyotypes of the pluripotential cell lines derived from them are also abnormal according to banding studies (Nicolas et al., 1976; E. P. Evans, personal communication). It is also known that many EC cell clones of independent origin are pluripotential and contain a variety of aneuploid karyotypes (e.g., Kahan and Ephrussi, 1970; Evans, 1972; Lehman et al., 1975; McBurney, 1976). The conclusion seems to be that EC cells with changed patterns of differentiation are always chromosomally abnormal, yet EC cells with abnormal karyotypes can remain pluripotential.

Why might an abnormal karyotype affect the observed differentiation of EC cells? First, it is certain that abnormal karyotypes by themselves do not prevent the expression of differentiated functions; in addition to the information given about EC cells above, it is known that the differentiated cell lines from teratocarcinomas are all chromosomally abnormal (usually hypotetraploid) and that other differentiated mouse lines are similar (Terzi, 1974; Lehman et al., 1975; Rheinwald and Green, 1975; Nicolas et al., 1976). Second, most of the EC cell karyotypes contain at least one copy of each pair of chromosomes and, therefore, it is unlikely that the cells lack any gene that is present in the normal diploid cell.

It seems more probable that unbalanced karyotypes slightly alter the rate of various processes inside the solid teratocarcinomas and change their apparent range of differentiation. The apparent range of differentiated tissues that EC cells form inside a large tumor depends on several factors, such as the number of histological sections cut and observed, the relative rates of multiplication of the EC cells and differentiated cells, and the rate of differentiation of the EC cells. Minor changes in either of these last two factors might completely alter the histology of a tumor, because EC cells dominate whenever they are rapidly dividing and slowly differentiating. This is almost certainly the correct explanation of the apparent restriction of differentiation in one of the teratocarcinomas with neural tendencies. This teratocarcinoma rarely formed other tissues, whereas cell lines grown from it differentiated into many cell types both in culture and after injection into a host; perhaps the tissue culture conditions altered the growth rate of the EC cells and thus increased the frequency at which differentiated cells appear in the tissue sections (compare Iles, 1977; McBurney, 1976). Similarly, manipulations of at least one "nullipotential" EC cell line in vitro causes the cells to differentiate (M. I. Sherman and R. A. Miller, in preparation).

Many of the abnormal karyotypes of transplantable teratocarcinomas are trisomies (Iles and Evans, 1977). Specific trisomies are associated with spontaneous leukemias of AKR mice, certain rat tumors, and a number of human hematological conditions (Mitelman, 1971; Dofuku et al., 1975; Wurster-Hill et al., 1976). Perhaps similar associations between chromosomal abnormalities and teratocarcinomas will be found, but presently the abnormalities appear to be a heterogeneous collection. As we have seen, teratocarcinomas can be transplantable without these abnormalities, and their appearance is presumably related to particular selection conditions inside the tumors (Iles and Evans, 1977).

7.3.2 Comparison with the Effect of Chromosomal Abnormalities on Normal Mouse Embryos

The chromosomal abnormalities of EC cells do not limit their ability to grow, but they do affect the frequency with which different cell types show up in the tumors; often the reverse seems to be the case for mouse embryos developing inside the uterus. Effects of chromosomal abnormalities on embryonic development are described below and in chapter 3.

Changes in the number of chromosome sets
Haploid and triploid embryos die soon after implantation, but tetraploid embryos sometimes survive for a day after birth [haploid embryos reviewed by Graham (1974); triploids by Wroblewska (1971); tetraploids by Snow (1973, 1975)]. The tetraploid embryos die at various stages of embryogenesis, but it is clear that tetraploid cells can differentiate into many of the tissues of an adult mouse. The death of haploid and triploid embryos precedes stages of extensive embryonic tissue differentiation; however, a haploid karyotype does not preclude subsequent cytodifferentiation, because originally haploid blastocysts can form teratomas with a range of cell types similar to those produced by fertilized diploid blastocysts (Iles et al., 1975). It is likely that triploid embryos could also form well-differentiated teratomas if they were transplanted to extrauterine sites because triploid rat (Bomsel-Helmreich, 1971) and human (Carr, 1969) embryos differentiate well. The conclusion is that changes in the number of chromosome sets reduce the ability of haploid, triploid, and tetraploid embryos to grow within the uterus. However, both haploid and tetraploid cells are able to differentiate into a wide range of tissues, and therefore the pluripotentiality of neartetraploid EC cells is not surprising (Kahan and Ephrussi, 1970; McBurney, 1976).

Changes in the sex chromosomes
Several teratocarcinomas and many EC cell clones become XO after progressive transplantation or cell culture; these XO cells have been derived from both XX and XY cells (Stevens and Bunker, 1964; Bunker, 1966; McBurney, 1976; Iles and Evans, 1977). These teratocarcinomas and EC cells remain pluripotential, and it is interesting, there-

fore, to find out if the same is true for mouse embryos bearing sex
chromosome abnormalities. Relatively normal mice have been born
with XO, XYY, XXY, and XXXY karyotypes. Clearly these abnormalities do not block differentiation of most tissues, although the
embryos can develop at a reduced frequency and all except the XO
individuals are infertile (reviewed by Ford, 1970; Cattanach, 1975;
Burgoyne and Biggers, 1976). The complete absence of an X chromosome stops preimplantation development, so it is likely that the X
chromosome contains genes required for cell multiplication and maintenance (Morris, 1968; Tarkowski and Rossant, 1976). No EC cells
lacking this chromosome have yet been found.

Monosomics and trisomics

Monosomic and trisomic embryos can be generated by breeding F_1
products of crosses between mice with metacentric chromosomes and
mice with normal karyotypes. Embryos containing monosomics of any
chromosome or trisomics of chromosomes 4, 8, 11, 15, and 17 all die
by the twelfth day of development (Ford, 1975; Gropp, 1976). Because
the embryonic part of the conceptus contains tissues that are immature
at this time, it is of interest that EC cells monosomic for chromosome
19 can multiply rapidly and form neural tissue [therefore, monosomy, is
not necessarily cell lethal (Iles and Evans, 1977)], and that EC cells
trisomic at 8 and 11 can form many mature tissues (McBurney, 1976).
Similarly, the tissue types produced by EC cells trisomic for 6 and 10
appear more advanced than the same tissues in dying embryos trisomic
for these chromosomes. It is probable, therefore, that the death of
most trisomic embryos is due not to their failure to produce a particular tissue type but rather to an asynchrony between their rate of development and progressive changes in the pregnant uterus that disturbs
the formation of particular regions of the embryo (reviewed by Ford,
1975; Gropp, 1976). Trisomy of chromosome 19 is the only trisomy that
allows survival to birth (White et al., 1974).

Reciprocal translocations and tertiary trisomics

It is clear that autosomal monosomy and trisomy are lethal to the
whole animal, but they do not block differentiation in teratomas. There
are, however, situations in which only part of a chromosome is trisomic (tertiary trisomics), and in some cases mice with this abnormality
do survive after birth and breed.

Chromosomal abnormalities can be induced by x-irradiation of adults, and the effect of these abnormalities has been studied in babies that they have produced after mating with normal mice (reviewed by Searle, 1974). Many of their offspring are heterozygous for reciprocal translocations, and during meiosis such karyotypes generate further abnormalities by nondisjunction; the majority of these embryos die after implantation, but those that contain a small region of chromosome in the trisomic state sometimes survive for some time after birth (Carter et al., 1955; Eicher and Green, 1972; Ford, 1975; Kaufman, 1976; de Boer and de Maar, 1976). In at least one instance, both male and female mice with a tertiary trisomy have been known to breed (de Boer, 1973; see also Lyon and Meredith, 1966). These animals are particularly interesting because, apart from XO females, they are the only mice that carry more or less than the normal diploid chromosomal material and still remain fertile. Thus, they are useful in defining the chromosomal abnormalities that might be tolerated in an EC cell if it were to colonize the germ line and form viable gametes (see section 7.4.3).

7.4 Differentiation of Embryonal Carcinoma Cells and Embryos Compared

7.4.1 Conditions That Allow Differentiation

Normal pluripotential stem cells of the embryo are not easy to grow in cell culture, and nobody has succeeded in obtaining a pluripotential cell line directly from the embryo or from PG cells. It is difficult even to sustain the growth and development of the blastocyst in culture, and special techniques have had to be employed to encourage 2–5 percent of blastocysts to form early somite embryos; the latter sometimes contain beating cardiac muscle surrounded by yolk sac with blood islands (Hsu and Baskar, 1974; Hsu et al., 1974; reviewed by McLaren and Hensleigh, 1975). Although the intact embryo is difficult to culture, other early embryonic cells are easier to grow because cell lines can be derived from blastocysts in culture (Sherman, 1975b) and egg cylinders routinely develop into nerve and beating muscle under ordinary culture conditions; egg cylinders also develop normally for extended periods in circulating culture medium (reviewed by Steele, 1975). In contrast to the stem cells of the embryo, the selected stem cells of teratocarcinomas are relatively easy to maintain and can be made to differentiate in cell culture (see, e.g., figure 7.3).

In comparing the differentiation of EC cells and the differentiation of normal embryonic stem cells, I shall refer mainly to the differentiation of EC cells in culture, where it is possible to see the progressive development of different cell types. In particular, I shall concentrate on experiments with cloned EC cells. In these experiments, differentiation is observed in a genetically homogeneous population, and so it is easier to analyze the process.

Selection of embryonal carcinoma cells for growth in culture

It is worth remembering that mammalian embryos have had at least 100 million years to adapt to life in the uterus, assuming that the common stock of the Eutheria and Marsupalia had a placenta (Clemens, 1971). The evolution of EC cells in culture has been underway for less than ten years; a longer period of selection will undoubtedly promote their adaptation to this novel environment.

When placed in culture, dispersed solid teratocarcinomas or embryoid bodies contain a mixture of EC cells and their differentiated progeny. A primary requirement for isolating EC cells is that they should grow faster than the differentiated cells and that they should not differentiate faster than they can divide; it is likely that culture selects for certain subpopulations of EC cells in the tumor that have these properties.

It is a general experience in my laboratory that it is difficult to isolate EC cells from teratocarcinomas that recently have been induced from an embryo. It is easier to do so when the tumor is in its fourth or subsequent transplant generation and probably contains more EC cells. The EC cells from teratocarcinomas must be nursed on nondividing feeder cells. A high serum concentration assists their growth. The main problem is to stop them from differentiating and thus disappearing from the cultures. The EC cells grown on feeder cells will subsequently be called "nursed" (reviewed by Martin, 1975).

In contrast, it is easier to isolate EC cells from embryoid bodies, probably because these structures have proportionately more EC cells than solid tumors. However, even these cells must be handled with care, and many laboratories avoid the use of trypsin during subculture. Frequently, EC cells can be grown without feeders; these cells will be referred to as "emancipated" EC cells (e.g., Jakob et al., 1973). It

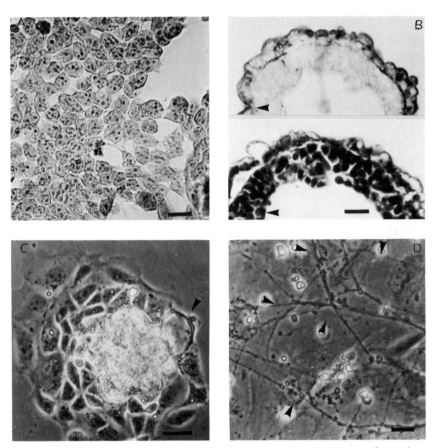

7.3 Carcinoma cells in culture. (a) Embryonal carcinoma cells on the bottom of a petri dish. Fixed formol saline and stained Mayer's hemalum. Scale bar = 25 μm. (b) Cystic embryoid body formed in culture. Top panel, stained with Schiff's reagent, shows the Schiff's positive endoderm layer (arrow). Bottom panel stained with Mayer's hemalum and Light Green to show the EC cells (arrow) beneath the endoderm layer. Scale bar = 25 μm. (c) Endoderm carcinoma in culture. The large refractile cells in the center of this colony are often vesiculated (arrow) and look like the outside layer of embryoid bodies. The endoderm carcinoma in this petri dish was synthesizing α-fetoprotein. Scale bar = 50 μm. (d) Nerve fibers (arrows) formed by EC cells in culture. The fibers are overlying a sheet of epithelial cells. Scale bar = 25 μm.

should be noted that emancipated EC cells can be derived from their nursed counterparts. It is clear that in isolating either nursed or emancipated EC cells in culture one is selecting stem cells that do not readily differentiate under the culture conditions used for their isolation and all procedures for inducing differentiation in these cells involve changing the culture conditions; therefore, the culture conditions for initiating development in nursed and emancipated EC cells are different because the cells have been selected in different environments. It is also clear that many of the EC cells used for differentiation studies have been chromosomally abnormal (E. P. Evans, personal communication; Lehman et al., 1975; Nicolas et al., 1976), and their differentiation should really be compared to that of aneuploid mouse embryos (section 7.3).

Differentiation of embryonal carcinoma cells
Nursed embryonal carcinoma cells   Studies by Martin and Evans 1975a,b,c; Evans and Martin, 1975; Martin, 1975, 1977; Evans, 1976b) have revealed that when single cell suspensions of nursed EC cells are plated on tissue culture dishes without feeders ("weaning"), they initially form small attached colonies. As they grow, they loosen their contact with the substratum and appear to form endoderm on their exposed surface. These in vitro embryoid bodies can be readily dislodged from the culture dish and subsequently grown in bacteriological dishes; they do not stick to these dishes and remain as balls of cells. These embryoid bodies sometimes are simple, with a core of EC cells surrounded by a rind of endoderm cells, but certain cell clones produce embryoid bodies that develop further into cystic or cavitated forms within seven days. A cystic embryoid body often develops cell layers with a dense extracellular matrix and a rare one will contract, indicating that muscle formation can occur. However, the greatest range of differentiated tissues forms after both simple and cystic embryoid bodies are plated out on tissue culture dishes, which need not be covered with feeder cells (McBurney, 1976). The embryoid bodies stick to the substratum, and first endoderm and then neural cells emerge from the multilayered cell mass. During the next two weeks, the cells reach high densities and many cell types can be observed: these include ectodermal derivatives, such as neural tissue, pigmented epithelium, and keratinizing epithelium; mesodermal derivatives, such as fibroblast connective tissue, muscle, and cartilage; and glandular epithelium, a defini-

tive endodermal derivative. There are variations of this procedure for
making nursed EC cells differentiate, but the pattern of differentiation
does not vary much.

In 1961, Pierce and Verney showed that embryoid bodies removed
from the peritoneal cavity develop a similar range of cell types during
in vitro differentiation. Their observations were repeated in 1974
(Teresky et al., 1974; Levine et al., 1974; Gearhart and Mintz, 1974;
Hsu and Baskar, 1974).

Emancipated embryonal carcinoma cells   Emancipated EC cells differentiate after they have reached confluence on the surface of a petri
dish (Kahan and Ephrussi, 1970; Rosenthal et al., 1970; Nicolas et al.,
1975). Flattened epithelial cells, which are probably endodermal cells,
appear at this time. During the next four days, there is considerable
cell death. Ten days later, areas of nerve cells appear. Subsequently,
the following cell types develop in turn: unicellular contractile cells
(cardiac muscle), cartilage, fat cells, keratinized cells, striated muscle,
and pigment cells (Nicolas et al., 1976). Frequently embryoid bodies
float free of the differentiating cells and differentiate in much the same
way as the embryoid bodies produced by nursed EC cells (Kahan and
Ephrussi, 1970; Rosenthal et al., 1970).

Some emancipated EC cells do not differentiate under these conditions, but these cells can be forced to differentiate if they are grown in
large aggregates (Sherman, 1975a). This technique of aggregation also
causes differentiation of cells that do not differentiate either under normal culture conditions or in adult hosts (M. I. Sherman and R. A.
Miller, in preparation).

The conclusion is that most EC cells that are either nursed or emancipated tend to differentiate when they are packed closely together.
Whether or not these conditions of close contact are necessary for
differentiation is discussed later (see section 7.5.1).

7.4.2 Changes in Gene Products with Differentiation

The appearance and disappearance of gene products in differentiating
EC cells and in embryos are discussed below. Note that the comparison is usually between cells differentiating in culture and cells differentiating inside the uterus. It must be determined whether differences are
due to inherent dissimilarities between the cells or whether culture
conditions by themselves account for the discrepancy. Embryonic bio-

chemical products and antigens that are not known to appear during EC cell differentiation are not discussed (see chapters 2 and 5).

Characteristics of embryonal carcinoma cells and embryonic stem cells
General properties   EC cells have the following features in common with PG cells and the embryonic ectoderm of the egg cylinder:

1. The nucleus contains no obvious condensed chromatin apart from the nucleoli, and the cytoplasm is rich is ribosomes but lacks a prominent Golgi apparatus or rough endoplasmic reticulum (RER) (Pierce and Beals, 1964; Pierce et al., 1967; Solter et al., 1970a; Damjanov et al., 1975).
2. The cells contain alkaline phosphatase, which is almost exclusively membrane bound (Damjanov and Solter, 1975; Clark and Eddy, 1975).
3. Membranes of embryonic ectoderm cells react with antibodies prepared against F9 EC cells (Nicolas et al., 1976; reaction of this antibody with PG cells is unknown).

It is conceivable that the EC cells are equivalent to the pluripotential cells of the preimplantation embryo (each cell of an 8-cell embryo and ICM cells of the blastocyst), but there are several reasons for believing otherwise:

1. During their early differentiation, EC cells never form giant trophoblast cells (see section 7.5.2) and, therefore, are not equivalent to the blastomeres of the 8-cell-stage mouse embryo, each of which can form this tissue (reviewed by Kelly, 1975).
2. It is known that EC cells express the lactate dehydrogenase-5 isozyme (E. Bernstine, personal communication; R. A. Miller, personal communication), yet the mouse embryo does not produce this gene product until the seventh day of development (Auerbach and Brinster, 1967).
3. They have a different pattern of glycosaminoglycan sulfation compared to the cells of the early embryo (Cantor et al., 1976).

X-chromosome inactivation   The difference between EC cells and the cells of the preimplantation embryo is reinforced by studies on X-chromosome inactivation. The inactivation of one of the female mouse embryo's X chromosomes after fertilization has been studied by various techniques (e.g., Gardner and Lyon, 1971). Both X chromosomes exhibit similar staining properties before the early blastocyst (39-cell stage). Subsequently, in some metaphase plates, one of the X chromosomes is heterochromatic; this X chromosome usually derives from the father (Takagi, 1974; Wake et al., 1976). Condensed sex chromatin appears first in some of the cells of the embryo at the 50-cell stage (De Mars, 1967). Although some asynchrony in X-chromosome replication has been reported in a small proportion of blastocyst cells (Mukherjee, 1976), the heterochromatic X is not initially late-labeling; rather, the heterochromatic X initiates replication later than the autosomes and completes replication earlier than the autosomes up to the eighth day of development, when the heterochromatic X is usually late-replicating (Takagi and Oshimura, 1973; Takagi, 1974). It seems likely that changes in the condensation and replication of the X chromosome occur independently in different parts of the embryo, because by the ninth day of development, 90 percent of the heterochromatic X chromosomes in the extraembryonic membranes are paternal, while in the embryonic part, only 60 percent are paternal (Takagi and Sasaki, 1975). Taken together, these studies of the female embryo suggest that the majority of cells have a condensed X chromosome from the blastocyst stage onwards, that asynchronous replication of the X chromosomes can be detected by reduced labeling of one X early and late in S phase by at least the sixth day onwards, but that X chromosome late replication does not occur in most cells of the embryo until the eighth day of development.

A study of the replication of X chromosomes in two female teratocarcinoma lines revealed that in one line, both X chromosomes initiated DNA synthesis about the time the autosomes did, but in the other line, there was a delayed initiation of replication in one of the X chromosomes and some metaphases had late-labeling X chromosomes (McBurney and Adamson, 1976). With these patterns of DNA synthesis, the first line is equivalent to cells in the embryo before the sixth day of development, whereas the second line is equivalent to a stage in development when asynchronous X replication has started (i.e., after the blastocyst stage).

X-chromosome inactivation can also be studied by observing the ratio of X-linked enzyme activity to autosome-linked enzyme activity in XX and XO cells. The ratio should be the same in cells of both karyotypes after X inactivation has occurred (Epstein, 1969, 1972). Within the embryo, X inactivation is reflected in the ratio of X-linked to autosomal enzyme activity between the blastocyst stage and the seventh-day egg cylinder (Kozak et al., 1974; Kozak and Quinn, 1975). In a study of XX and XO EC cell lines (McBurney and Adamson, 1976), the ratio of specific activities of X-linked and autosomal enzymes (Chapman and Shows, 1976; Chapman, 1975) was found to be similar, suggesting that genetic X inactivation had occurred in both the XX EC cell lines. In other EC cell lines, enzyme activities are more consistent with the activity of both X chromosomes (G. R. Martin, personal communication). Again this evidence suggests that EC cell lines are equivalent to cells at different stages of embryogenesis.

Antigens of embryonal carcinoma cells   A variety of EC cell lines with different properties have been used to raise antibodies against early embryonic cells (table 7.1; reviewed by Jacob, 1975; Gachelin, 1976; Edidin, 1976; see also Stern et al., 1975). These antibodies were raised in syngeneic hosts, which greatly increases the probability that they are directed only toward determinants specific to the early embryonic cells and which are not present in the adult. In one case, however, antibodies directed against EC cells have been shown to cross-react with antigens in the kidney and brain of adult mice (Stern et al., 1975).

Antibodies raised against EC cells of one line often cross-react with other EC cell lines. Thus, EC cells can have common surface properties. The anti-EC-cell antibodies do not cross-react with cell lines that have been transformed with onocogenic viruses. Therefore, their reactivity, is not directed to other known carcinoembryonic antigens (e.g., Alexander, 1972; Coggin and Anderson, 1974). They appear to be directed against a wild-type product of the T complex (see chapter 4), so it is particularly interesting that these antigens are also expressed in the early embryo (see table 7.1). The antibodies prepared against F9 EC cells cross-react with 2-cell and later preimplantation embryos (Babinet et al., 1975), but the antibodies specific to PCC4 EC cells do not react with the cells of the cleaving embryo until the blastocyst stage (Gachelin, 1976). Consequently, different EC cells could be equivalent to different stages of embryogenesis, as studies on X inactivation have suggested. The antigens recognized by the anti-EC-cell antibodies are often present on only a proportion of the EC cells, and they tend to disappear from the surface of the differentiated products of EC cells.

The aggregation of embryonal carcinoma cells   The confluence or aggregation of EC cells at the start of their differentiation slows the rate of cell multiplication (Martin and Evans, 1975b), and similar changes occur in the derivatives of the embryonic ectoderm developing in the

Table 7.1. Expression of Antigens on the Surface of Embryonal Carcinoma Cells and Their Differentiated Derivatives and on the Surface of Embryos

Cell Type	Antigens Expressed				
	F9	PCC4	Endo	H-2	Ia
EC cells					
F9, EC cells, nullipotential or limited differentiation [1]	+	−	−	−	−
PCC1, EC cells, multipotential in vivo [2]	+	+	?	−	?
PCC3, EC cells, multipotential in vivo [2] and in vitro [3]	+	+	traces	−	−
PCC4, EC cells, multipotential in vivo [2] and in vitro [4]	+	+	±	−	−
LT1, EC cells from LT mice [5, 6]	+	+	?	?	?
EC cells differentiating in culture					
NDI, after 4 days in culture	+	+	+ (5–10 % of cells)	traces (5% of cells)	−
Differentiated cell lines from teratocarcinomas					
PCD1, cardiac myoblasts [7]	−	−	+	±	?
PCD2, striated myoblasts [7]	−	−	−	±	?
PCD3, fibroblasts [5]	−	−	−	+	?
Endo, endoderm [8]	−	−	+	+	?
PYS2, parietal endoderm [9]	−	−	+	+	?
Embryo					
1-cell, unfertilized	−	−	−	−	−
1-cell, fertilized	?	−	−	−	?
16-cell	+	−	+	−	?
ICM, blastocyst	+	+	?	?	?
Trophectoderm, blastocyst	+	−	?	−	−

Table 7.1 (continued). Expression of Antigens on the Surface of Embryonal Carcinoma Cells and Their Differentiated Derivatives and on the Surface of Embryos

Cell Type	Antigens Expressed				
	F9	PCC4	Endo	H-2	Ia
8th- to 9th-day embryos	+	?	?	−	?
10th-day embryos	−	?	?	±	?
14th-day embryos	−	?	+	+	?
16th-day embryos	−	−	?	?	?
Embryonic fibroblasts	−	−	−	+	?
Adult sperm	+	+	+	+	+

Data from studies at the Pasteur Insitute (reviewed by Gachelin, 1976). + indicates the presence of surface antigens, − indicates their absence, and ? means that results are not available or are uncertain. The studies on antigens are as follows: Artzt et al. (1973, 1974, 1976), Artzt and Jacob (1974), Babinet et al. (1975), Buc-Caron et al. (1974), Fellous et al. (1974), and Gachelin (1976). Parallel studies by other groups are described in the text, and the expression of H-2 and F9 is also discussed in chapters 4 and 5.

Cells used in these studies can be traced from the following references: [1] Artzt et al. (1973); [2] Jakob et al. (1973); [3] Nicolas et al. (1975); [4] Sherman (1975a); [5] Gachelin (1976); [6] Stevens and Varnum (1974); [7] Boon et al. (1974); [8] Artzt et al. (1976); [9] Lehman et al. (1974).

uterus (Snow, 1976). Aggregation of EC cells is an active process, and as EC cells become confluent, their adhesiveness appears to increase. The adhesive tendency is probably due to the increased production of carbohydrate-protein complexes, which appear to mediate cell adhesion (Oppenheimer et al., 1969; Oppenheimer and Humphreys, 1971; Oppenheimer, 1973, 1974, 1975). Ascitic EC cells used in these studies required glutamine for cell adhesion, and it is thought that this amino acid is used in the transamination of fructose-6-phosphate in the formation of the adhesion factor. It is interesting, therefore, that as the EC cells reach confluence, there is a transitory increase in glutamine synthetase that could reflect an increased demand for glutamine and its incorporation into the adhesive factor (Connolly and Oppenheimer, 1975).

Endoderm formation
As soon as a discrete cell layer has formed on the blastocoel side of the ICM, its cells are distinct: they contain more rough endoplasmic

reticulum than the cells from which they formed (Nadijcka and
Hillman, 1974), and they subsequently develop into cells that resemble
secretory cells in a number of other structural features (see Siekevitz
and Palade, 1959). It is necessary to recapitulate briefly the available
information about endoderm in the normal embryo before proceeding
to an analysis of endoderm formation from EC cells.

The first endoderm to be formed is the primitive endoderm, or hypoblast. Histological and transplantation studies suggest that these cells
subsequently associate and possibly interact with a variety of other
tissues to form the endoderm of all extraembryonic structures. It is
argued that the endoderm of the late fetus (definitive endoderm) must
originate from the embryonic ectoderm (see chapter 1). Proceeding
from the mouth to the anus, the definitive endoderm derivatives are
believed to include part of the Eustachian tube, tonsil, thyroid, para-
thyroid, thymus, ultimobranchial bodies, bronchi and lungs, stomach,
pancreas, liver, gut, cloaca, and urogenital sinus (Rugh, 1968). The
development of many of these organs depends on endoderm-mesoderm
interactions (reviewed by Lash, 1974), and endoderm products whose
synthesis depends on cell interactions for initiation or maintenance are
not likely to be expressed by the haphazard tissue arrays of teratocarcinomas. For this reason the present section focuses on the characters
of the various types of primitive endoderm.

Table 7.2 illustrates the changes in gene products that occur during
the formation of endoderm. Several points emerge from the comparison
of these events in embryos and in teratocarcinomas:

1. The endoderm associated with embryonic ectoderm changes its
characteristics between the seventh and eighth day of development.
2. The definitive endoderm of the fetus and the adult shares many
characteristics with the visceral and parietal endoderm of the embryo.
3. Endoderm carcinoma has many of the characteristics of other types
of endoderm.

It is important to assess the extent to which cell-culture conditions
interfere with the development of primitive endoderm characters. Morphological studies suggest that blastocysts can form both parietal and
visceral endoderm in culture and that this endoderm is slightly
disorganized (Solter et al., 1974; McLaren and Hensleigh, 1975). Biochemical evidence supports this view: the parietal endoderm is characterized by the high activity of plasminogen activator, and the endoderm

Table 7.2. Endoderm Development from Embryonal Carcinoma Cells and Embryonic Stem Cells

Characteristic	EC Cells	Endoderm from EC Cells		
		Cloned EC	Embryoid Bodies from Peritoneum	Solid Tumors
Ultrastructure				
Microvilli	− [6, 7, 8, 10, 22, 30, 34, 43]	+ [19, 22]	+ [43]	?
Prominent Golgi, RER, polysomes	− [6, 7, 8, 10, 22, 30, 34, 43]	+ [19, 22]	+ [43]	?
Vesicles and lysosomes	− [6, 7, 8, 10, 22 30, 34, 43]	+ [19, 22]	+ [43]	?
Basement membrane				
Readily seen under microscope	− [6, 7, 8, 10, 22, 30, 34, 43]	+ [19, 22]	+ [43]	?
PAS positive	− [22]	+ [19, 22]	+ [43]	?
Reacts with anti-epithelial BM antibody	−	?	?	?
Collagen or collagenlike material	+ [1]	+ [1]	?	?
Synthesis or localization				
α-Fetoprotein	− [1, 18]	+ [1]	+ [24]	+ [12, 42]
Albumin	+? [24]	?	+? [24]	?
Enzymes				
Alkaline phosphatase (membrane bound)	High [3, 22]	Low [22]	Low [3]	Low [7]
Acid phosphatase	Low [9]	?	?	High [9]
Plasminogen activator	− [36]	+ [36]	+ [21]	?

Notes: [a]Associated with the trophoblast layer. [b]Associated with the extraembryonic ectoderm. [c]Associated with the embryonic ectoderm. [d]Associated with the extraembryonic mesoderm. [e]Associated with the embryonic mesoderm and ectoderm.

References (data from species other than the mouse are noted): [1] E. D. Adamson and

Parietal[a]	Visceral Extraembryonic[b]	Visceral Embryonic[c]
?	+ [26, 35, 37, 39]	+ [26, 35, 37, 39]
?	+ [37, 39]	+ [37, 39]
?	+ [37]	+ [37]
+ [35]	?	?
?	?	?
+ [28]	− [28]	− [28]
− [28]	− [28]	− [28]
− [12]	+ [12]	+ [12]
?	?	?
?	Low [38]	Low [38]
?	High [38]	High [3]
?	?	?

M.J. Evans (personal communication); [2] Adinolfi et al. (1975); [3] Bernstine et al. (1973); [4] Cade-Treyer (1973, 1975) (AFP synthesis found in bovine kidney in culture); [5] Clark et al. (1975) (rat); [6] Damjanov and Solter (1973); [7] Damjanov and Solter (1975); [8] Damjanov et al. (1971a); [9] Damjanov et al. (1971c); [10] Damjanov et al. (1975); [11] Deren et al. (1966) (rat); [12] Engelhardt et al. (1973); [13] Everett (1935)

Table 7.2 (continued). Endoderm Development from Embryonal Carcinoma Cells and Embryonic Stem Cells

| | Endoderm in 8th-Day Embryos | | | |
Characteristic	Parietal[a]	Visceral Extraembryonic[b]	Visceral Extraembryonic[d]	Visceral Embryonic[e]
Ultrastructure				
Microvilli	?	+ [35, 37, 39]	+ [37, 39]	− [37, 39]
Prominent Golgi, RER, polysomes	?	+ [37, 39]	+ 37, 39	− [37, 39]
Vesicles and lysosomes	?	+ [37, 39]	+ [37, 39]	− [37, 39]
Basement membrane				
Readily seen under microscope	+ [35]	+ [35]	− [37, 39]	− [37, 39]
PAS positive	?	?	?	?
Reacts with anti-epithelial BM antibody	+ [28]	+ [28]	+ [28]	+ [28]
Collagen or collagenlike material	− [28]	− [28]	− [28]	− [28]
Synthesis or localization				
α-Fetoprotein	?	+ [12]	+ [12]	+ [12]
Albumin	?	?	?	?
Enzymes				
Alkaline phosphatase (membrane bound)	?	Low, [38], but high in PG cells (see section 7.2.1)		
Acid phosphatase	?	High [38]	High [38]	Low [38]
Plasminogen activator	?	?	?	?

(rat); [14] Gitlin and Perricelli (1970) (human); [15] Gitlin et al. (1972) (human); [16] Gustine and Zimmerman (1973); [17] Jollie (1968) (rat); [18] Kahan and Levine (1971) (claimed AFP production by EC cells; possibly endoderm present); [19] Cudennec and Nicolas (1977); [20] Lee et al. (1969); [21] Linney and Levinson (1977); [22] Martin and Evans (1975b); [23] N. A. McGregor, H. Dziadek, and E. D. Adamson (personal communication); [24] Mintz et al. (1975) (evidence of presence but not synthesis AFP and albumin); [25] Mukerjee et al. (1965); [26] Nadijcka and Hillman (1975); [27] Padykula et

Endoderm in 10th–16th-Day Embryos			
Parietal[a]	Visceral Extraembryonic[d]	Definitive Fetal	Endoderm Carcinoma
− [10]	+ [11, 12, 27, 45]	+ (e.g., gut)	+ [32]
+ [17, 32]	+ [11, 27, 45]	+ (e.g., secretory epithelia)	+ [32]
+ [17]	+ [11, 27]	+ (e.g., gut)	+ [32]
+ [32, 45]	+ [45]	+ (e.g., gut)	+ [6, 32]
+ [32, 45]	+ [45]	?	+ [6, 32]
+ [28, 32]	+ [33]	+ [28]	+ [6, 25, 29, 31]
+ [5]	?	?	+ [20, 25, 29, 31]
?	+ [16, 23, 44]	+ [2, 15, 16, 23]	+ [1, 2, 4]
?	+ [14, 16]	+ [14, 16, 46]	+? [24]
?	?	?	?
?	?	?	?
+ [36, 40]	− [36, 40]	− [36, 40]	?

al. (1966) (rat); [28] Pierce (1966); [29] Pierce (1970); [30] Pierce and Beals (1964); [31] Pierce and Nakane (1967); [32] Pierce et al. (1962); [33] Pierce et al. (1964); [34] Pierce et al. (1967); [35] Reinius (1965); [36] Sherman et al. (1976); [37] Solter et al. (1970a); [38] Solter et al. (1973); [39] Solter et al. (1974); [40] Strickland et al. (1976); [41] Tamaoki et al. (1974); [42] Teilum et al. (1974, 1975) (human); [43] Teresky et al. (1974); [44] Wilson and Zimmerman (1975); [45] Wislocki and Padykula (1953) (rat); [46] Yeoh and Morgan (1974) (rat).

produced by cultured ICMs also synthesizes this protease (Strickland et al., 1976); the visceral endoderm is characterized by a high rate of alpha fetoprotein (AFP) synthesis (Engelhardt et al., 1973; N. A. McGregor, M. A. Dziadek, and E. D. Adamson, personal communication), and this protein is secreted by the derivatives of blastocysts in culture (Hensleigh, 1976). The conclusion is that culture conditions by themselves should not inhibit the development of at least some parietal and visceral endoderm characters.

EC cells are able to form different types of endoderm; some of these resemble the parietal endoderm and others the visceral endoderm of the embryo (G. R. Martin, L. Wiley, and I. Damjanov, personal communication). In the endoderm formed by cloned EC cells and by embryoid bodies there are columnar cells with microvilli on their exposed surface. These cells have enlarged Golgi apparatuses, many free and membrane bound ribosomes, and pinocytotic and lysosome vesicles; they resemble the cells of the visceral endoderm. There are also flattened cells that lack dense microvilli, have fewer vesicles, and are associated with a thick extracellular matrix; these cells resemble parietal endoderm. As a whole, the endoderm has high acid phosphatase activity, low alkaline phosphatase activity, and a thick basement membrane, which is eosinophilic and periodic acid Schiff (PAS) positive, separating it from the EC cells. Like parietal endoderm, this endoderm produces plasminogen activator, but it is probable that it also contains visceral endoderm because some differentiating EC cell lines produce substantial amounts of α-fetoprotein (AFP, see Table 7.2 for references).

It is worth commenting on the production of AFP because it illustrates some of the difficulties involved in interpreting biochemical data of this kind. AFP appears in the primitive endoderm of seventh-day egg cylinders; later in embryogenesis, however, AFP is synthesized not only in the visceral endoderm (derived from the primitive endoderm), but also in the parenchymal cells of the liver, which are definitive endoderm derivatives. It is also known that its production is diminished shortly after the birth of mice and that it is observed again in substantial quantities only when mice develop hepatomas. There is some evidence that serum factors control its synthesis (Tumyan et al., 1975), so there is a possibility that the serum used in cell culture could alter its synthesis. It is also known that a culture of kidney cells that contain no derivatives of either the primitive or the definitive endoderm can switch on AFP synthesis (Cade-Treyer, 1973, 1975). The conclusion is that the character of AFP synthesis by itself should not be used to identify endoderm formation.

The endoderm formed on the outside of aggregated cloned EC cells does not cross-react with anti-EC-cell antibodies or with antibodies directed against H-2 or the thymocyte antigen [Thy-1 (Θ) antigen; Stern et al., 1975]. Similarly, H-2 has not been detected in the endoderm layer of embryoid bodies taken from the peritoneum (Edidin et al., 1971). Some of the endodermlike cells that are formed by confluent emancipated EC cells during differentiation have antigens that cross-react with antibodies prepared against transplantable endoderm cells derived from teratocarcinomas (Gachelin, 1976). These observations on the endoderm antigens suggest that EC-cell-derived endoderm developing early in the differentiation of EC cells has lost EC cell antigens and gained endoderm-associated antigens, but has not acquired the H-2 antigens of fully differentiated cells.

If we consider as a group the characters that do develop in the early differentiated derivatives of EC cells, it is permissible to say that these cells form endoderm that in both morphology and biochemical features can resemble both the parietal and the visceral endoderm of the embryo. It is still too early to decide whether they form the definitive endoderm tissues of the fetus.

Mesoderm formation

Extraembryonic mesoderm: hemopoietic tissue Hemoglobin synthesis, which begins in the extraembryonic mesoderm on the eighth-day embryo, is the first biochemical product of the mesoderm that is readily recognized (reviewed by Cole, 1975). The process is sustained by contact with the visceral endoderm in some species, and there is evidence that the microenvironment of the cells is important in the development and maturation of the hemoglobin-producing cells (Miura and Wilt, 1970; Cole, 1975). Blastocysts and EC cells developing in culture might be expected to produce hemoglobin only rarely, both because of the environmental requirements mentioned above and because special culture conditions are required for the development and maturation of hemopoietic cells of the normal embryo (Metcalfe and Moore, 1971); nevertheless it is known that embryonic erythrocytes, once formed, can maintain a high rate of hemoglobin synthesis in culture (Cole and Paul, 1966). It is not surprising that there has been limited success in obtaining hemopoiesis from the derivatives of EC cells grown in cell culture, because it has recently been shown that exposing these cells to 5 percent CO_2 in air on an organ culture raft stimulates hemopoiesis; this condition has not generally been used [compare Nicolas et al., (1975) and Cudennec and Nicolas (1977)]. In cultures of differentiated EC cells and of embryoid bodies taken from the peritoneal cavity, hemoglobin has been detected cytochemically in nucleated erythrocytes (Hsu and Baskar, 1974; Teresky et al., 1974; Cudennec and Nicolas, 1977; Martin, 1977; B. Hogan, personal communication). Because fetal hemopoiesis in the liver, spleen, and bone marrow produces non-nucleated erythrocytes, the observation of nucleated erythrocytes suggests that the hemopoiesis observed in these cells is equivalent to that observed in the yolk sac mesoderm of the embryo.

Embryonic mesoderm: muscle   Unfortunately, much of the information about normal muscle development has come from experiments with the chick embryo (e.g., Hauschka, 1968), and this discussion must emphasize observations on the markers of normal mouse muscle development, which have also been studied in the derivatives of EC cells.

The high specific activity of the enzymes acetylcholinesterase (EC 3.1.1.7), creatine phosphokinase (EC 2.7.3.2), phosphoglycerate mutase (EC 2.7.5.3), and myokinase (adenylate kinase, EC 2.7.4.3) characterize adult mouse muscle (Adamson, 1976). The first three of these enzymes have been thought to have muscle-specific forms, but the validity of this conclusion depends on what is meant by "specificity" and which type of muscle is studied.

Acetylcholinesterase activity reportedly exists in different molecular species in various tissues at progressive stages of development (reviewed by Silver, 1974). However, studies of the forms of this enzyme that are present in adult mouse brain, muscle, and erythrocytes suggest that all the forms result from the aggregation of a single monomer (Adamson et al., 1975; Adamson, 1977). Aggregation of a monomer probably accounts also for the appearance of different forms during muscle development in the chick (Wilson et al., 1975), so it will henceforth be assumed that the tissues of the mouse contain the same acetylcholinesterase polypeptide with a molecular weight of about 80,000 daltons (Adamson et al., 1975).

Other enzymes that have been used to follow the development of muscle have more distinct forms. Thus creatine phosphokinase has a

form that is thought to be characteristic of muscle. This enzyme is a
dimer that can be composed of M (muscle) and B (brain) subunits as
homodimers or heterodimers. These subunits in the chick are anti-
genically distinct and have molecular weights close to 42,000 daltons
(reviewed by Turner and Eppenberger, 1973; Turner, 1975), and it will
be assumed that in the mouse the M and B subunits are products of
different structural genes. The M form appears during the fourteenth
and fifteenth days of development in the muscles of the limbs, tongue,
heart, and bladder, but it is not found in the adult muscle of the
stomach and the uterus. The M form of the enzyme, then, can be used
to follow the development of only certain types of muscle by EC cells
(Adamson, 1976). The enzyme phosphoglycerate mutase resembles creatine phosphokinase in that it also has homodimers and heterodimers of
M and B subunits; in man the subunits differ in heat stability and
amino acid composition and each subunit has a molecular weight of
about 30,000 daltons (reviewed by Omenn and Hermodson, 1975). In
some other species the distribution of the M and B subunits of phosphoglycerate mutase is similar to that of creatine phosphokinase, but
although we know that the M form appears during limb development in
the mouse, its tissue distribution in this species has not been studied in
detail (Adamson, 1976).

Most studies of enzyme changes during muscle development have
been with myoblasts determined to form striated muscle, rather than
cardiac or smooth muscle. In cultures of chick myoblasts, the development of striated muscle is promoted by a collagen, or gelatin, substrate
and a rich culture medium containing embryo extract (e.g., O'Neill and
Stockdale, 1972). Acetylcholinesterase is present in the chick myoblasts before they fuse into myotubes, and enzyme activity dramatically increases during fusion (e.g., Wilson et al., 1973; Fluck and
Strohman, 1973). Much of the enzyme is released into the tissue culture medium, and the rise in enzyme activity that accompanies fusion
does not continue unless an acetylcholine analog is present in the culture medium (Goodwin and Sizer, 1965). Under rich culture conditions,
studies with mammalian myoblasts have shown that there is a sudden
rise in creatine phosphokinase and myokinase activity as fusion proceeds,
but the final specific activity of creating phosphokinase also depends
on the concentration of calcium and creatine in the culture medium
(rat: Shainberg et al., 1971; Yaffe and Dym, 1972; Delain et al., 1973;

Ingwall and Wildenthall, 1976; mouse: Morris et al., 1972, 1976). In culture, the M subunit of creatin phosphokinase appears after fusion (Morris et al., 1972).

Thus, there are several reasons why EC cells might form muscle without showing the characteristic range of enzyme changes that accompany muscle formation in the whole animal or in culture:

1. The EC cells form a variety of cell types in addition to muscle, so changes in specific activity of enzymes cannot be used to follow the development of this single tissue.
2. The presence of these other cell types might interfere with muscle formation (e.g., Kagen et al., 1976).
3. If the EC cells form smooth muscle and cardiac muscle as well as striated muscle, then one would expect a low proportion of the M form of creatine phosphokinase to develop even though mature muscle was present.
4. The special culture conditions used in myoblast fusion studies generally have not been employed in work with EC cells.

Despite these difficulties, it is clear that as cloned EC cells differentiate into muscle and other cell types, there is a large rise in acetylcholinesterase activity (E. D. Adamson and M. J. Evans, personal communication), and sometimes creatine phosphokinase activity increases as well. The rise in creatine phosphokinase activity is transitory [in only one case did the differentiated EC cells develop some of the M form on this enzyme; none showed a transition to the muscle form of phosphoglycerate mutase (Adamson, 1976). Studies on embryoid bodies developing in culture after removal from the peritoneal cavity have shown similar changes in the specific activity of acetylcholinesterase and creatine phosphokinase (Gearhart and Mintz, 1974, 1975; Mintz et al., 1975; Levine et al., 1974; Hall et al., 1975). One of these studies demonstrated that culture conditions and the other difficulties mentioned above were responsible for the failure of EC cells to regularly display the normal characteristics of striated muscle; embryoid bodies that formed morphologically recognizable striated muscle in culture failed to show transition to the M form of creating phosphokinase, although the same embryoid bodies formed solid teratocarcinomas in adult hosts in which the muscle had the M form of this enzyme (Gearhart and Mintz, 1975).

The formation of other mesoderm derivatives, such as connective tissue and cartilage, have not been studied biochemically, but it is known that some fibroblasts express the Thy-1 antigen (Stern et al., 1975) and that teratocarcinoma-derived cell lines of fibroblasts express H-2 antigens, as do myoblast lines (Gachelin, 1976).

Ectoderm formation: nerve development
Neurons are the only ectoderm derivative whose biochemical differentiation has been followed during EC cell development. In cultures of EC cells and in early embryos, the neurons appear to cytodifferentiate early. Because the brain has been studied in detail, this tissue can be taken as an example of the biochemical changes that occur during normal neuron development, despite the fact that the brain also contains a variety of other cell types.

Brain formation is accompanied by a substantial rise in acetylcholinesterase activity. Unfortunately, this rise in activity either does not occur or does not persist unless dissociated embryonic or early postnatal brain cells are subsequently grown as aggregates [compare surface cultures of Wilson et al., (1972) and aggregate cultures of Seeds (1971)]. After newborn mouse brain cells have been cultured on the surface of petri dishes for thirty days, the specific activity of this enzyme falls to 4 percent of its original level, and because this activity is similar to that in other cultured nonneuronal cells, such as fibroblasts (Glinos and Bartos, 1974), it cannot be used to recognize neuron differentiation in cell culture (Wilson et al., 1972). It should not be concluded that none of the acetylcholinesterase activity in differentiating EC cell cultures is contributed by neurons because neuroblastomas, for instance, have high levels of activity when grown in particular conditions (Blume et al., 1970); it is not certain, however, that neurons seen in culture of EC cells contain large quantities of acetylcholinesterase.

Fortunately, neuron development can be recognized in another way. The forms of the enzyme fructose diphosphate aldolase (aldolase, EC 4.1.2.13) have been used to follow nerve differentiation. Aldolase is a tetramer containing combinations of subunits A, B, and C (reviewed by Masters, 1968). In the mouse, the brain contains C_4 and various mixtures of the C and A subunits; adult muscle from the limbs, heart, uterus, and stomach contains A_4 with traces of A_3C_1, and liver, kidney, and spleen contain the A and B subunits (Adamson, 1976). Because early embryonic cells and EC cells produce only the A subunit, the formation of tetramers containing a high proportion of C subunits is a good marker of neuron development. High proportions of C subunits in the tetramers are regularly observed when EC cells differentiate in culture (Adamson, 1976).

Another way of studying neuron differentiation is to look for the expression of the Thy-1 antigen. The cells of the peripheral and central nervous system express this antigen (Reif and Allen, 1964), as do thymocytes, thymus-derived lymphocytes, epidermal cells, and fibroblasts (Reif and Allen, 1966; Raff, 1969; Schlesinger and Yron, 1969; Scheid et al., 1972; Stern, 1973). However, neurons from fetal brain cultures do not express this antigen until they have been in vitro for a long period (Mirsky and Thompson, 1975), and it is not surprising, therefore, that EC-cell-derived neuronlike cells also do not express Thy-1 (Stern et al., 1975).

Conclusions

These studies provide a clear demonstration that EC cells display a varying range of gene products as they differentiate in culture: EC cell characteristics disappear and products that are characteristic of mature differentiated tissues appear. In many cases, the cultured derivatives of EC cells do not produce the same quantities of differentiated products as do similar cells developing in the intact embryo. This discrepancy can be shown to be caused by the limitations of current culture techniques, and it is probably not due to inherent differences between the differentiated cells produced by EC cells and by the stem cells of the embryo. Those biochemical products of EC cells differentiating in culture that have been analyzed all appear in the embryo before birth and, in this sense, are embryonic. Cultured EC cells then, provide an excellent system for studying the control of gene expression in defined conditions, but one must turn to studies on the interaction of EC cells with normal cells in order to observe their full expression of adult mouse differentiated characters.

7.4.3 Differentiation in Association with Normal Embryonic Cells

Totipotency of embryonal carcinoma cells

For some time there have been clues that the EC cells and their derivatives participate in developmental interactions with normal cells; they have been shown to induce kidney tubule formation in culture and they have shown signs of repopulating the hemopoietic system of x-irradiated adults (Auerbach, 1972a, b). It was first shown by Brinster (1974, 1975) that cells taken from dissociated embryoid bodies and injected into the blastocyst could form part of the coat of a mouse. More important was the finding that EC cells could contribute to *most* of the tissues of the mouse after they had been injected into the blastocyst (Mintz et al., 1975; Mintz and Illmensee, 1975; Papaioannou et al., 1975; Illmensee and Mintz, 1976). Showing that the EC cells could form viable sperm completed the proof of their genetic totipotency (Mintz et al., 1975; Mintz and Illmensee, 1975).

The EC cells are exposed to all the tissue interactions of normal embryogenesis after injection into the blastocyst and the range of different tissues that they subsequently form is much greater than when they differentiate by themselves either in solid teratocarcinomas or in culture. The experiments to test the developmental potential of EC cells in the blastocyst are still in progress because many of the resulting EC cell ↔ embryo chimeras are being mated to try to detect a contribution of the EC cells to the germ line, and so the following is a "halftime" review.

In these experiments, the EC cells were from whole embryoid bodies taken from the peritoneal cavity (Brinster, 1974, 1975), from the center of simple embryoid bodies obtained from the peritoneal cavity (Mintz et al., 1975; Illmensee and Mintz, 1975; Mintz and Illmensee, 1975), or from cultured and sometimes cloned EC cells (Papaioannou et al., 1975). In most cases 2–40 EC cells were injected, and in such experiments it is not possible to test the developmental potential of each EC cell. The injection of single EC cells, however, gave similar results (Illmensee and Mintz, 1976). The results from both types of experiment will be treated together. The differentiation of EC cells in chimeras has been recognized by several techniques, as outlined below.

Direct recognition of differentiated products   In situations in which the strain of the EC cells and the strain of the recipient blastocyst have different alleles of genes that code for differentiated products, it is possible to demonstrate the differentiation of the EC cells directly (see, e.g., figure 7.4). Thus, it has been shown that these cells can form melanocytes of the skin and retina, the dermal elements of the skin follicle, adult hemoglobin, immunoglobulins (IgG1, IgG2a), major urinary protein (from parenchyma of the liver), and sperm cells.

7.4 Embryonal carcinoma cell ↔ embryo chimera. The EC cells were from C3H embryos. Their pigmented melanocytes show up on the head and flank of the white skin formed by the cells of the recipient blastocyst. Photograph kindly provided by V. E. Papaioannou.

Recognition of glucose phosphate isomerase isozymes   In most experiments, the EC cells and the recipient blastocyst differed at the locus for glucose phosphate isomerase (GPI–1; Carter and Parr, 1967), a dimer enzyme that is present in most mouse cells. In embryo aggregation chimeras between strains having different electrophoretic variants of this enzyme, the only tissue reported to show hybrid enzyme between the electrophoretic variants is skeletal muscle; this finding is consistent with the observations suggesting that skeletal muscle is formed by cell fusion (Mintz and Baker, 1967; Chapman et al., 1972; Gearhart and Mintz, 1972). The discovery of hybrid enzyme in the skeletal muscle of EC cell ↔ embryo chimeras is, therefore, evidence that the EC cells are participating in the normal cell fusions of muscle development. It is also possible to use this enzyme to show that the EC cells often contribute to the blood. This phenomenon raises a problem because almost all tissues are perfused by blood and therefore any apparent EC cell contribution to a tissue might be due to blood contamination. Luckily, in some of the EC cell ↔ embryo chimeras, the EC cells had not formed the blood cells but had contributed to other tissues, and in these cases the enzyme variant could be used to follow the EC cells and their products. In mice without the EC cell GPI isozyme in the blood, the EC cell products formed a substantial part of the following tissues and organs: brain, heart, kidney, liver, salivary gland, thymus, lung, intestine, pancreas, stomach, placenta, reproductive tract, and gonads (Mintz et al., 1975; Mintz and Illmensee, 1975; Illmensee and Mintz, 1976; Papaioannou et al., 1975). It is likely, therefore, that EC cells can form any cell type in an adult mouse.

Colonization of the embryo

An important question is: why have very few EC cell ↔ embryo chimeras produced EC-cell-derived viable gametes? This question is important because it is the hope of all workers in this field to introduce mutants selected in cultures of EC cells into the germ line of mice after injection into the blastocyst. It is likely that several factors are responsible for the current low rate of germ-line colonization.

Pattern of colonization   The colonization of the embryo by EC cells is different from that of injected ICM cells (see chapter 1): the latter are widespread and evenly distributed in the fetus as are the cells of each embryo in an aggregation chimera (injection chimeras: Ford et al., 1975b; aggregation chimeras: e.g., Mintz, 1971). In contrast, the distribution of EC cell products is uneven and they are often found in only a few organs of the chimera. One possibility is that the cells do not mix with the ICM cells; certainly the cells from postimplantation embryos do not readily colonize the blastocyst (Moustafa and Brinster, 1972a,b).

Despite the uneven distribution of EC cell products, they have been found within the gonads many times. Unfortunately, it is not possible to tell whether they were in the germ line or soma of the gonads by observing the GPI types in homogenates of this tissue. Even if they did form part of the germ line, there is no guarantee that they could produce viable gametes because XX cells in XX ↔ XY chimeras with a male phenotype do not produce sperm and, with one exception, XY cells in XX ↔ XY chimeras with a female phenotype do not form eggs (reviewed by Ford et al., 1975a). The conclusion is that the chances of EC cells reaching the gonads is smaller than that of ICM cells, and even if they do form part of the germ line, then they, like ICM cells, have only a 50 percent chance of residing in a blastocyst of the same sex and thus producing viable gametes.

Chromosomal abnormalities   In addition, EC cells might have to have a normal karyotype in order to produce viable gametes. A wide range of chromosomal abnormalities are associated with male sterility in mice. The list includes X-autosome translocations, Y-autosome transloca- tions, insertional translocations, tertiary trisomy, failure of X and Y association in meiotic prophase, sex chromosome trisomies, double heterozygosity for certain reciprocal translocations involving a common chromosome, spontaneous univalence, and Robertsonian translocations with homologous arms (reviewed by Evans, 1976a). Many of these abnormalities are also associated with sterility in female mice. In rare instances these animals are fertile, but it is clear that any cell with a chromosomal abnormality has problems at meiosis. In mice, the only karyotypes that have more or less than the normal amount of chromo- some material and are still fertile are XO mice and mice with one of the tertiary trisomies (see section 7.3.2).

The effects of the above abnormalities have all been studied in mice that are not chimeras. In chimeras it is possible that cells with abnor- mal karyotypes might be helped through meiosis by interactions with normal cells; certainly the somatic cells of mutant embryos can be sustained by such interactions (Eicher and Hoppe, 1973). Unfortu- nately, there is no evidence that this is the case; instead, cells with abnormal karyotypes, such as XO embryos, have failed so far to colo- nize the embryo at all when placed in chimeric associations with nor- mal embryos (Burgoyne and Biggers, 1976; C. F. Graham, unpublished observations).

It is now possible to assess the EC cell injection experiments. The cells used by Mintz and Illmensee came from the cores of embryoid bodies taken from the peritoneum, in which 40 percent of the meta- phases contained more or less than the normal number of mouse chro- mosomes. Banding studies on the cells are not yet complete, so it is impossible to tell whether the metaphases with the correct number of chromosomes also contained abnormalities. In the experiments of these authors, therefore, it is possible that the low rate of germ-line coloniza- tion was due to chromosomal abnormalities in the EC cells at the time of injection. This conclusion is consistent with the observations of Papaioannou et al. (1975) that cultured EC cells with abnormal karyotypes do not colonize the germ line. Thus, it will be impossible to obtain an estimate of the potential to form germ-line chimeras until a cell injection study is conducted with cloned EC cells containing normal karyotypes.

Tumor formation
In section 7.8, it will be noted that most of the differentiated products of EC cells are unable to form tumors. The same situation occurs in EC cell ↔ embryo chimeras, with some exceptions. In chimeras obtained by injecting cells from the center of embryoid bodies into blastocysts, only one tumor has been reported (Illmensee and Mintz, 1976). This tumor, which resembled a pancreatic adenocarcinoma, was shown to be derived partly from the EC cells. It would be interesting to know whether this tumor arose by a change in the products of the EC cell after normal differentiation. In contrast to these results, EC cells obtained from prolonged cell culture were found to frequently form tumors in the EC cell ↔ embryo chimeras (Papaioannou et al., 1975). The tumors, which grew rapidly after birth, were poorly differentiated teratocarcinomas. In this case, some EC cells appeared to have survived unchanged through the process of normal development. It is still not clear why EC cells obtained from cell culture should give results different from those obtained when EC cells from embryoid bodies are injected into the blastocyst.

7.5 The Mechanism of Development

The mechanism by which stem cells generate a variety of cell types is one of the great unsolved problems of modern biology. This section examines whether or not a comparison of differentiation from EC cells and from embryonic stem cells will provide a solution.

7.5.1 Critical Mass for Differentiation?

Histological observations of solid teratocarcinomas show that the EC cells often grow in aggregates or nests, and this property has led to the idea that the aggregation of EC cells might be necessary for subsequent cytodifferentiation (Solter et al., 1975a). Certainly EC cells in culture differentiate readily after they have been aggregated, and it has been shown that a particular cell mass must be exceeded in aggregates of emancipated EC cells if cytodifferentiation is to occur (Nicolas et al., 1975; Sherman, 1975b). This situation must be contrasted with that in the normal embryo, where the differentiation into trophectoderm and primitive endoderm depends on the exposure of the morula and ICM cells, respectively, to outside conditions; the total cell mass is irrelevant to these differentiations (chapter 1). In the selection of emancipated EC cells during cloning, any initially pluripotential cell that differentiates when exposed to outside conditions is discarded because it does not grow as an EC cell; consequently the EC cells that do grow will have been selected not to differentiate when exposed to outside conditions. On the other hand, nursed EC cells, which are cloned in contact with other cells, are never so severely exposed to outside conditions and will not be selected to suppress differentiation in these circumstances. It is not surprising, therefore, that nursed EC cells, like the isolated ICM, regularly form endoderm on the outside of aggregates (Martin and Evans, 1975a,b), and endoderm formation does not appear to depend on any particular number of cells in the aggregates (C. F. Graham, unpublished observations). Clearly, further studies are required to resolve the matter of whether or not a critical mass is required for determination and differentiation.

7.5.2 Sequence of Differentiation in Teratocarcinomas and in Embryos

It is one of the obvious features of normal embryonic development that cell types appear in a fixed sequence. As the mouse embryo develops, determined cells are sequentially thrown off from the pluripotent stem cells: in turn, trophectoderm, primitive endoderm, extraembryonic mesoderm, embryonic mesoderm, definitive endoderm, and ectoderm are delimited [see chapter 1 and compare with amphibian embryos (Horder, 1976)]. This sequence might be necessary for subsequent gene expression. One could envisage a situation in which trophectoderm, once formed, would interact with the stem cells so that they would produce primitive endoderm at the next determination, and this tissue might in its turn interact with the stem cells so that they would produce mesoderm. If this sequence is necessary for the expression of advanced differentiated functions, then it should be repeated every time the EC cells differentiate.

It is clear that trophectoderm formation is not required for subsequent differentiation of the stem cells, because this tissue is not regu-

larly formed during the initiation of spontaneous testicular teratocarcinomas (Stevens, 1959) or during the early differentiation of any EC cell
in any circumstance (see section 7.4). The fact that spontaneous ovarian teratocarcinomas proceed through a stage of trophectoderm formation is simply a consequence of their origin from parthenogenetically
activated eggs; the EC cells of these tumors subsequently differentiate
without the immediate formation of trophectoderm both in vivo
(Stevens and Varnum, 1974; Damjanov et al., 1975) and in vitro (G. R.
Martin, personal communication).

It has been suggested that endoderm formation might be required for
the subsequent cytodifferentiation of the stem cells (discussed by
Martin and Evans, 1975a, b; Martin, 1975). There are two reasons for
thinking that this might be the case:

1. The first cell type formed by cultured EC cells under particular
conditions resembles the endoderm of the early embryo.
2. EC cells that cannot differentiate (nullipotential) are unable to form
endoderm.

Neither of these arguments is compelling. In histological section, EC
cells in solid teratocarcinomas appear to be transforming directly into
both neural tissue and serous glands (Damjanov et al., 1973; Solter et
al., 1975a), and fibroblastic cells seem to be the first cell type formed
by certain EC cells in culture (Nicolas et al., 1975; Sherman, 1975b);
there seems to be no absolute requirement for endoderm formation.
More experiments are required to settle the matter.

Even if it were shown that endoderm cells are not required for the
development of other cell types from EC cells, it is worth bearing in
mind that there is extensive evidence that cell interactions in the embryo are required at least for the sustenance of many differentiated
functions, and that these interactions are particularly necessary when
parts of the embryo are grown in culture (reviewed by Lash, 1974;
Saxén and Wartiovaara, 1976).

7.5.3 Developmental Mutants

Mutants of embryonal carcinoma cells
So far in this book, several classes of mutants that affect development
have been described. Some mutants and chromosomal abnormalities
kill the embryo, and their deleterious effects sometimes appear to be

confined to a particular tissue in the embryo (see chapters 3 and 4 and section 7.3). Others affect the time of expression, tissue levels, and inducibility of particular enzymes in viable embryos (chapter 3).

It has already been noted that chromosomal abnormalities affect the frequency with which EC cells express differentiated functions. In a more systematic attempt to understand the genetics of EC cell differentiation (Boon et al., 1975), a pluripotential EC cell clone was mutagenized. The cells that survived were cloned and injected into adult hosts so that their differentiation could be studied. Five of the 53 clones that grew in the host had a greatly reduced ability to differentiate. This restriction was not absolute since sublines of each of the five clones sometimes formed derivatives of all three germ layers. The clones obtained after mutagenesis were often aneuploid. One possible conclusion from these experiments is that any mutation in an EC cell that is not cell lethal cannot block the formation of any particular cell type. It is perhaps permissible to go further and argue that these studies support the view that mutant EC cells that cannot form a particular cell type are also unable to divide and survive and that this relationship is the consequence of the mechanism of differentiation involving some general cell function which is required for cell maintenance and division. This view is at variance with current interpretations of mutations that affect mouse development, but interpretations of such mutations might change (see chapter 4).

Cell fusion

Another way of studying the genetic and epigenetic control of differentiation is to fuse cells together with the Sendai virus and then to observe the expression of differentiated functions in the resulting hybrid cells (reviewed by Davidson, 1974). In evaluating hybrids between EC cells and differentiated cells, the goal is to determine whether the characters of the EC cell or of the differentiated cell dominate the characters of the hybrid.

Fusions of EC cells and fibroblasts produce hybrid cells that are fibroblastlike in cell culture and produce fibrosarcomas when injected into adult hosts (Finch and Ephrussi, 1967; Jami et al., 1973). The apparent domination of the hybrid cells by fibroblast characters could be partly the consequence of the damaging conditions of cell fusion and of the selection system used to isolate the hybrid cells, because hybridization of EC cells with other EC cells can also produce hybrids that give rise to fibrosarcomas [Sit (1973), quoted in Evans (1975)]. There is, however, evidence that at least some nonfibroblast differentiated characters can survive the fusion conditions. In neuroblastoma × EC cells hybrids, the enzyme monoamine oxidase has been detected at high concentration, and this is a neuroblastoma cell characteristic [Sit (1973), quoted in Evans (1975)].

In contrast, hybrids of EC cells and mouse thymocytes and hybrids of EC cells and Friend leukemia cells are pluripotential, and there is no obvious influence of the differentiated cells' characteristics on the properties of the hybrid (Miller and Ruddle, 1976). It is possible that technical differences between these experiments and the cell fusions mentioned earlier account for the differences in result.

The most important problem now is to discover whether the EC cell genome itself has been directed to code for the synthesis of differentiated products in hybrids that express differentiated functions; such an observation would provide evidence that some part of the original differentiated cell is responsible for controlling the activity of the EC cell's genome. And in pluripotential cell hybrids it could be interesting to know whether the EC cell, like the frog's egg (reviewed by Gurdon, 1976), can direct the differentiated cell components in the hybrid cell to completely revert to the embryonic state. Further studies are necessary to answer these questions.

7.5.4 Conclusions

It is depressing to conclude this section by noting that none of these studies has yet shown how cells become different from each other. However, it is now practical to attempt to answer the following questions:

1. What significant biochemical changes accompany EC cell aggregation?
2. What are the environmental differences between cells on the outside (endoderm forming cells) and cells on the inside of an aggregate?
3. What cell receptors detect these environmental differences?
4. How do the cell receptors communicate with the processes that control gene-product formation?
5. Does the process of determination involve single or multiple changes in cell state?

It is clearly impossible to answer these questions by studying the mouse embryo because it is so small; EC cells provide the biological material, and persistent research is required.

7.6 Comparison of Embryonal Carcinoma Cells and Other Tumor Cells

It has been pointed out that EC cells can become the stem cells of the embryo (section 7.4.3) and are also the stem cells of teratocarcinomas that kill their host. Consequently, they have in common the properties of normal and malignant cells, and it is important to understand this bewildering duality. It may be explained in two ways. First, it is conceivable that the stem cells of an embryo are inherently capable of neoplastic growth in adult animals and, as a corollary, that all tumors form from cells that have remained in, or have returned to, a partly embryonic state (e.g., Markert, 1968). There is certainly evidence that many other tumors express fetal antigens and secrete fetal products (e.g., Alexander, 1972; Coggin and Anderson, 1974; review of AFP by Adinolfi et al., 1975). Second, rapidly dividing embryonic cells might be readily transformed to neoplastic growth by unknown factors. The correct explanation might become evident if one could readily recognize the difference between embryonic stem cells and transplantable tumor cells (see Damjanov and Solter, 1974b). The following discussion distinguishes between EC cells with a normal karyotype and those which are obtained after repeated passage and usually have a grossly abnormal karyotype; the latter have undergone secondary changes and are not informative about the primary properties of EC cells (see section 7.3).

7.6.1 Growth of Embryonic Stem Cells and Tumor Cells

Growth in adult hosts
Both embryonic stem cells and tumor cells are transplantable. The transplantability of tumor cells is well known. Ehrlich ascites cells hold the record, for they have been maintained for at least eighty years by repeated passage in adult hosts. Embryonic stem cells do not have an indefinite lifespan, but their transplantability is impressive: skin and bone-marrow stem cells have continued to grow for up to seven transplantations over five years, a period that is two or three times the lifespan of the donor animal (Krohn, 1962; Micklem and Loutit, 1966). Experiments of shorter duration have also shown that embryonic stem

cells have an inherently longer lifespan than their adult counterparts. Thus, the ability of hemopoietic stem cells to form spleen colonies after repeated transplants decreases with age; the stem cells of the yolk sac (ninth day of development) can be transplanted for longer periods than similar cells in the fetal liver (fifteenth day of development), which in turn can be transplanted for longer periods than bone-marrow hemopoietic stem cells from newborn or adult mice (Metcalf and Moore, 1971). Hemopoietic stem cells originate later in development than EC cells, and on this basis alone EC cells can be expected to have an inherently longer lifespan after transplantation. It is known that some EC cells can be maintained for at least twenty-two years through repeated transplantation (L. C. Stevens, personal communication). In this way they resemble other tumor cells; however, most of these long-lived tumors have abnormal karyotypes and might have been secondarily changed. Chromosomally normal EC cells have been maintained for two and a quarter years, a period that is clearly well within the range of transplantability of normal embryonic stem cells (Iles and Evans, 1977). The conclusion is that the transplantability of EC cells is not a reason for believing that they are other than normal stem cells.

It is usually thought that the growth of embryonic stem cells is controlled, whereas that of other tumor cells is not, but again this distinction is difficult to apply to EC cells. On the one hand, many physiological substances can affect the growth rate of tumor cells in their host, but on the other hand, embryonic stem cells seem to be remarkably resistant to humoral mechanisms of growth control that are supposed to operate in adult animals (see papers in Goss, 1972). Thus, transplanted embryonic guts form giant intestines in hosts that already have an alimentary canal (Zinzar et al., 1971), and young bones continue to grow in skeletally complete adults [Green (1955), reviewed by Goss (1964)]. The conclusion must be that although growth regulation can occur inside an intact embryo enclosed in the reproductive tract (e.g., Buehr and McLaren, 1974) and although the uterus can control both implantation and trophoblast invasiveness (e.g., Kirby and Cowell, 1968; McLaren, 1973), the adult has rather restricted powers for limiting the growth and development of early embryonic cells by soluble factors (see also section 7.7.2). One can expect that adults would lack soluble factors to block the growth of embryonic cells in viviparous animals, and it is clear that both EC cells and other embryonic stem cells can grow for many years in adult animals and that this ability is not peculiar to cells that have been deliberately transformed by oncogenic viruses.

Growth in culture

In culture, epithelial cells and fibroblasts multiply to a particular density and then stop growing, but if the same cells are transformed by a tumor virus, they multiply to much higher densities (reviewed by Rubin, 1974; see also Dulbecco, 1970, and Stoker and Piggott, 1974). This phenomenon of density-dependent growth might serve to distinguish normal and transformed cells. Density-dependent growth of normal cells is not directly related to cell contact, for in many instances the membranes of adjacent cells overlap long before the cells stop growing, and both embryos and EC cells grow in multilayers in culture. It could be that it is very difficult to demonstrate density-dependent growth in EC cells; the cells grow rapidly when packed together (Martin and Evans, 1974), and at high density the cells quickly die (e.g., Nicolas et al., 1975). It is therefore impossible to apply this criterion to find out whether the cells are normal or transformed. It would be particularly interesting to find out whether density-dependent death is a method of growth control of EC cells because it is an odd phenomenon in itself. All that is known at present is that protease inhibitors do not prevent it (C. F. Graham, unpublished observations). It is also important that the differentiated products of EC cells do show density-dependent growth in culture, and that if the rapid growth of EC cells in culture is considered to be transformed growth, then one must say that they quickly "untransform" (e.g., Evans, 1972).

It was once thought that normal cells and cells transformed by tumor viruses always differ by the degree to which they can be agglutinated by plant lectins such as concanavalin A; the transformed cells are readily agglutinated, but the normal cells are not. The difference is not related to the ability of concanavalin A to bind to the cell surface, and it was thought that the cells must differ either in the fluidity of their membranes or in the distribution of the lectin receptors; agglutinability could be induced in normal cells by protease treatment (Cline and Livingston, 1971; Ozanne and Sambrook, 1971). However, it is now clear that the agglutinability of normal cells changes with the cell cycle, and in mitosis and the early G_1 phase they are as agglutinable as transformed cells; indeed, most of the surface distinctions of transformed cells appear on normal cells during these stages of the·cell cycle

(Burger and Noonan, 1970; Mannino and Burger, 1975). Consequently, rapidly dividing normal cells might be expected to agglutinate more readily than slower dividing cells, and they do; neural retina cells divide more slowly as development proceeds, and their agglutinability with concanavalin A also declines (Kleinschuster and Moscona, 1972). A rapid division rate is not a complete explanation of the agglutinability or embryonic cells, however. The experiment described above showed that when cell multiplication was blocked with cytosine arabinoside (a DNA synthesis inhibitor), agglutinability did not change. It is also known that 4-cell mouse embryos are agglutinated by concanavalin A at 10μg/ml (Pienkowski, 1974) and yet they lack a long G_1 phase and spend only 5 percent of a 13-hr cell cycle in mitosis (Barlow et al., 1972). Indeed, the mouse embryo provides a good example of changes in agglutinability of cells with development which is unrelated to rates of cell division (Rowinski et al., 1976). As development proceeds, the embryo requires greater concentration of concanavalin A for agglutination: 4- and 8-cell stages require 10μg/ml, morulae require 100μg/ml, but blastocysts fail to agglutinate even when the concentration is raised to 5000μg/ml. In contrast to the whole blastocyst, the ICM resembles the early cleavage stages in that it agglutinates at 10μg/ml.

It is difficult to compare the agglutinability of EC cells with that of preimplantation embroys or cells transformed by tumor viruses. Agglutination depends on the chance of two or more cells colliding and on the concentration of lectin per cell, and these conditions often vary among laboratories. Conseqeuntly, some workers consider EC cells readily agglutinable and similar to tumor cells (Oppenheimer and Odencrantz, 1972), whereas others report that they are poorly agglutinable and are therefore dissimilar to many tumor cells (Gachelin et al., 1976). All that can be said is that the agglutinability of EC cells falls within the range of agglutinability of embryonic, transformed cells and normal cells at particular stages of the cell cycle.

7.6.2 Embryonal Carcinoma Cells and Viral Transformation

The subject here is the susceptibility of EC cells to infection by viruses and whether or not viral effects are responsible for the special features of EC cells. The EC cells are resistant to infection by DNA viruses such as polyoma, simian virus 40 (SV40), and minute virus of mice (MVM), but are susceptible to infection by adenovirus-2 and vaccinia virus (Lehman et al., 1975; Kelly and Boccara, 1976; R. A. Miller, personal communication). Rarely do EC cells that have been exposed to polyoma become T or V antigen positive (less than one in 10^4 cells), and SV40-exposed EC cells never display the T antigen. It appears that the polyoma and SV40 viruses can penetrate the EC cells, but are blocked early in the lytic cycle. As the EC cells differentiate, their cells become susceptible to infection by these two viruses; bromodeoxyuridine appears to convert EC cells to a cell type that is susceptible to SV40 and polyoma infection (Speers and Lehman, 1976). Similar observations have been made in the early embryo (see chapter 6).

It is known that EC cells are susceptible to and lysed by the RNA virus mengovirus. The production of mengovirus both by the EC and by the early mouse embryo is less than that produced by an equivalent volume of differentiated cells (Lehman et al., 1975; see chapter 6). Another RNA virus, the vesicular stomatitis virus (VSV), also infects and lyses the EC cells (N. Teich, G. R. Martin, and R. A. Wiess, personal communication); this characteristic contrasts with the nonlytic infection of the 2- and 4-cell embryo by this virus (chapter 6).

The RNA tumor virus murine leukemia virus (MuLV) appears to be able to penetrate EC cells, but it does not replicate in them. In contrast, the virus establishes a productive infection in the differentiated products of EC cells. It remains to be seen whether the virus can integrate into the EC cells before differentiation (N. Teich, G. R. Martin, and R. A. Weiss, personal communication).

There is no compelling evidence that normal EC cells have been transformed by external agents and that their ability to grow for long periods is shared with other embryonic stem cells. However, it is possible that EC cells are particularly susceptible to "internal transformation." Several virologists believe that all mouse embryonic cells contain within themselves the seeds of transformation to malignant growth. The so-called oncogene theory (Huebner and Todaro, 1969; Nowinski et al., 1970; Todaro and Huebner, 1972) proposes that all mouse cells contain a transforming gene (the oncogene), which can be attached to, or is part of, virogenes that code for endogenous viruses. The oncogene and virogene are thought to be integral parts of the chromosomal DNA and to be held inactive by repressor substances in untransformed cells. Because cells can produce endogenous virus without transformation and vice versa, it must be supposed that the virogene and oncogene are expressed (or act) separately. Many normal and malignant cells contain within them numerous endogenous viruslike particles. These particles are classified either as "intracisternal A-type particles," which have no known biological activity, or as "C-type particles," which are similar to RNA tumor viruses and might transform cells. The latter are favored as the usual carriers of the oncogene. There is no doubt that the blastocysts of several mouse strains contain these particles, and that they are present throughout embryogenesis and are therefore available to transform cells (Todaro, 1974; reviewed by Piko, 1975). The EC cells either lack or do not express C-type particles, but they do possess A-type particles; the EC cells of embryoid bodies in the peritoneum have a few particles, but many more of these are in the endoderm layer; A-type particles are found in both EC cells and differentiated cells of ovarian teratomas (Teresky et al., 1974; Damjanov et al., 1975; Damjanov and Solter, 1975). The A-type particles are also abundant in endoderm carcinoma (Damjanov and Solter, 1973). So far, C-type particles have been reported only in the endoderm layer of in vivo embryoid bodies (Lehman et al., 1974). These observations do not allow us to decide whether or not an oncogene is active in the EC cells, but this theory should be kept in mind when attempting to decide whether EC cells are normal stem cells or not.

7.7 The Induction of Teratocarcinomas

The conditions that accompany EC cell formation and multiplication are reviewed here because they might resemble conditions that promote or accompany the growth of other tumors.

7.7.1 Evidence for Chromosome Segregation

A large number of tumors contain chromosomal abnormalities (German, 1974). There is evidence that in tumor cells gene segregation can occur and lead to homozygosity for particular loci (e.g., Chasin and Urlaub, 1975).

Evidence exists also that some teratocarcinomas form after chromosome segregation in a progenitor cell, and this event might initiate EC cell growth. The segregation of chromosomes can sometimes be observed by examining karyotypes.

Teratocarcinomas from female cells

The spontaneous ovarian teratocarcinomas of LT mice develop parthenogenetically. Within the ovary, oocytes can be seen at different stages of meiosis and polar bodies are sometimes visible. The activated ovulated eggs from such ovaries can form haploid, diploid, or polyploid embryos (Stevens and Varnum, 1974), but they usually have diploid chromosome counts (P. Hoppe, personal communication). The presence of the haploid embryos demonstrates that complete meiosis can occur, but the chromosomal origin of the others is less certain. Diploid and polyploid embryos could be formed either as a result of the failure of meiotic reduction divisions or by the completion of meiosis followed by subsequent chromosome replication without cell division; both processes are known to occur in artificially activated mouse eggs (reviewed by Tarkowski, 1975).

However, chromosome segregation by itself does not create neoplastic cells. Thus human ovarian teratomas are formed by chromosome segregation, but they are rarely malignant in adult women. The chromosome complement of these tumors is always diploid, and segregation has been detected by studying the inheritance by teratomas of enzyme loci that are heterozygous in the female who bears them (Linder et al., 1975). In addition, large numbers of artificially activated haploid mouse embryos can form teratomas, but fail to produce teratocarcinomas (Iles et al., 1975).

Teratocarcinomas from male cells

Spontaneous testicular teratocarcinomas and about half the embryo-derived teratocarcinomas are XY normal diploid cells (Stevens and Bunker, 1964; Dunn and Stevens, 1970; E. P. Evans and C. F. Graham, unpublished). This observation demonstrates that the sex chromosomes do not segregate and therefore have not passed through a regular first meiotic division. It is still possible that autosomes segregate, but this possibility has not yet been studied with enzyme variants as human ovarian teratomas have been.

Adding this information to what is known about EC cells during progressive transplantation (section 7.3), it is clear that chromosome segregation by itself does not cause EC cell formation. At present there is no evidence that segregation is a necessary condition for EC cell formation.

7.7.2 Permissive and Nonpermissive Strains

Genetics of permissive hosts
Embryo-derived teratocarcinomas   In syngeneic hosts, seventh- and eighth-day embryos of strains 129/J, A/He, C3H, BALB/c, and CBA readily form teratocarcinomas, but similar embryos of strains C57BL and AKR rarely do so (reviewed by Solter et al., 1975b, and personal communication). Strain C57BL and AKR embryos are, however, able to form teratocarcinomas in F_1 hosts, and this ability suggests that it is the response of the host that controls EC cell growth. The observation that these teratocarcinomas are transplantable only in F_1 hosts confirms this view. It is legitimate, therefore, to describe AKR and C57BL as nonpermissive strains for EC cell multiplication, while other strains are described as permissive. The difference between the two groups is not absolute, but it is obvious: in strain C3H, up to 67 percent of transplanted embryos will form teratocarcinomas, but in strains C57BL and AKR, only 4–7 percent will do so. The effect of the host is not immediate, for both AKR and C57BL embryos in syngeneic hosts do contain undifferentiated cells for up to a fortnight after transplantation. The difference between these hosts and permissive hosts is that in the former all the cells differentiate.
Teratoma formation: spontaneous and from grafted genital ridges   The distinction between permissive and nonpermissive hosts is not so clearcut for spontaneous teratomas and those obtained from grafted genital ridges. This fact is not surprising because these teratomas are rarely transplantable, unlike the embryo-derived tumors. The host genotype does not have a large effect on the incidence of these teratomas, but the genotype of the teratoma progenitor cells is very important. Strain 129/J is the only mouse strain to have regular spontaneous testicular teratomas, and only three other tumors of this kind have been reported in other strains of mice. The base incidence of teratoma formation is below 1 percent in 129/J, but the gene Steel (Sl^J) increases the incidence to about 5 percent in heterozygotes, and a mutation within a subline of the strain (129/ter Sv) has raised the incidence to 32 percent (Stevens and Mackensen, 1961; Stevens, 1967b; Meier et al., 1971; Stevens, 1973). In this strain the incidence of teratoma formation depends on the genotype of the embryo and not of the mother as has been elegantly shown by ovary transplant. The genetic effect on the em-

bryo is effectively recessive because F_1 crosses between 129 and other strains do not have spontaneous teratomas. In contrast, the genetic factors causing spontaneous ovarian teratomas in LT mice are not completely recessive, for F_1 hybrids with other strains show about 10 percent of the incidence of the LT strain (Stevens and Varnum, 1974).

Sublines of both 129 and strain AL/KS are able to support teratoma formation in more than 40 percent of transplanted genital ridges (Stevens, 1964), and it is the donor genotype that seems to be important. The most interesting points to emerge from such studies are that strains permissive for embryo-derived teratocarcinomas do not necessarily support teratoma formation from transplanted genital ridges (strain C3H: Stevens, 1970) and that the gene Steel, which in the heterozygous state increases teratoma formation in the testis, in the homozygous state almost completely suppresses teratoma formation from genital ridges (Stevens, 1967c). Mouse embryos homozygous for Steel are deficient in PG cells, and the latter observation supports the view that the PG cells in grafted genital ridges are the progenitors of teratomas.

Immunology of the nonpermissive hosts

Embryo-derived teratomas induce splenomegaly, which is more pronounced when teratocarcinomas are formed (Damjanov and Solter, 1974a). This observation and the fact that EC cells bear antigens that are recognized as foreign by syngeneic hosts (e.g., Artzt et al., 1973) have directed attention to the immune responses of permissive and nonpermissive hosts (Solter et al., 1975a,b).

Various abuses of the immune system, including splenectomy and treatment with antithymocyte and antilymphocyte serum, did not increase the incidence of teratocarcinoma formation from embryos in the nonpermissive strain C57BL; the most noticeable effect was that mild treatment with antithymocyte serum reduced teratocarcinoma formation in the permissive strain C3H. Despite this failure to identify the host factor involved in teratocarcinoma formation, there was considerable evidence that the immune system affected tumor growth. Thus, neither C57BL nor C3H embryos were able to form teratocarcinomas in athymic hosts (nu/nu), and the tumor weights of C57BL embryos were reduced in C57BL hosts that had been either splenectomized or treated with antithymocyte serum.

7.8 The Cessation of Embryonal Carcinoma Cell Proliferation by Their Differentiation

7.8.1 Teratocarcinomas

Differentiation in the animal
It is clear that inside the intact embryo something happens by the ninth day of development to block teratocarcinoma formation and that by the sixteenth day another event occurs within the genital ridge to stop the PG cells from forming teratomas (section 7.2.1). It is not unreasonable to suppose that this restriction of tumor formation is related to the differentiation of the embryo and the early maturation of PG cells. However, the clearest evidence that differentiation suppresses the transplantable growth of cells comes from the classic studies by Pierce and his colleagues (1960) on embryoid bodies. They found that 64 percent of cystic embryoid bodies lacked EC cells and that after transfer of single embryoid bodies to adult hosts, 69 percent formed well-differentiated benign teratomas. This close correlation suggests that EC cells lost neoplastic growth potential as they differentiated in the cystic embryoid bodies; subsequent observations showing that the cystic embryoid bodies must have been previously formed by EC cells (section 7.4) support this conclusion.

It has also been possible to isolate differentiated cell lines directly from solid teratocarcinomas. Clones of a keratinizing epithelial cell have been shown to be unable to form tumors in both isologous and athymic mice (Rheinweld and Green, 1975), and because it is clear that they must originally have arisen from EC cells, the capacity for neoplasia has been lost.

Differentiation in culture
Viable stem cells, which are determined to form particular tissues, are retained during the differentiation of embryoid bodies inside the animal, but the differentiation of these embryoid bodies in culture blocks not only their capacity to form transplantable teratocarcinomas but also their ability to form teratomas. The explanation must be that both the totipotent EC cells and stem cells determined to form particular tissues rapidly disappear in culture. When embroid bodies are plated out in overcrowded cultures, their growth and differentiation is accompanied

by cell death and the EC cells seem to be preferentially killed. After embryoid bodies have been in culture for only two weeks, it is possible to show that their ability to form tumors is reduced (Hall et al., 1975). It is also the case that differentiated (non-EC) cell lines from embryoid bodies are usually unable to form tumors: thus, fibroblastlike, epitheloid, endodermal cells and cardiac and skeletal myoblasts are unable to form tumors (Evans, 1972; Martin and Evans, 1974; Bernstine et al., 1973; Boon et al., 1974; Lehman et al., 1974).

Taken together with the studies of tumors inside the animal, these observations clearly demonstrate that the immediate differentiated products of EC cells are unable to form transplantable tumors.

Transformation in culture
When the differentiated cells obtained from teratocarcinomas are propagated for long periods in culture, it is possible to recover cells that can form tumors. The long time required for this change in their properties suggests that the cells are transformed in culture, but there is no evidence that this explanation is correct. The clearest facts about this transition in culture came from Martin and Evans (1974). Cultures of teratocarcinoma cells contained fibroblastic "E" cells that looked quite different from the EC cells; subclones of the E cells initially showed density-dependent growth inhibition and formed monolayers, but after many passages, the cloning efficiency of the cells increased and they were able to overgrow one another. Before the transition none of the E cells could form tumors, but afterward they formed rapidly dividing fibrosarcomas; the modal chromosome count did not change during the transition.

After prolonged culture of cells from embryoid bodies, differentiated cells that can form tumors can be isolated: these cells are epithelial, stellate, or spindle-shaped in culture, and these cell types form, respectively, parietal endoderm carcinoma, neural tissue, and fibrosarcomas when reinjected into adult hosts (Lehman et al., 1975). These cells are hypotetraploid and therefore have undergone some change in culture during or after their formation from hypodiploid EC cells.

It is clear that differentiation of EC cells can suppress their ability to grow progressively in adult hosts until some secondary event occurs. However, these studies also show that the expression of differentiated functions and transplantable growth are not incompatible, as previous studies on many other tumors (e.g., neuroblastomas) indicated.

7.8.2 Other Transplantable Tumors

It is important to find out whether other tumor cells become benign as they differentiate. Pierce and his colleagues have clearly shown that other tumors consist of rapidly dividing stem cells and more slowly dividing cells, which are more obviously differentiated (reviewed by Pierce, 1974).

The transition from malignant to benign growth was demonstrated with a squamous cell carcinoma, namely, a transplantable tumor of the rat lip (Pierce and Wallace, 1971). The tumor consisted of a large mass of undifferentiated cells that surrounded pearls of differentiated squamous cells. It could be shown that the undifferentiated cells contributed to the growth of the pearls and that within the pearls cell division was rare: tritiated thymidine was pulsed into the host; two hours later the labeled cells were almost exclusively confined to the undifferentiated zone, and it was some time before cells labeled during the thymidine pulse had differentiated and become part of a pearl. To discover whether the squamous cells formed by the undifferentiated cells were benign, 78 were transplanted to syngeneic hosts. None formed tumors, but about one-third of control transplanted undifferentiated cells formed squamous cell carcinoma.

The transition from malignant to benign growth was not proved for two other tumors studied, but there are good reasons for believing that a similar transition could have occurred. In a chondrosarcoma of the hamster, the cells surrounded by the most mucoprotein do not incorporate thymidine after short time intervals, but the less differentiated cells do (Pierce, 1974). The situation is more complex in the case of primary adenocarcinomas of the breast in mice. In this case, both differentiated cells and some undifferentiated cells do not label after the host has been perfused with tritiated thymidine for several days, and there appears to be a nondividing population of stem cells (Wylie et al., 1973). In both the chondrosarcoma and the adenocarcinoma, it is not certain whether the most fully differentiated cells do not divide because they are fully differentiated or whether slowly dividing cells just happen to look more fully differentiated. In any event, these studies make a strong case for studying the conditions that promote the differentiation of neoplastic cells.

It is reasonable to accept the view that the stem cells of teratocarcinomas and many other tumors resemble embryonic stem cells. The

growth and differentiation of embryonic stem cells are not always controlled by humoral mechanisms operating in the adult host, and the
mechanisms that control the differentiation of EC cells still are not
known. After these mechanisms have been discovered, mammalian embryology will be more understandable, and it will be possible to offer
advice on the clinical control of cancer.

Acknowledgments   My colleagues in Oxford have clarified several of
the uncertain regions of teratocarcinoma biology and have helped me to
write this review. I would like to thank them (Eileen Adamson, Marie
Dziadek, Ted Evans, Richard Gardner, Susan Iles, Mike McBurney,
Nicola McGregor, and Virginia Papaioannou) and those who have critically checked the manuscript for me (Charles Babinet, Martin Evans,
Gail Martin, Karl Illmensee, Richard Miller, Davor Solter, and Roy
Stevens). They eliminated many errors. Numerous other scientists gen-
erously sent me preprints of their work. The Medical Research Council
and Cancer Research Campaign have kindly supported the studies of
myself and my colleagues in Oxford.

References

ADAMSON, E. D., AYERS, S. E., DEUSSEN, Z. A., and GRAHAM, C. F. (1975).
Analysis of the forms of acetylcholinesterase from adult mouse brain. Biochem. J. *147*, 205–214.

ADAMSON, E. D. (1976).
Isoenzyme transitions of creatine phosphokinase, aldolase, and phosphoglycerate mutase in differentiating mouse cells. J. Embryol. Exp. Morph. *35*, 355–367.

ADAMSON, E. D. (1977).
Acetylcholinesterase in mouse brain, erythrocytes, and muscle. J. Neurochem. *26*, 605–615.

ADINOLFI, A., ADINOLFI, M., and LESSOF, M. H. (1975).
Alpha-feto-protein during development and in disease. J. Med. Genet. *12*, 138–151.

ALEXANDER, P. (1972).
Foetal "antigens" in cancer. Nature *235*, 137–140.

ARTZT, K., BENNETT, D., and JACOB, F. (1974).
Primitive teratocarcinoma cells express a differentiation antigen specified by a gene at the T-locus in the mouse. Proc. Nat. Acad. Sci. USA *71*, 811–814.

ARTZT, K., DUBOIS, P., BENNETT, D., CONDAMINE, H., BABINET, C., and JACOB, F. (1973).
Surface antigens common to mouse cleavage embryos and primitive teratocarcinoma cells in culture. Proc. Nat. Acad. Sci. USA *70*, 2988–2992.

ARTZT, K., HAMBURGER, L., JAKOB, H., AND JACOB, F. (1976).
Embryonic surface antigens: a "quasi endodermal" teratoma antigen. Dev. Biol. *51*, 152–157.

ARTZT, K., and JACOB, F., (1974).
Absence of serologically detectable H-2 on primitive teratocarcinoma cells in culture. Transplantation *17*, 633–634.

AUERBACH, R. (1972a).
The use of tumors in the analysis of inductive tissue interactions. Dev. Biol. *28*, 304–309.

AUERBACH, R. (1972b).
Controlled differentiation of teratoma cells. *In* Cell Differentiation, R. Harris, P. Alin, and D. Viza, eds. (Copenhagen: Munksgaard), pp. 119–123.

AUERBACH, S., and BRINSTER, R. L. (1967).
Lactate dehydrogenase isozymes in the early mouse embryo. Exp. Cell. Res. *46*, 89–92.

AVERY, G. B., and HUNT, C. V. (1968).
The survival and differentiation of fetal membranes grafted into the peritoneal cavity in mice. Anat. Rec. *160*, 751–758.

BABINET, C., CONDAMINE, H., FELLOUS, M., GACHELIN, G., KEMLER, R., and JACOB, F. (1975).
Expression of a cell surface antigen common to primitive mouse teratocarcinoma cells and cleavage embryos during embryogensis. *In* Teratomas and Differentiation, M. I. Sherman and D. Solter, eds. (New York: Academic Press), pp. 101–107.

BALLS, M., and WILD, A. E., eds. (1975).
The Early Development of Mammals. London: Cambridge University Press.

BARLOW, P. W., OWEN, D. A. J., and GRAHAM, C. F. (1972).
DNA synthesis in the preimplantation mouse embryo. J. Embryol. Exp. Morph. 27, 431–445.

BENNETT, D. (1956).
Developmental analysis of a mutation with pleiotropic effects in the mouse. J. Morph. 98, 199–234.

BERNSTINE, E. G., HOOPER, M. L., GRANDCHAMP, S., and EPHRUSSI, B. (1973).
Alkaline phosphatase activity in mouse teratoma. Proc. Nat. Acad. Sci. USA 70, 3899–3902.

BILLINGTON, W. D. (1965).
The invasiveness of transplanted mouse trophoblast and the influence of immunological factors. J. Reprod. Fertil. 10, 343–352.

BILLINGTON, W. D., GRAHAM, C. F., and MCCLAREN, A. (1968).
Extra-uterine development of mouse blastocysts cultured in vitro from early cleavage stages. J. Embryol. Exp. Morph. 20, 391–400.

BLUME, A., GILBERT, F., WILSON, S., FARBER, J., ROSENBERG, R., and NIRENBERG, M. (1970).
Regulation of acetylcholinesterase in neuroblastoma cells. Proc. Nat. Acad. Sci. USA 67, 786–792.

BOMSEL-HELMREICH, O. (1971).
Fate of heteroploid embryos. Advan. Biosciences 6, 381–400.

BOON, T., BUCKINGHAM, M. E., DEXTER, D. L., JAKOB, H., and JACOB, F. (1974).
Tératocarcinome de la souris: isolement et propriétés de deux lignées de myoblastes. Ann. Microbiol. (Inst. Pasteur) 125B, 13–28.

BOON, T., KELLERMANN, O., MATHY, E., and GAILLARD, J. A. (1975).
Mutagenized clones of a pluripotent teratoma cell line: variants with decreased differentiation or tumor-formation ability. In Teratomas and Differentiation, M. I. Sherman and D. Solter, eds. (New York: Academic Press), pp. 161–166.

BRINSTER, R. L. (1974).
The effect of cells transferred into the mouse blastocyst on subsequent development. J. Exp. Med. 140, 1049–1056.

BRINSTER, R. L. (1975).
Can teratocarcinoma cells colonize the mouse embryo? In Teratomas and Differentiation, M. I. Sherman and D. Solter, eds. (New York: Academic Press), pp. 51–58.

BUC-CARON, M. H., GACHELIN, G., HOFNUNG, M., and JACOB F. (1974).
Presence of a mouse embryonic antigen on human spermatozoa. Proc. Nat. Acad. Sci. USA 71, 1730–1733.

BUEHR, M., and MCCLAREN, A. (1974).
Size regulation in chimaeric mouse embryos. J. Embryol. Exp. Morph. 31, 229–234.

BUNKER, M. C. (1966).
Y-chromosome loss in transplanted testicular teratomas of mice. Can. J. Genet. Cytol. 8, 312–327.

BURGER, M. M., and NOONAN, K. D. (1970).
Restoration of normal growth by covering of agglutinin sites on tumour cell surface. Nature 228, 512–515.

BURGOYNE, P. S., and BIGGERS, J. D. (1976).
The consequence of X dosage deficiency in the germ line: impaired development in vitro of preimplantation embryos from XO mice. Dev. Biol. 51, 109–117.

CADE-TREYER, D. (1973).
Neosynthesis of an α-fetoprotein in calf kidney cells cultured in vitro. Ann. Immunol. (Inst. Pasteur) 124C, 27–43.

CADE-TREYER, D. (1975).
In vitro culture of the proximal tubule of the bovine nephron: the fate of the histiospecific antigens and neosynthesis of an α-fetoprotein. Ann. Immunol. (Inst. Pasteur) 126C, 201–218.

CANTOR, J., SHAPIRO, S. S., and SHERMAN, M. I. (1976).
Chondroitin sulfate synthesis by mouse embryonic, extraembryonic, and teratoma cells in vitro. Dev. Biol. 50, 367–377.

CARR, D. H. (1969).
Chromosomal errors and development. Am. J. Obstet. Gynec. 104, 327–347.

CARTER, N. D., and PARR, C. W. (1967).
Isoenzymes of phosphoglucose isomerase in mice. Nature 216, 511.

CARTER, T. C., LYON, M. F., and PHILLIPS, R. J. S. (1955).
Gene-tagged chromosome translocations in eleven stocks of mice. J. Genetics 53, 154–166.

CATTANACH, B. M. (1975).
Sex reversal in the mouse and other mammals. In The Early Development of Mammals, M. Balls and A. E. Wild, eds. (London: Cambridge University Press), pp. 305–317.

CHAPMAN, V. M. (1975).
6-Phosphogluconate dehydrogenase (PGD) genetics in the mouse: linkage with metabolically related enzyme loci. Biochem. Genet. 13, 849–856.

CHAPMAN, V. M., ANSELL, J. D., and MCLAREN, A. (1972).
Trophoblast differentiation in the mouse: expression of glucose phosphate isomerase (GPI-1) electrophoretic variants in transferred and chimeric embryos. Dev. Biol. 29, 48–54.

CHAPMAN, V. M., and SHOWS, T. B. (1976).
X-chromosome linkage of three enzymes in the mouse. Nature 259, 665–667.

CHASIN, L. A., and URLAUB, G. (1975).
Chromosome-wide event accompanies the expression of recessive mutations in tetraploid cells. Science 187, 1091–1093.

CHIQUOINE, A. D. (1954).
The identification, origin, and migration of the primordial germ cells of the mouse. Anat. Rec. 118, 135–146.

CLARK, C. C., TOMICHEK, E. A., KOSZALKA, T. R., MINOR, R. R., and KEFALIDES, N. A. (1975).
The embryonic rat parietal yolk sac: the role of parietal endoderm in the biosynthesis of basement membrane

collagen and glycoprotein *in vitro*. J. Biol. Chem. *250*, 5259–5267.

CLARK, J. M., and EDDY, E. M. (1975).
Fine structure observations on the origin and associations of primordial germ cells of the mouse. Dev. Biol. *47*, 136–155.

CLEMENS, W. A. (1971).
Mammalian evolution in the Cretaceous. *In* Early Mammals, D. M. Kermack and K. A. Kermack, eds. (New York: Academic Press), pp. 165–180.

CLINE, M. J., and LIVINGSTON, D. C. (1971).
Binding of concanavalin A by normal and transformed cells. Nature New Biol. *232*, 155–156.

COGGIN. J. H., and ANDERSON, N. G. (1974).
Embryonic and fetal antigens in cancer cells. *In* Developmental Aspects of Carcinogenesis and Immunity, T. J. King, ed. (New York: Academic Press), pp. 173–185.

COLE, R. J. (1975).
Regulatory functions of micro-environmental and hormonal factors in pre-natal haemopoietic tissues. *In* The Early Development of Mammals, M. Balls and A. E. Wild, eds. (London: Cambridge University Press), pp. 335–358.

COLE, R. J., and PAUL, J. (1966).
The effects of erythropoietin on haem synthesis in mouse yolk-sac and cultured foetal liver cells. J. Embryol. Exp. Morph. *15*, 245–260.

CONNOLLY, D. T., and OPPENHEIMER, S. B. (1975).
Cell density-dependent stimulation of

glutamine synthetase activity in cultured mouse teratoma cells. Exp. Cell. Res. *94*, 459–464.

CUDENNEC, C. A., and NICOLAS, J. F. (1977).
Blood formation in a clonal cell line of mouse teratocarcinoma. J. Embryol. Exp. Morph., *38*, 203–210.

DAMJANOV, I., KATIC, V., and STEVENS, L. C. (1975).
Ultrastructure of ovarian teratomas in LT mice. Z. Krebsforsch. *83*, 261–267.

DAMJANOV, I., and SOLTER, D. (1973).
Yolk sac carcinoma grown from explanted egg cylinder. Arch. Path. *95*, 182–184.

DAMJANOV, I., and SOLTER, D. (1974a).
Host-related factors determine the outgrowth of teratocarcinomas from mouse egg cylinders. Z. Krebsforsch. *81*, 63–69.

DAMJANOV, I., and SOLTER, D. (1974b).
Experimental teratoma. Current Topics Path. *59*, 69–130.

DAMJANOV, I., and SOLTER, D. (1975).
Ultrastructure of murine teratocarcinomas. *In* Teratomas and Differentiation, M. I. Sherman and D. Solter, eds. (New York: Academic Press), pp. 209–220.

DAMJANOV, I., SOLTER, D., BELICZA, M., and ŠKREB, N. (1971a).
Teratomas obtained through extrauterine growth of seven-day old mouse embryos. J. Nat. Cancer Inst. *46*, 471–480.

DAMJANOV, I., SOLTER, D., and ŠERMAN, D. (1973).
Teratocarcinoma with the capacity for differentiation restricted to neuro-ectodermal tissue. Virchows. Arch. Abt. B *13*, 179–195.

DAMJANOV, I., SOLTER, D., and ŠKREB, N. (1971b).
Teratocarcinogenesis as related to the age of embryos grafted under the kidney capsule. Wilhelm Roux Arch. *167*, 288–290.

DAMJANOV, I., SOLTER, D., and ŠKREB, N. (1971c).
Enzyme histochemistry of experimental embryo-derived teratocarcinomas. Z. Krebsforsch. *76*, 249–256.

DAVIDSON, R. L. (1974).
Control of expression of differentiated functions in somatic cell hybrids. *In* Somatic Cell Hybridization, R. L. Davidson and F. F. de la Cruz, eds. (New York: Raven Press), pp. 131–146.

DE BOER, P. (1973).
Fertile tertiary trisomy in the mouse. Cytogent. Cell. Genet. *12*, 435–442.

DE BOER, P., and DE MAAR, P. H. M. D. (1976).
A histological study of embryonic death caused by heterozygosity for the T26H reciprocal mouse transloca-tion. J. Embryol. Exp. Morph. *35*, 595–606.

DELAIN, D., MEINHOFER, M. C., PROUX, D., and SCHAPIRA, F. (1973).
Studies on myogenesis *in vitro:* changes of creatine kinase, phospho-rylase, and phosphofructokinase iso-zymes. Differentiation *1*, 349–354.

DE MARS, R. (1967).
The single active-X: functional differ-entiation at the chromosome level. Nat. Cancer. Inst. Monograph *26*, 327–351.

DEREN, J. J., PADYKULA, H. A., and HASTINGS WILSON, T. (1966).
Development of structure and func-tion in mammalian yolk sac. Dev. Biol. *13*, 349–369.

DIWAN, S., and STEVENS, L. C. (1976).
Development of teratomas from the ectoderm of mouse egg cylinders. J. Nat. Cancer Inst. *57*, 937–942.

DOFUKU, R., BIEDLER, J. L., SPENGLER, B. L., and OLD, L. J. (1975).
Trisomy of chromosome 15 in sponta-neous leukemia of AKR mice. Proc. Nat. Acad. Sci. USA *72*, 1515–1517.

DOMINIS, M., DAMJANOV, I., and SOLTER, D. (1975).
Cytology of experimental teratomas and teratocarcinomas. Experientia *31*, 107–108.

DULBECCO, R. (1970).
Topoinhibition and serum requirement of transformed and untransformed cells. Nature *227*, 802–806.

DUNN, G. R., and STEVENS, L. C. (1970).
Determination of sex of teratomas derived from early mouse embryos. J. Nat. Cancer Inst. *44*, 99–105.

EDIDIN, M. (1976).
The appearance of cell-surface antigens in the development of the mouse embryo: a study of cell surface differentiation. *In* Embryogenesis in Mammals, K. Elliott and M. O'Connor, eds., (Amsterdam: Associ-ated Scientific Publishers), pp. 177–194.

EDIDIN, M., PATTHEY, H. L., MCGUIRE, E. J., and SHEFFIELD, W. D. (1971).
An antiserum to embryoid body tumor cells that reacts with normal mouse embryos. *In* Embryonic and Fetal Antigens in Cancer, N. G., Anderson and J. H. Coggin, eds. (Oak Ridge, Tenn.: Oak Ridge National Laboratory), pp. 239–248.

EICHER, E. M., and GREEN, M. C. (1972).
The T6 translocation in the mouse: its use in trisomy and mapping, centromere localization, and cytological identification of linkage group III. Genetics *71*, 621–632.

EICHER, E. M., and HOPPE, P. C. (1973).
Use of chimeras to transmit lethal genes in the mouse and to demonstrate allelism of the two X-linked male lethal genes *jp* and *msd*. J. Exp. Zool. *183*, 181–184.

ENGELHARDT, N. V., POLTORANINA, V. S., and YAZOVA, A. K. (1973).
Localization of alpha-fetoprotein in transplantable murine teratocarcinomas. Int. J. Cancer *11*, 448–459.

EPSTEIN, C. J. (1969).
Mammalian oocytes: X-chromosome activity. Science *163*, 1078–1079.

EPSTEIN, C. J. (1972).
Expression of the mammalian X-chromosome before and after fertilization. Science *175*, 1467–1468.

EVANS, E. P. (1976a).
Male sterility and double heterozygosity for Robertsonian translocations in mouse. *In* Chromosomes Today, P. L. Pearson and K. R.

Lewis, eds., vol. 5 (New York: Wiley), pp. 75–81.

EVANS, M. J. (1972).
The isolation and properties of a clonal tissue culture strain of pluripotent mouse teratoma cells. J. Embryol. Exp. Morph. *28*, 163–176.

EVANS. M. J. (1975).
Studies with teratoma cells *in vitro*. *In* The Early Development of Mammals, M. Balls and A. E. Wild, eds. (London: Cambridge University Press), pp. 265–284.

EVANS, M. J. (1976b).
Totipotency of animal cells. *In* The Development of Plants and Animals, C. F. Graham and P. F. Wareing, eds. (Oxford: Blackwell's Scientific Publications), pp. 64–72.

EVANS, M. J., and MARTIN, G. R. (1975).
The differentiation of clonal teratocarcinoma cell cultures *in vitro*. *In* Teratomas and Differentiation, M. I. Sherman and D. Solter, eds. (New York: Academic Press), pp. 237–250.

EVERETT, J. W. (1935).
Morphological and physiological studies on the placenta of the albino rat. J. Exp. Zool. *70*, 243–286.

EVERETT, N. B. (1943).
Observational and experimental evidences relating to the origin and differentiation of definitive germ cells in mice. J. Exp. Zool. *92*, 49–91.

FELLOUS, M., GACHELIN, G., BUC-CARON, M. H., DUBOIS, P., and JACOB, F. (1974).
Similar location of an early embryonic antigen on mouse and human spermatozoa. Dev. Biol. *41*, 331–337.

FINCH, B. W., and EPHRUSSI, B. (1967).
Retention of multiple developmental potentialities by cells of a mouse testicular teratocarcinoma during prolonged culture *in vitro* and their extinction upon hybridization with cells of a permanent cell line. Proc. Nat. Acad. Sci. USA *57*, 615–621.

FLUCK, R. A., and STROHMAN, R. C. (1973).
Acetylcholinesterase activity in developing skeletal muscle cells *in vitro*. Dev. Biol. *33*, 417–428.

FORD, C. E. (1970).
Cytogenetics and sex determination in man and mammals. J. Biosoc. Sci. Suppl. *2*, 7–30.

FORD, C. E. (1975).
The time in development at which gross genome unbalance is expressed. *In* The Early Development of Mammals, M. Balls, and A. E. Wild, eds. (London: Cambridge University Press), pp. 285–304.

FORD, C. E., EVANS, E. P., BURTENSHAW, M. D., CLEGG, H. M., TUFFREY, M., and BARNES, R. D. (1975a).
A functional 'sex reversed' oocyte in the mouse. Proc. Roy. Soc. B *190*, 187–197.

FORD, C. E., EVANS, E. P., and GARDNER, R. L. (1975b).
Marker chromosome analysis of two mouse chimeras. J. Embryol. Exp. Morph. *33*, 447–457.

GACHELIN, G. (1976).
Le tératocarcinome expérimental de la souris: une système modele pour l'étude des relations entre antigènes des surfaces cellulaires et différencia-tion embryonnaire. Bull. Cancer *63*, 95–110.

GACHELIN, G., BUC-CARON, M. H., LIS, H., and SHARON, N. (1976).
Saccharides on teratocarcinoma cell plasma membranes: their investigation with radioactively labelled lectins. Biochim. Biophys. Acta *436*, 825–832.

GARDNER, R. L., and LYON, M. F. (1972).
X-chromosome inactivation studied by injection of a single cell into the mouse blastocyst. Nature New Biol. *231*, 383–386.

GARTLER, S. M., ANDINA, R., and GANT, N. (1975).
Ontogeny of X-chromosome inactivation in the female germ line. Exp. Cell Res. *91*, 454–457.

GARTLER, S. M., LISKAY, R. M., and GANT, N. (1973).
Two functional X-chromosomes in human fetal oocytes. Exp. Cell Res. *82*, 464–466.

GEARHART, J. D., and MINTZ, B. (1972).
Glucose phosphate isomerase subunit reassociation tests for maternal-fetal and fetal-fetal cell fusion in mouse placenta. Dev. Biol. *29*, 55–64.

GEARHART, J. D., and MINTZ, B. (1974).
Contact-mediated myogenesis and increased acetylcholinesterase activity in primary cultures of mouse teratocarcinoma cells. Proc. Nat. Acad. Sci. USA *71*, 1734–1738.

GEARHART, J. D., and MINTZ, B. (1975).
Creatine kinase, myokinase, and acetylcholinesterase activities in muscle-

forming primary cultures of mouse teratocarcinoma cells. Cell 6, 61–66.

GERMAN, J. (1974).
Chromosomes and Cancer. New York: Wiley.

GITLIN, D., and PERRICELLI, A. (1970).
Synthesis of serum albumin, prealbumin, α-fetoprotein, α_1-antitrypsin, and transferrin by human yolk sac. Nature 228, 995–997.

GITLIN, D., PERRICELLI, A., and GITLIN, G. M. (1972).
Synthesis of α-foetoprotein by liver, yolk sac, and gastrointestinal tract of the human conceptus. Cancer Res. 32, 979–982.

GLINOS, A. D., and BARTOS, E. H. (1974).
Density dependent regulation of growth in L cell suspension cultures. III. Elevation of the specific activity of acetylcholinesterase. J. Cell. Physiol. 83, 131–139.

GOLD, P., and FREEDMAN, S. O. (1965).
Specific carcinoembryonic antigens of human digestive tract. J. Exp. Med. 122, 467–481.

GOODWIN, B. C., and SIZER, I. W. (1965).
Effect of spinal cord and substrate on acetylcholinesterase in chick embryonic skeletal muscle. Dev. Biol. 11, 136–153.

GOSS, R. J. (1964).
Adaptive Growth. London: Logos Press.

GOSS, R. J., ed. (1972).
Regulation of Organ and Tissue Growth. New York: Academic Press.

GRAHAM, C. F. (1974).
The production of parthenogenetic mammalian embryos and their use in biological research. Biol. Rev. 49, 399–422.

GREENE, H. S. N. (1955).
Compatibility and non-compatibility in tissue transplantation. In Biological Specificity and Growth, E. G. Butler, ed. (Princeton, N. J.: Princeton University Press), pp. 177–194.

GROPP, A. (1976).
Morphological consequences of trisomy in mammals. In Embryogenesis in Mammals, K. Elliott and M. O'Connor, eds. (Amsterdam: Associated Scientific Publishers), pp. 155–175.

GUENET, J. L., JAKOB, H., NICOLAS, J. F., and JACOB, F. (1974).
Tératocarcinome de la souris: étude cytogenetique de cellules à potentialitiés multiples. Ann. Microbiol. (Inst. Pasteur) 125A, 135–151.

GURDON, J. B. (1976).
Pluripotentiality of cell nuclei. In The Developmental Biology of Plants and Animals, C. F. Graham and P. F. Wareing, eds. (Oxford: Blackwell's Scientific Publications), pp. 55–63.

GUSTINE, D. L., and ZIMMERMAN, E. F. (1973).
Developmental changes in microheterogeneity of fetal glycoproteins of mice. Biochem. J. 132, 541–555.

HALL, J. D., MARSDEN, M., RIFKIN, D., TERESKY, A. K., and LEVINE, A. J. (1975).
The in vitro differentiation of embryoid bodies produced by a transplantable teratoma of mice. In Teratomas and Differentiation, M. I.

Sherman and D. Solter, eds. (New York: Academic Press), pp. 251–269.

HAUSCHKA, S. D. (1968).
Clonal aspects of muscle development and the stability of the differentiated state. *In* The Stability of the Differentiated State, J. Ursprung, ed. (Berlin: Springer-Verlag), pp. 37–57.

HENSLEIGH, H. C. (1976).
Synthesis of α-fetoprotein and transferrin by yolk sac-like vesicles *in vitro*. Anat. Rec. *184*, 424–425.

HORDER, T. (1976).
Pattern formation in animal embryos. *In* The Developmental Biology of Plants and Animals, C. F. Graham and P. F. Wareing, eds. (Oxford: Blackwell's Scientific Publications), pp. 169–197.

HSU, Y. C., and BASKAR, J. (1974).
Differentiation *in vitro* of normal mouse embryos and mouse embryonal carcinoma. J. Nat. Cancer Inst. *53*, 177–185.

HSU, Y. C., BASKAR, J., STEVENS, L. C., and RASH M. E. (1974).
Development *in vitro* of mouse embryos from the two-cell egg stage to the early somite stage. J. Embryol. Exp. Morph. *31*, 235–245.

HUEBNER, R. J., and TODARO, G. J. (1969).
Oncogenes of RNA tumor viruses as determinents of cancer. Proc. Nat. Acad. Sci. USA *64*, 1087–1094.

ILES, S. A. (1977).
Mouse teratomas and embryoid bodies: their induction and differentiation. J. Embryol. Exp. Morph., *38*, 63–75.

ILES, S. A., and EVANS, E. P. (1977).
Karyotype analysis of teratocarcinomas and embryoid bodies of C3H mice. J. Embryol. Exp. Morph., *38*, 77–92.

ILES, S. A., MCBURNEY, M. W., BRAMWELL, S. R., DEUSSEN, Z. A., and GRAHAM, C. F. (1975).
Development of parthenogentic mouse embryos in the uterus and in extra-uterine sites. J. Embryol. Exp. Morph. *34*, 387–405.

ILLMENSEE, K., and MINTZ, B. (1976).
Totipotency and normal differentiation of single teratocarcinoma cells cloned by injection into blastocysts. Proc. Nat. Acad. Sci. USA *73*, 549–553.

INGWALL, J. S., and WILDENTHAL, K. (1976).
Role of creatine in the regulation of cardiac protein synthesis. J. Cell. Biol. *68*, 159–163.

JACOB, F. (1975).
Mouse teratocarcinomas as a tool for the study of the mouse embryo. *In* The Early Development of Mammals, M. Balls and A. E. Wild, eds. (London: Cambridge University Press), pp. 233–241.

JAKOB, H., BOON, T., GAILLARD, J., NICOLAS, J. F., and JACOB, F. (1973).
Tératocarcinome de la souris: isolement, culture et propriétés de cellules à potentialités multiples. Ann. Microbiol. (Inst. Pasteur) *124B*, 269–282.

JAMI, J., FAILLY, C., and RITZ, E. (1973).
Lack of expression of differentiation

in mouse teratoma-fibroblast somatic cell hybrids. Exp. Cell Res. *76*, 191–199.

JENKINSON, J. W. (1913). Vertebrate Embryology. Oxford: Clarendon Press.

JEON, K. W., and KENNEDY, J. R. (1973). The primordial germ cells in early mouse embryos: light and electron microscopic studies. Dev. Biol. *31*, 275–284.

JOLLIE, W. P. (1968). Changes in the fine structure of the parietal yolk sac of the rat placenta with increasing gestational age. Am. J. Anat. *122*, 513–532.

KAGEN, L. J., COLLINS, K., ROBERTS, L., and BUTT, A. (1976). Inhibition of muscle cell development in culture by cells from spinal cord due to production of low molecular weight factor. Dev. Biol. *48*, 25–34.

KAHAN, B. W., and EPHRUSSI, B. (1970). Developmental potentialities of clonal *in vitro* cultures of mouse testicular teratoma. J. Nat. Cancer Inst. *44*, 1015–1036.

KAHAN, B. W., and LEVINE, L. (1971). The occurrence of a serum fetal protein in developing mice and murine hepatomas and teratomas. Cancer Res. *31*, 930–936.

KAUFMAN, M. H. (1976). The incidence of chromosomally unbalanced gametes in T (14;15) 6 Ca heterozygote mice. Genet. Res. *27*, 77–84.

KELLY, F., and BOCCARA, M. (1976). Susceptibility of teratocarcinoma cells to adenovirus type 2. Nature *262*, 409–411.

KELLY, S. J. (1975). Studies of the potency of the early cleavage blastomeres of the mouse. *In* The Early Development of Mammals, M. Balls and A. E. Wild, eds. (London: Cambridge University Press), pp. 97–105.

KIRBY, D. R. S. (1963). The development of mouse blastocysts transplanted to the scrotal and cryptorchid testis. J. Anat. *97*, 119–130.

KIRBY, D. R. S. (1965). The role of the uterus in early stages of mouse development. *In* Preimplantation Stages of Pregnancy, G. E. W. Wolstenholme and M. O'Connor, eds. (London: Churchill), pp. 325–339.

KIRBY, D. R. S., and COWELL, T. P. (1968). Trophoblast-host interactions. *In* Epithelial-Mesenchymal Interactions, R. Fleischmajer and R. E. Billingham, eds. (Baltimore: Williams & Wilkins), pp. 64–77.

KLEINSCHUSTER, S. J., and MOSCONA, A. A. (1972). Interactions of embryonic and fetal neural retina cells with carbohydrate-binding phytoagglutinins: cell surface changes with differentiation. Exp. Cell. Res. *70*, 397–410.

KLEINSMITH, L. J., and PIERCE, G. B. (1964). Multipotentiality of single embryonal carcinoma cells. Cancer Res. *24*, 1544–1552.

KOZAK, C., NICHOLS, E., and
RUDDLE, F. H. (1975).
Gene linkage analysis in the mouse by
somatic cell hybridization: assignment
of adenine phosphoribosyltransferase
to chromosome 8 and α-galactosidase
to the X-chromosome. Somatic Cell.
Genet. *1*, 371–383.

KOZAK, L. P., MCLEAN, G. K.,
and EICHER, E. M. (1974).
X-Linkage of phosphoglycerate kinase
in the mouse. Biochem. Genet. *11*,
41–48.

KOZAK, L. P., and QUINN, P. J.
(1975).
Evidence for dosage compensation of
an X-linked gene in the 6-day mouse
embryo. Dev. Biol. *45*, 65–73.

KROHN, P. L. (1962).
Heterochronic transplantation in the
study of ageing. Proc. Roy. Soc. B
157, 128–147.

LASH, J. (1974).
Tissue interactions and related sub-
jects. *In* Concepts of Development, J.
Lash and J. R. Whittaker, eds.
(Stamford, Conn.: Sinauer Associ-
ates), pp. 197–212.

LEE, P. A., BLASEY, K.,
GOLDSTEIN, I. J., and PIERCE,
G. B. (1969).
Basement membrane: carbohydrates
and X-ray diffraction. Exp. Mol.
Pathol. *10*, 323–330.

LEHMAN, J. M., KLEIN, I. B., and
HACKENBERG, R. M. (1975).
The response of murine teratocarci-
noma cells to infection with DNA and
RNA viruses. *In* Teratomas and Dif-
ferentiation, M. I. Sherman and D.
Solter, eds. (New York: Academic
Press), pp. 289–301.

LEHMAN, J. M., SPEERS, W. C.,
SWARTZENDRUBER, D. E., and
PIERCE, G. B. (1974).
Neoplastic differentiation: characteristics of cell lines derived from a murine teratocarcinoma. J. Cell. Physiol.
84, 13–28.

LEVINE, A. J., TOROSIAN, M.,
SAROKHAN, A. J., and TERESKY,
A. K. (1974).
Biochemical criteria for the *in vitro*
differentiation of embryoid bodies
produced by a transplantable teratoma
of mice. The production of acetylcholine esterase and creatine
phosphokinase by teratoma cells. J.
Cell. Physiol. *84*, 311–318.

LINDER, P., MCCAW, K. K., and
HECHT, F. (1975).
Parthenogenetic origin of benign ovarian teratomas. New England J. Med.
292, 63–66.

LINNEY, E., and LEVINSON, B. B.
(1976).
Teratocarcinoma differntiation: plasminogen activator activity associated
with embryoid body formation. Cell
10, 297–304.

LYON, M. F., and MEREDITH, R.
(1966).
Autosomal translocations causing
male sterility and viable aneuploidy in
the mouse. Cytogenetics *5*, 335–354.

MANNINO, R. J., and BURGER,
M. M. (1975).
Growth inhibition of animal cells by
succinylated concanavalin A. Nature,
256, 19–22.

MARKERT, C. L. (1968).
Neoplasia: a disease of cell differentiation. Cancer Res. *24*, 1544–1551.

MARTIN, G. R. (1975).
Teratocarcinomas as a model system

for the study of embryogenesis and neoplasia. Cell 5, 229–243.

MARTIN, G. R. (1977).
The differentiation of teratocarcinoma cells in vitro: parallels to normal embryogenesis. In Cell Interactions in Differentiation, M. Karhinen-Jaaskelainen, L. Saxén, and L. Weiss, eds. (New York: Academic Press), in press.

MARTIN, G. R., and EVANS, M. J. (1974).
The morphology and growth of pluripotent teratocarcinoma cell line and its derivatives in tissue culture. Cell 2, 163–172.

MARTIN, G. R., and EVANS, M. J. (1975a).
Differentiation of clonal lines of teratocarcinoma cells: formation of embryoid bodies in vitro. Proc. Nat. Acad. Sci. USA 72, 1441–1447.

MARTIN, G. R., and EVANS, M. J. (1975b).
The formation of embryoid bodies in vitro by homogeneous embryonal carcinoma cell cultures derived from isolated single cells. In Teratomas and Differentiation, M. I. Sherman and D. Solter, eds. (New York: Academic Press), pp. 169–187.

MARTIN, G. R., and EVANS, M. J. (1975c).
Multiple differentiation of teratocarcinoma stem cells following embryoid body formation in vitro. Cell 6, 467–474.

MASTERS, C. J. (1968).
The ontogeny of mammalian fructose-1,6-diphosphate aldolase. Biochim. Biophys. Acta 167, 161–171.

MCBURNEY, M. W. (1976).
Clonal lines of teratocarcinoma cells in vitro: differentiation and cytogenetic characteristics. J. Cell. Physiol. 89, 441–456.

MCBURNEY, M. W., and ADAMSON, E. D. (1976).
Studies on the activity of the X-chromosome in female teratocarcinoma cells in culture. Cell 9, 57–70.

MCCOSHEN, J. A., and MCCALLION, D. J. (1975).
A study of primordial germ cells during their migratory phase in steel mutant mice. Experientia 31, 589–590.

MCLAREN, A. (1973).
Blastocyst activation. In The Regulation of Mammalian Reproduction, R. Crozier and P. Corfman, eds. (Springfield, Ill.: C. C Thomas), pp. 321–328.

MCLAREN, A., and HENSLEIGH, H. C. (1975).
Culture of mammalian embryos over the implantation period. In The Early Development of Mammals, M. Balls and A. E. Wild, eds. (London: Cambridge University Press), pp. 45–60.

MEIER, H., MYERS, D. D., FOX, R. R., and LAIRD, C. W. (1970).
Occurrence, pathological features, and propagation of gonadal teratomas in inbred mice and rabbits. Cancer Res. 30, 31–34.

METCALF, D., and MOORE, M. A. S. (1971).
Haemopoietic Cells. Amsterdam: North-Holland.

MICKLEM, H. S., and LOUTIT, J. F. (1966).
Tissue Grafting and Radiation. New York: Academic Press.

MILLER, R. A., and RUDDLE, F H. (1976).
Pluripotent teratocarcinoma-thymus somatic cell hybrids. Cell 9, 45–55.

MINTZ, B. (1957).
Embryological development of primordial germ-cells in the mouse: influence of a new mutation, W^j. J. Embryol. Exp. Morph. 5, 396–403.

MINTZ, B. (1971).
Clonal basis of mammalian differentiation. In Control Mechanisms of Growth and Differentiation, D. D. Davies and M. Balls, eds. (London: Cambridge University Press), pp. 345–370.

MINTZ, B., and BAKER, W. W. (1967).
Normal mammalian muscle differentiation and gene control of isocitrate dehydrogenase synthesis. Proc. Nat. Acad. Sci. USA 58, 592–598.

MINTZ, B., and ILLMENSEE, K. (1975).
Normal genetically mosaic mice produced from malignant teratocarcinoma cells. Proc. Nat. Acad. Sci. USA 72, 3585–3589.

MINTZ, B., ILLMENSEE, K., and GEARHART, J. D. (1975).
Developmental and experimental potentialities of mouse teratocarcinoma cells from embryoid body cores. In Teratomas and Differentiation, M. I. Sherman and D. Solter, eds. (New York: Academic Press), pp. 59–82.

MINTZ, B., and RUSSELL, E. S. (1957).
Gene-induced embryological modifications of primordial germ cells in the mouse. J. Exp. Zool. 134, 207–229.

MIRSKY, R., and THOMPSON, E. J. (1975).
Thy-1 (theta) antigen on the surface of morphologically distinct brain cell types. Cell 4, 95–101.

MITELMAN, F. (1971).
The chromosomes of fifty primary Rous rat sarcomas. Hereditas 69, 155–186.

MIURA, Y., and WILT, F. H. (1970).
The formations of blood islands in dissociated-reaggregated chick embryo yolk sac cells. Exp. Cell. Res. 59, 217–226.

MORRIS, G. E., COOKE, A., and COLE, R. J. (1972).
Isoenzymes of creatine phosphokinase during myogenesis in vitro. Exp. Cell Res. 74, 582–585.

MORRIS, G. E., PIPER, M., and COLE, R. J. (1976).
Differential effects of calcium ion concentration on cell fusion, cell division and creatine kinase activity in muscle cell cultures. Exp. Cell Res. 99, 106–114.

MORRIS, T. (1968).
The XO and OY chromosome constitution in the mouse. Genet. Res. 12, 125–136.

MOUSTAFA, L. A., and BRINSTER, R. L. (1972a).
The fate of transplanted cells in mouse blastocysts in vitro. J. Exp. Zool. 181, 181–192.

MOUSTAFA, L. A., and BRINSTER, R. L. (1972b).
Induced chimaerism by transplanting embryonic cells into mouse blastocysts. J. Exp. Zool. 181, 193–202.

MUKERJEE, H., SRI RAM, J., and PIERCE, G. B. (1965).
Basement membranes. V. Chemical composition of neoplastic basement membrane mucoprotein. Am. J. Path. *46*, 49–59.

MUKHERJEE, A. B. (1976).
Cell cycle analysis and X-chromosome inactivation in the developing mouse. Proc. Nat. Acad. Sci. USA *73*, 1608–1611.

NADIJCKA, M., and HILLMAN, N. (1974).
Ultrastructural studies of mouse blastocyst substages. J. Embryol. Exp. Morph. *32*, 675–695.

NADIJCKA, M., and HILLMAN, N. (1975).
Studies of t^6/t^6 mouse embryos. J. Embryol. Exp. Morph. *33*, 697–713.

NICOLAS. J. F. AVNER, P., GAILLARD, J. GUENET, J. L., JAKOB, H., and JACOB, F. (1976).
Cell lines derived from teratocarcinomas. Cancer Res. *36*, 4224–4231.

NICOLAS, J. F., DUBOIS, P., JAKOB, H., GAILLARD, J., and JACOB, F. (1975).
Tératocarcinome de la souris: differenciation en culture d'une lignée de cellules primitives à potentialités multiples. Ann. Microbiol. (Inst. Pasteur) *126A*, 3–22.

NOWINSKI, R. C., OLD, L. J., SARKHAR, N. H., and MOORE, D. H. (1970).
Common properties of the oncogenic RNA viruses (oncornaviruses). Virology *42*, 1152–1157.

OMENN, G. S., and HERMODSON, M. A. (1975).
Human phosphoglycerate mutase: isozyme marker for muslce differentiation and neoplasia. *In* Isozymes C. L. Markert, ed., vol. 3 (New York: Academic Press), pp. 1005–1017.

O'NEILL, M. C., and STOCKDALE, F. E. (1972).
A kinetic analysis of myogenesis *in vitro*. J. Cell. Biol. *52*, 52–65.

OPPENHEIMER, S. B. (1973).
Utilization of L-glutamine in intercellular adhesion: ascites tumor and embryonic cells. Exp. Cell Res. *77*, 175–182.

OPPENHEIMER, S. B. (1974).
Functional carbohydrate in teratoma cell adhesion factor. J. Cell. Biol. *63*, 251a.

OPPENHEIMER, S. B. (1975).
Functional involvement of specific carbohydrate in teratoma cell adhesion factor. Exp. Cell Res. *92*, 122–126.

OPPENHEIMER, S. B., EDIDIN, M., ORR, C. W., and ROSEMAN, S. (1969).
An L-gutamine requirement for intercellular adhesion. Proc. Nat. Acad. Sci. USA *63*, 1395–1402.

OPPENHEIMER, S. B., and HUMPHREYS, T. (1971).
Isolation of specific macromolecules required for adhesion of mouse tumour cells. Nature *232*, 125–127.

OPPENHEIMER, S. B., and ODENCRANTZ, J. (1972).
A quantitative assay for measuring cell agglutination: agglutination of sea urchin embryo and mouse teratoma cells by concanavalin A. Exp. Cell Res. *73*, 475–480.

OZANNE, B., and SAMBROOK, J. (1971).
Binding of radioactively labelled concanavalin A and wheat germ agglutinin to normal and transformed cells. Nature New Biol. *232*, 156–160.

OZDZENSKI, W. (1967).
Observations on the origin of primordial germ cells in the mouse. Zool. Pol. *17*, 65–78.

OZDZENSKI, W., (1969).
Fate of primordial germ cells in the transplanted hind gut of mouse embryos. J. Embryol. Exp. Morph. *22*, 505–510.

PADYKULA, H. A., DEREN, J. J., and HASTINGS WILSON, T. (1966).
Development of structure and function in the mammalian yolk sac. Dev. Biol. *13*, 311–348.

PAPAIOANNOU, V. E., MCBURNEY, M. W., GARDNER, R. L., and EVANS, M. J., (1975).
Fate of teratocarcinoma cells injected into early mouse embryos. Nature *258*, 70–73.

PAYNE, J. M., and PAYNE, S. (1961).
Placental grafts in rats. J. Embryol. Exp. Morph. *9*, 106–116.

PETERS, H. (1970).
Migration of gonocytes into mammalian gonads and their differentiation. Phil. Trans. Roy. Soc. B *259*, 91–101.

PIENKOWSKI, M. (1974).
Study on the growth regulation of preimplantation mouse embryos using concanavalin A. Proc. Soc. Exp. Biol. *145*, 464–469.

PIERCE, G. B. (1966).
The development of basement membranes of the mouse embryo. Dev. Biol. *13*, 231–249.

PIERCE, G. B. (1970).
Eipthelial basement membranes: origin, development and role in disease. *In* Chemistry and Molecular Biology of the Intercellular Matrix, E. A. Balazs, ed. (New York: Academic Press), pp. 471–510.

PIERCE, G. B. (1974).
The benign cells of malignant tumors. *In* Developmental Aspects of Carcinogenesis and Immunity, T. J. King, ed. (New York: Academic Press), pp. 3–22.

PIERCE, G. B., and BEALS, T. F. (1964).
The ultrastructure of primordial germinal cells of the foetal testis and of embryonal carcinoma cells in mice. Cancer. Res. *24*, 1553–1567.

PIERCE, G. B., BEALS, T. F., SRI RAM, J., and MIDGLEY, A. R. (1964).
Basement membranes. IV. Epithelial origins and immunological cross reactions. Am. J. Path. *45*, 929–941.

PIERCE, G. B., BULLOCK, W. K., and HUNTINGDON, R. W. (1970).
Yolk sac tumors of the testis. Cancer *25*, 644–658.

PIERCE, G. B., and DIXON, F. J. (1959).
Testicular teratomas. II. Teratocarcinoma as an ascitic fluid. Cancer *12*, 584–589.

PIERCE, G. B., DIXON, F. J., and VERNEY, E. L. (1960).
Teratocarcinogenic and tissue-forming potentialities of the cell types comprising neoplastic embryoid bodies. Lab. Invest. *9*, 583–602.

PIERCE, G. B., and JOHNSON, L. D. (1971).
Differentiation and cancer. In Vitro 7, 140–145.

PIERCE, G. B., MIDGLEY, A. R., SRI RAM, J., and FELDMAN, J. D. (1962).
Parietal yolk sac carcinoma: clue to the histogenesis of Reichert's membrane in the mouse. Am. J. Path. 41, 549–557.

PIERCE, G. B., and NAKANE, P. K. (1967).
Antigens of epithelial basement membranes of mouse, rat, and man. Lab. Invest. 17, 499–514.

PIERCE, G. B., STEVENS, L. C., and NAKANE, P. K. (1967).
Ultrastructural analysis of the early development of teratocarcinoma. J. Nat. Cancer Inst. 39, 755–773.

PIERCE, G. B., and VERNEY, E. L. (1961).
An in vitro and in vivo study of differentiation in teratocarcinomas. Cancer 14, 1017–1029.

PIERCE, G. B., and WALLACE, C. (1971).
Differentiation of malignant to benign cells. Cancer Res. 31, 127–134.

PIKÓ, L. (1975).
Expression of mitochondrial and nuclear genes during early development. In The Early Development of Mammals, M. Balls and A. E. Wild, eds. (London: Cambridge University Press), pp. 167–187.

RAFF, M. C. (1969).
Theta isoantigen as a marker of thymus-derived lymphocytes in mice. Nature 224, 378–379.

REIF, A. E., and ALLEN, J. M. V. (1964).
The AKR thymic antigen and its distribution in leukemias and nervous tissue. J. Exp. Med. 120, 413–433.

REIF, A. E., and ALLEN, J. M. V. (1966).
Mouse thymic isoantigen. Nature 209, 521–523.

REINIUS, S. (1965).
The morphology of the mouse embryo from the time of implantation to mesoderm formation. Z. Zellforschung 68, 711–723.

RHEINWALD, J. G., and GREEN, H. (1975).
Formation of keratinizing epithelium in culture by a cloned cell line derived from a teratoma. Cell 6, 317–330.

ROSENTHAL, M. D., WISHNOW, R. M., and SATO, G. H. (1970).
In vitro growth and differentiation of clonal populations of multipotential mouse cells derived from a transplantable testicular teratocarcinoma. J. Nat. Cancer. Inst. 44, 1001–1014.

ROWINSKI, J., SOLTER, D., and KOPROWSKI, H. (1976).
Changes in concanavalin A induced agglutinability during preimplantation mouse development. Exp. Cell Res. 100, 404–408.

RUBIN, H. (1974).
Regulation of animal cell growth. In Cell Communication, R. P. Cox, ed. (New York: Wiley), pp. 127–146.

RUGH, R. (1968).
The Mouse: Its Reproduction and Development. Minneapolis, Minn: Burgess.

SAXÉN, L., and WARTIOVAARA, J. (1976).
Embryonic induction. *In* The Developmental Biology of Plants and Animals, C. F. Graham and P. F. Wareing, eds. (Oxford: Blackwell's Scientific Publications), pp. 127–140.

SCHEID, M. P., BOYSE, E. A., CARSWELL, E. A., and OLD, L. J. (1972).
Serologically demonstrable alloantigens of mouse epidermal cells. J. Exp. Med. *135*, 938–955.

SCHLESINGER, M., and YRON, I. (1969).
Antigenic changes in lymph node cells after administration of antiserum to thymus cells. Science *164*, 1412–1413.

SEARLE, A. G. (1974).
Nature and consequence of induced chromosome damage in mammals. Genetics *78*, 173–186.

SEEDS, N. W. (1971).
Biochemical differentiation in reaggregating brain cell culture. Proc. Nat. Acad. Sci. USA *68*, 1858–1861.

SHAINBERG, A., YAGIL, G., and YAFFE, D. (1971).
Alterations of enzymatic activities during muscle differentiation *in vitro*. Dev. Biol. *25*, 1–29.

SHARON, N., and LIS, H. (1972).
Lectins: cell agglutinating and sugar specific proteins. Science *177*, 949–959.

SHERMAN, M. I. (1975a).
Differentiation of teratoma cell line PCC4.aza1 *in vitro*. *In* Teratomas and Differentiation, M. I. Sherman and D. Solter, ed. (New York: Academic Press), pp. 189–205.

SHERMAN, M. I. (1975b).
The culture of cells derived from mouse blastocysts. Cell *5*, 343–349.

SHERMAN, M. I., and SOLTER, D. eds., (1975).
Teratomas and Differentiation. New York: Academic Press.

SHERMAN, M. I., STRICKLAND, S., and REICH, E. (1976).
Differentiation of early mouse embryonic and teratocarcinoma cells *in vitro*: plasminogen activator production. Cancer Res. *36*, 4208–4216.

SIEKEVITZ, P., and PALADE, G. E. (1959).
A cytochemical study on the pancreas of the guinea pig. IV. Chemical and metabolic investigation of ribonucleoprotein particles. J. Biophys. Biochem. Cytol. *5*, 1–10.

SILVER, A. (1974).
The Biology of Cholinesterases. Amsterdam: North-Holland.

SIMMONS, R. L., and RUSSELL, P. S. (1962).
The antigenicity of mouse trophoblast. Ann. N. Y. Acad. Sci. *99*, 717–732.

SIMMONS, R. L., and WEINTRAUB, J. (1965).
Transplantation experiments on placental ageing. Nature *208*, 82–83.

SLYE, M., HOLMES, H. F., and WELLS, H. G. (1920).
Primary spontaneous tumors of the ovary in mice. Studies on the incidence and inheritability of spontaneous tumors in mice. J. Cancer Res. *5*, 205–226.

SNOW, M. H. L. (1973).
Tetraploid mouse embryos produced by cytochalasin B during cleavage. Nature *244*, 513–515.

SNOW, M. H. L. (1975).
Embryonic development of tetraploid mice during the second half of gestation. J. Embryol. Exp. Morph. *34*, 707–721.

SNOW, M. H. L. (1976).
Embryo growth during the immediate postimplantation period. *In* Embryogenesis in Mammals, K. Elliott and M. O'Connor, eds. (Amsterdam: Associated Scientific Publishers), pp. 53–70.

SOBIS, H., and VANDEPUTTE, M. (1975).
Sequential morphological study of teratomas from displaced yolk sac. Dev. Biol. *45*, 276–290.

SOLTER, D., ADAMS, N., DAMJANOV, I., and KOPROWSKI, H. (1975a).
Control of teratocarcinogenesis. *In* Teratomas and Differentiation, M. I. Sherman and D. Solter, eds. (New York: Academic Press), pp. 139–159.

SOLTER, D., BICZYSKO, W., PIENKOWSKI, M., and KOPROWSKI, H. (1974).
Ultrastructure of mouse egg cylinders developed *in vitro*. Anat. Rec. *180*, 263–280.

SOLTER, D., and DAMJANOV, I. (1973).
Explantation of extraembryonic parts of 7-day-old mouse egg cylinder. Experientia *29*, 701.

SOLTER, D., DAMJANOV, I., and KOPROWSKI, H. (1975b).
Embryo derived teratoma: a model system in developmental and tumor biology. *In* The Early Development of Mammals, M. Balls and A. E. Wild, eds. (London: Cambridge University Press), pp. 243–264.

SOLTER, D., DAMJANOV, I., and ŠKREB, N. (1970a).
Ultrastructure of mouse egg cylinder. Z. Anat. Entwicklungsgesch. *132*, 291–298.

SOLTER, D., DAMJANOV, I., and ŠKREB, N. (1973).
Distribution of hydrolytic enzymes in early rat and mouse embryos: a reappraisal. Z. Anat. Entwicklungsgesch. *139*, 119–126.

SOLTER, D., ŠKREB, N., and DAMJANOV, I. (1970b).
Extra-uterine growth of mouse egg cylinders results in malignant teratoma. Nature *227*, 503–504.

SPEERS, W. C., and LEHMAN, J. M. (1976).
Increased susceptibility of murine teratocarcinoma cells to simian virus 40 and polyoma virus following treatment with 5-bromodeoxyuridine. J. Cell. Physiol. *88*, 297–306.

SPIEGELMAN, M., and BENNETT, D. (1973).
A light- and electron-microscopic study of primordial germ cells in the early mouse embryo. J. Embryol. Exp. Morph. *30*, 97–118.

STEELE, C. E. (1975).
The culture of post-implantation mammalian embryos. *In* The Early Development of Mammals, M. Balls and A. E. Wild, eds. (London: Cambridge University Press), pp. 61–79.

STERN, P. L. (1973).
Alloantigen on rat and mouse fibroblasts. Nature New Biol. *246*, 76–78.

STERN, P. L., MARTIN, G. R., and EVANS, M. J. (1975).
Cell surface antigens of clonal terato-

carcinoma cells at various stages of differentiation. Cell *6*, 455–465.

STEVENS, L. C. (1959).
Embryology of testicular teratomas in strain 129 mice. J. Nat. Cancer Inst. *23*, 1249–1295.

STEVENS, L. C. (1960).
Embryonic potency of embryoid bodies derived from a transplantable testicular teratoma of the mouse. Dev. Biol. *2*, 285–297.

STEVENS, L. C. (1962).
Testicular teratomas in fetal mice. J. Nat. Cancer Inst. *28*, 247–268.

STEVENS, L. C. (1964).
Experimental production of testicular teratomas in mice. Proc. Nat. Acad. Sci. USA *52*, 654–661.

STEVENS, L. C. (1966).
Development of resistance to teratocarcinogenesis by primordial germ cells in mice. J. Nat. Cancer Inst. *37*, 859–867.

STEVENS, L. C. (1967a).
The development of teratomas from intratesticular grafts of 2-cell mouse eggs. Anat. Rec. *157*, 328.

STEVENS, L. C. (1967b).
The biology of teratomas. Advan. Morphogenesis *6*, 1–81.

STEVENS, L. C. (1967c).
Origin of testicular teratomas from primordial germ cells in mice. J. Nat. Cancer Inst. *38*, 549–552.

STEVENS, L. C. (1968).
The development of teratomas from intratesticular grafts of tubal mouse eggs. J. Embryol. Exp. Morph. *20*, 329–341.

STEVENS, L. C. (1970a).
Experimental production of testicular teratomas in mice of strains 129, A/He, and their F_1 hybrids. J. Nat. Cancer Inst. *44*, 923–929.

STEVENS, L. C. (1970b).
Environmental influences on experimental teratocarcinogenesis in testes of mice. J. Exp. Zool. *174*, 407–414.

STEVENS, L. C. (1970c).
The development of transplantable teratocarcinomas from intertesticular grafts of pre- and post-implantation mouse. Dev. Biol. *21*, 364–382.

STEVENS, L. C. (1973).
A new inbred subline of mice (129/terSv) with a high incidence of spontaneous congenital testicular teratoma. J. Nat. Cancer Inst. *50*, 235–242.

STEVENS, L. C. (1974).
Teratocarcinogenesis and spontaneous parthenogenesis in mice. *In* The Developmental Biology of Reproduction, C. L. Markert and J. Papaconstantinou, eds. (New York: Academic Press), pp. 93–106.

STEVENS, L. C., and BUNKER, M. C. (1964).
Karyotype and sex of primary testicular teratomas in mice. J. Nat. Cancer Inst. *33*, 65–78.

STEVENS, L. C., and MACKENSEN, J. A. (1961).
Genetic and environmental influences on teratocarcinogenesis in the mouse. J. Nat. Cancer Inst. *27*, 443–453.

STEVENS, L. C., and PIERCE, G. B. (1975).
Teratomas: definitions and terminology. *In* Teratomas and Differentiation, M. I. Sherman and D. Solter, eds. (New York: Academic Press), pp. 13–14.

STEVENS, L. C., and VARNUM, D. S. (1974).
The development of teratomas from parthenogenetically activated ovarian mouse eggs. Dev. Biol. *37*, 369–380.

STOKER, M., and PIGGOTT, D. (1974).
Shaking 3T3 cells: further studies of diffusion boundary effects. Cell *3*, 207–215.

STRICKLAND, S., REICH, E., and SHERMAN, M. I. (1976).
Plasminogen activator in early embryogenesis: enzyme production by trophoblast and parietalendoderm. Cell *9*, 231–240.

TAKAGI, N. (1974).
Differentiation of X-chromosomes in early female mouse embryos. Exp. Cell Res. *86*, 127–135.

TAKAGI, N., and OSHIMURA, M. (1973).
Fluorescence and Giemsa banding studies of the allocyclic X-chromosome in embryonic and adult mouse cells. Exp. Cell Res *78*, 127–135.

TAKAGI, N., and SASAKI, M. (1975).
Preferential inactivation of paternally derived X-chromosome in the extraembryonic membranes of the mouse. Nature *256*, 640–642.

TAMAOKI, T., THOMAS, K., and SCHINDLER, I. (1974).
Cell-free studies of developmental changes in synthesis of α-fetoprotein and albumin in mouse liver. Nature *249*, 269–271.

TARKOWSKI, A. K. (1975).
Induced parthenogenesis in the mouse. *In* The Developmental Biology of Reproduction, C. L. Markert and J. Papaconstantinou, eds. (New York: Academic Press), pp. 107–129.

TARKOWSKI, A. K., and ROSSANT, J. (1976).
Haploid blastocysts from bisected zygotes. Nature *259*, 663–665.

TEILUM, G., ALBRECHTSEN, R., and NORGAARD-PEDERSEN, B. (1974).
Immunofluorescent localization of alpha-fetoprotein synthesis in endodermal sinus tumor (yolk sac tumor). Acta. Path. Microbiol. Scand. A *82*, 586–588.

TEILUM, G., ALBRECHTSEN, R., and NORGAARD-PEDERSEN, B. (1975).
The histogenetic-embryologic basis for reappearance of alpha-fetoprotein in endodermal sinus. Acta. Path. Microbiol. Scand. A *83*, 80–86.

TERESKY, A. K., MARSDEN, M., KUFF, E. L., and LEVINE, A. R. (1974).
Morphological criteria for the *in vitro* differentiation of embryoid bodies produced by a transplantable teratoma in mice. J. Cell. Physiol. *84*, 319–332.

TERZI, M. (1974).
Genetics and the Animal Cell. New York: Wiley.

TODARO, G. J. (1974).
Endogenous viruses in normal and transformed cells. *In* Developmental Aspects of Carcinogenesis and Immunity, T. J. King, ed. (New York: Academic Press), pp. 145–158.

TODARO, G. J., and HUEBNER, R. J. (1972)
The viral oncogene hypothesis: new evidence. Proc. Nat. Acad. Sci. USA *69*, 1009–1015.

TOPP, W. D., HALL, J. D.,
MARSDEN, M., TERESKY, A. K.,
RIFKIN, D., LEVINE, A. J., and
POLLACK, R. (1976).
In vitro differentiation of teratomas:
the distribution of creatine phospho-
kinase and plasminogen activator in
teratocarcinoma-derived cells. Cancer
Res. *36*, 4217–4223.

TUMYAN, B. G., SVET-
MOLDAVSKY, G. G., and
KARMANOVA, N. V. (1975).
Factor suppressing α-fetoprotein pro-
duction in newborn mice. Nature *255*,
244–245.

TURNER, D. C. (1975).
Isozyme transitions of creatine kinase
and aldolase during muscle differenti-
ation *in vitro*. *In* Isozymes, C. L.
Markert, ed., vol. 3 (New York:
Academic Press), pp. 145–158.

TURNER, D. C., and
EPPENBERGER, H. M. (1973).
Developmental changes in creatine
kinase and aldolase isoenzymes and
their possible function in association
with contractile elements. Enzyme *15*,
224–238.

WAKE, N., TAKAGI, N., and
SASAKI, M. (1976).
Non-random inactivation of
X-chromosome in the rat yolk sac.
Nature *262*, 580–581.

WHITE, B. J., TJIO, J. H., VAN DE
WATER, L. C., and GRANDALL, C.
(1974).
Trisomy 19 in the laboratory mouse.
I. Frequency in different crosses at
specific developmental stages and
relationship of trisomy to cleft palate.
Cytogenet. Cell Genet. *13*, 217–231.

WILSON, B. W., LINKHART, T. A.,
WALKER, C. R., and YEE, G. W.
(1975).
Acetylcholinesterase isozymes and
muscle development: newly synthesized enzymes and cellular site of
action of dystrophy of the chicken. *In*
Isozymes, C. L. Markert, ed., vol. 3.
(New York: Academic Press),
pp. 739–752.

WILSON, B. W., NIEBERG, P. S.,
WALKER, C. R., LINKHART,
T. A., and FRY, D. M. (1973).
Production and release of acetylcho-
linesterase by cultured chick embryo
muscle. Dev. Biol. *33*, 285–299.

WILSON, J. R., and ZIMMERMAN,
E. F. (1975).
Yolk sac: site of developmental
microheterogeneity of mouse alpha-
fetoprotein. Fed. Proc. *34*, 679.

WILSON, S. H., SCHRIER, B. K.,
FARBER, J. L., THOMPSON, E. J.,
ROSENBERG, R. N., BLUME, A. J.,
and NIRENBERG, M. W. (1972).
Markers for gene expression in cul-
tured cells from the nervous system.
J. Biol. Chem. *247*, 3159–3169.

WISLOCKI, G. B., and
PADYKULA, H. A. (1953).
Reicherts membrane and yolk sac of
the rat investigated by histochemical
means. Am. J. Anat. *92*, 117–152.

WROBLEWSKA, J. (1971).
Developmental anomaly in the mouse
associated with triploidy.
Cytogenetics *10*, 199–207.

WURSTER-HILL, D., WHANG-
PENG, J., MCINTYRE, O. R., HSU,
L. Y. F., HIRSCHKORN, K.,
HODAN, B., PISCIOTTA, A. V.,
PIERRE, R., BALCERZAK, S. P.,

WEINFELD, A., and MURPHY, S. (1976).
Cytogenetic studies in polycythemia vera. Seminars in Haematology *13*.

WYLIE, C. V., NAKANE, P. K., and PIERCE, G. B. (1973).
Degrees of differentiation of non-proliferating cells of mammary carcinoma. Differentiation *1*, 11–20.

YAFFE, D., and DYM, H. (1972).
Gene expression during differentiation of contractile muscle fibres. Cold Spring Harbor Symp. Quant. Biol. *37*, 543–547.

YEOH, G. C. T. and MORGAN, E. H. (1974).
Albumin and transferrin synthesis during development in the rat. Biochem. J. *144*, 215–224.

ZAMBONI, I., and MERCHANT, H. (1973).
The fine structure of mouse primordial germ cells in extragonadal locations. Am. J. Anat. *137*, 299–336.

ZINZAR, S. N., SVET-MOLDAVSKY, G. J., LATINA, B. I., and TUMYAN, B. G. (1971).
Enormous organ-like growth of transplant of fetal digestive tract. Transplantation *11*, 499–502.

MILLER, R. A., and RUDDLE, F. H. (1977).
Teratocarcinoma × Friend erythroleukemia cell hybrids resemble their pluripotent embryonal carcinoma cell parent. Dev. Biol. *56*, 157–173.

PERIES, J., ALVES-CARDOSO, E., CANIVERT, M., DEBENS-GUIL-LEMIN, M. C. and LASNERET, J. (1977)
Lack of multiplication of ectotropic murine C type viruses in murine teratocarcinoma primitive cells. J. Nat. Cancer Inst., in press.

SWARTZENDRUBER, D. E. (1976).
Squamous cell differentiation in a clonal teratocarcinoma cell line. Differentiation *7*, 7–12.

SWARTZENDRUBER, D. E., CRAM, L. S., and LEHMAN, J. M. (1976)
Microfluorometric analysis of DNA content changes in murine teratocarcinoma. Cancer Res. *36*, 1894–1899.

WADA, H. G., VANDENBERG, S R., SUSSMAN, H. H., GROVE, W. E., and HERMAN, M. M. (1976).
Characterization of two different alkaline phosphatases in mouse teratoma: partial purification, electrophoretic, and histochemical studies. Cell *9*, 37–44.

Further useful references (added in proof):

LITTLE, C. D., CHURCH, R. L., MILLER, R. A. and RUDDLE, F. H. (1977).
Procollagen and collagen produced by a teratorcarcinoma-derived cell line, TSD-4: evidence for a new molecular form of collagen. Cell *10*, 287–296.

INDEX